# From Combinatorics to Philosophy

Ernesto Damiani · Ottavio D'Antona ·
Vincenzo Marra · Fabrizio Palombi
Editors

# From Combinatorics to Philosophy

## The Legacy of G.-C. Rota

Springer

*Editors*
Ernesto Damiani
Università Milano
Dipto Tecnologie Informatiche
Via Bramante, 65
26013 Crema
Italy
damiani@dti.unimi.it

Ottavio D'Antona
Università Milano
Dipto. Informatica e Comunicazione
Via Comelico, 39
20135 Milano
Italy
dantona@dti.unimi.it

Vincenzo Marra
Università Milano
Dipto. Informatica e Comunicazione
Via Comelico, 39
20135 Milano
Italy
marra@dti.unimi.it

Fabrizio Palombi
Università della Calabria
Dipto. Filosofia
Ponte Bucci Cubo 18/c
87036 Arcavacata di Rende (Cs)
Italy
fabriziopalombi@yahoo.it

ISBN 978-1-4899-8298-8      ISBN 978-0-387-88753-1 (eBook)
DOI 10.1007/978-0-387-88753-1
Springer Dordrecht Heidelberg London New York

Printed on acid-free paper

Springer is part of Springer Science+Business Media (www.springer.com)

# Foreword

Even those who were fortunate enough to know Gian-Carlo Rota, or perhaps to match wits with him, would be hard put to answer the question: "Are you sure you know who Gian-Carlo Rota was?" The point underlying this question is well illustrated by the Latin saying, *Si duo dicunt aliud, non est idem.* In other words, although each of them may bear within their memory the image and the feeling of Gian-Carlo, it is also true that none of them knows Gian-Carlo Rota the figure.

Like it or not, every such fortunate contemporary has built his/her own personal Mr. Rota. Indeed each has painted a portrait using her/his own canvas and own palette. To display it, each has then chosen the frame deemed more appropriate. This portrait has been painted in dozens of faithful copies – each different from one another.

This book does not attempt to answer the question that opens these introductory remarks. Rather, it provides a number of selected contributions by authors who have interacted either with Rota himself or with his work. We trust that Rota's intellectual footprint will thereby emerge more vividly than through an attempt at yet another portrait.

## Rota, the man.

The book begins with two authoritative, first-hand accounts. Ester Gasperoni Rota brings to bear the special intimacy of her and his brother's childhood. The fascinating story she tells ends with the young Rota's departure for Princeton in 1950. Eight years later, Joseph Kohn met Rota in Cambridge. His contribution first connects with the shared energy of youth and early career, and then with the shared halls of the academia and of off-campus sommeliers.

## Rota, the mathematician.

The second part of the volume collects papers of a mathematical nature.

Andrea Brini's contribution is an exposition of some of the contemporary developments in invariant theory. Here we see how Rota has his place among the several mathematicians who contributed to the modern rebirth of the subject.

Pietro Codara introduces and investigates three different notions of partitions of a finite partially ordered set, each induced by a different sort of morphism between them.

Henry Crapo's paper discusses Hermann Grassmann's seminal contributions to mathematics, and Rota's life-long intellectual interaction with his profound ideas. As a feature of this essay to be particularly treasured, let us mention the quotations from the mathematical correspondence between Rota and the author in the section on the Whitney algebra of a matroid.

Rota's Fubini Lectures (Crapo, Senato, 2001, pp. 27-93) feature prominently in the paper by Elvira Di Nardo and Domenico Senato. Here the authors fruitfully apply the techniques of the modern umbral calculus – the evolution of the symbolic method, that "old war-horse of nineteenth-century mathematics" (D'Antona, Rota, 1978) – to a variety of statistical problems.[1]

The first part of the essay by Joseph Kung is devoted to the theory of Baxter algebras, here surveyed in considerable detail. The second part of the paper offers a revisitation of the theory of random variables, after ideas that Rota expounded in a course he delivered at the Massachusetts Institute of Technology in 1977.

Daniele Mundici surveys a non-idempotent, many-valued generalization of a wide range of concepts and results from classical probability theory. To mention only one example, he shows how de Finetti's celebrated coherence criterion for probability assignments can be extended from yes/no, classical events to continuously-valued ones. The author's line of thought is interspersed with quotations from the aforementioned Fubini Lectures.

Emanuele Munarini characterizes classes of finite distributive lattices in terms of their Euler characteristic. In several interesting cases he obtains an explicit expression for the values of the characteristic.

Again applying umbral methods, Pasquale Petrullo shows how moments and cumulants are representable by generalized, umbral, Abel polynomials.

Each paper featured in the second part of the book has its own autonomous motivations and goals, but there is at least one important sense in which all these contributions fully relate to Gian-Carlo Rota's intellectual legacy. While any sound proof can vouch for the truth of a mathematical statement, not all mathematical proofs are created equal. In Rota's own words, "the mere logical truth of a statement does not enlighten us to the sense of the statement". But in fact, "enlightenment is a

---

[1] A similar sentence is attributed (with no reference) to Hermann Weyl in Kung, Rota (1984), p. 28.

fundamental, essential feature of mathematics". (Rota, 1997a, p. 181). By way of example, let us consider two celebrated theorems, and their proofs.

In 1995, after over three centuries of efforts by generations of mathematicians, Fermat's last theorem was finally proved by Andrew Wiles and his collaborators. Now, "no one will seriously entertain the notion that the statement of Fermat's last theorem is of any interest whatsoever". Clearly, "what is remarkable about Fermat's last theorem is the proof", for it is the proof that "opens up new possibilities for mathematics". (Rota, 1997b, pp. 188-190).

In 1976, Kenneth Appel and Wolfgang Haken gave a computer-assisted proof of the four color theorem, thereby solving a mathematical problem that had resisted all assaults for over a century. To say the very least, their full argument is impractical to check by hand, and so are the subsequent simplified proofs along the same lines. Nowadays the correctness of Appel's and Haken's proof seems to be acknowledged by the majority of mathematicians. Nonetheless, it may be argued that there is a sense in which their remarkable achievement does not compare to Wiles' proof of Fermat's last theorem. The proof of the four color theorem is "the glamorous instance of a verification"; and while "verification is proof", it "may not give the reason". Characteristically, Rota goes all the way: "every mathematician knows that the computer verification of the four color conjecture is of considerably lesser value than Wiles's proof, *because it fails to open up any significant mathematical possibilities*". (Rota, 1997b, pp. 187-191; the emphasis is ours).

Only an inspired choice of the right mathematical tools can support the crystal-clear reasoning that may open up new mathematical vistas through enlightening proofs. We hope readers will share our view that all mathematical contributions to this volume strive to honor Gian-Carlo's intellectual legacy in at least one respect. That is, they endeavor to *prove*, not to merely *verify*.

## Rota, the philosopher.

The third part of the book is devoted to Rota's philosophical reflection. It is framed around two approaches. The first one, consisting of the contributions of Carlo Cellucci and Massimo Mugnai, presents a general picture of Rota's philosophical work. The second approach, consisting of texts written by Francesca Bonicalzi, Albino Lanciani with Claudio Majolino, and Fabrizio Palombi, examines from different perspectives the antiscientistic and antireductionist value of Rota's phenomenology.

Carlo Cellucci analyzes four epistemological problems: the ontological status of mathematical objects, the function of mathematical definitions, the concept of demonstration and, more generally, the complex relationship between mathematics and philosophy. The examination of such questions underlines qualities as well as some limits of Rota's philosophical reflection and, above all, it points out its unconventional nature. As a matter of fact, according to Cellucci, Rota is a nonconformist scholar who interprets mathematics as a dynamic form of knowledge and studies the problems connected with its growth.

Massimo Mugnai seems to share this opinion and describes Rota as an outsider to philosophy, who is capable of individuating new problems from unsuspected perspectives and of criticizing old paradigms. For example, Rota's criticism of analytical philosophy constitutes a "legitimate reaction" to the scholastic attitude adopted by some of its exponents. Nonetheless, according to Mugnai, Rota's work presents some historiographic inaccuracies and theoretic limits, as in the case of the interpretation of reductionism and psychologism.

The authors adopting the second kind of approach agree that Rota's philosophical research has its own personality and independence. In fact, Francesca Bonicalzi lingers over some well-known articles written by Rota on the difference between mathematical and philosophical truth, which caused an uproar when they were published. This is the starting point for a reflection that proceeds from an historical contextualization of the torsion in thought made by Descartes and Leibniz to demonstrate the impossibility of reducing the philosophical reflection to a sort of sophisticated logical, linguistic or methodological analysis. The fundamental Greek etymology of the word "method" is recalled by Bonicalzi to exhort contemporary philosophy to be aware of Rota's monitions and to recover its own way, a way that is made of richness of sense and not of exactness of calculation.

Fabrizio Palombi draws on Bonicalzi's considerations in order to attempt an examination of the peculiar phenomenological realism that characterizes Rota's reflection. The Husserlian theory of *Fundierung* and the Heideggerian "State-of-Mind" are used to explain certain rules of Rota's phenomenological description and of his peculiar mereology. These considerations are summarised in the suggestion of defining Rota's philosophical perspective as a form of "phenomenological surrealism".

The originality of Rota's reflection may also be valued through his courage, that, according to Albino Lanciani and Claudio Majolino, is a distinctive feature of phenomenological thought. Inspired by Ortega y Gasset, Rota is not afraid of facing the "constant shipwreck" of philosophy, that the two authors examine to describe some of its historical features. Their contribution is completed by a description of the "three senses of phenomenology" with relation to temporality, to an "act of resistance" and to a "readjustment of perception".

**Acknowledgments.**

This book is based on the talks delivered at the Rota Memorial Conference held at the Università degli Studi di Milano, Italy, from the 16th through the 18th of February 2009. We would like to take this occasion to thank all speakers and participants. The event would not have been possible without the financial help of several companies and institutions. First, we wish to acknowledge the generous support of our main sponsor ARIA S.p.A. We also record our gratitude to the Dipartimento di Informatica e Comunicazione, the Dipartimento di Tecnologie dell'Informazione, and the Dipartimento di Matematica of the Università degli Studi di Milano, and to the

Rector of the latter university; to the Dipartimento di Filosofia and the Facoltà di Lettere e Filosofia of the Università della Calabria; and, finally, to the Italian branch of Springer Verlag. Finally, one cannot forget the help given to us by S. Aguzzoli, M. Barnabei, P. Codara, G. Dalmasso, A. Ferrara, F. Frati, O. Scotti, M. Torelli, and G. Zampieri.

Milano, 16.IV.2009                                                               *Ernesto Damiani*
                                                                                    *Ottavio M. D'Antona*
                                                                                    *Vincenzo Marra*
                                                                                    *Fabrizio Palombi*

# References

D'Antona, O. M., Rota, G.-C. (1978), *Two Rings Connected with the Inclusion-Exclusion Principle*, in "J. Combin. Theory Ser. A", 24/3, pp. 395-402.

Crapo, H., Senato, D. (eds.) (2001), *Algebraic Combinatorics and Computer Science. A tribute to Gian-Carlo Rota*, Milan, Springer.

Kung, J. P. S., Rota, G.-C. (1984), *The Invariant Theory of Binary Forms*, in "Bull. Amer. Math. Soc.", 10, pp. 27-85.

Rota, G.-C. (1997a), *The Phenomenology of Mathematical Beauty*, in "Synthese", 111, pp. 171-182.

Rota, G.-C. (1997b), *The Phenomenology of Mathematical Proof*, in "Synthese", 111, pp. 183-196.

# List of talks

List of talks delivered at the Rota Memorial Conference held in Milan on February 16/17/18, 2009.

Francesca Bonicalzi: *Ethics in thought. Gian-Carlo Rota and philosophy*

Andrea Brini: *A glimpse on vector invariant theory. The points of view of Weyl, Rota, De Concini and Procesi, and Grosshans*

Carlo Cellucci: *Indiscrete variations on Gian-Carlo Rota's themes*

John H. Conway: *On the Free Will Theorem*

Henry Crapo: *Gian-Carlo Rota's work in geometry*

Elio Franzini: *Fenomenologia, fondazione e incrocio di saperi: a partire da Gian-Carlo Rota*

Ester Gasperoni Rota: *Gianco, my brother*

Luca Giudici: *von Neumann's coordinatization and some of Rota's problems in lattice theory*

Joseph J. Kohn: *Remembering Gian-Carlo Rota and some problems in complex analysis*

Massimo Mugnai: *Rota's philosophical insights*

Daniele Mundici: *Rota, probability and logic*

Fabrizio Palombi:
*"A minority view". Gian-Carlo Rota's phenomenological realism*

Domenico Senato: *The eleventh problem of Fubini Lectures: cumulants*

# Contents

# Part I
# Gian-Carlo Rota, the Man

# Chapter 1
# Gianco, my Brother
## Invited Chapter

Ester Gasperoni Rota

As a child, Gianco (the nickname by which he was known to family and friends) wanted to be a cook. Since cooking has never been my "cup of tea", I gladly allowed him to experiment on a stove that had been given to me for Christmas, a little electric marvel equipped with a real oven. He concocted sauces, baked cookies and cakes, which were eminently edible. Perhaps thanks to his childhood practice, Gianco's cooking was always far better than mine.[1]

Having mastered cooking, my brother decided to try surgery. For lack of patients, he practised on my dolls. I had one doll I particularly cherished: a sailor with a porcelain face and a limp body stuffed with wool. One day I found my beloved sailor with a gaping wound that revealed his woollen innards. "He had appendicitis and I had to operate", explained my brother lamely. I burst into tears: Gianco never touched my dolls again and directed his curiosity towards mechanics, a science in which he probably perceived certain similarities with surgery. He took apart and re-assembled toys, played with hammers and screw-drivers and nuts and bolts. Father, who had frowned on his cooking and laughed at his surgical skills, encouraged this new interest. He bought an electric train, complete with a set of railways, passenger and freight cars, level crossings, tunnels made of papier mch and train stations with red roofs. This German made masterpiece attracted dozens of Gianco's friends. My brother, a magnanimous host even as a child, asked me to make ice-cream for this boisterous crowd. I protested. In the forties making ice-cream was quite a feat, since the ice-cream machine had to be operated manually by means of a crank. Gianco offered a deal: ice-cream against admission into his newly founded Secret Society. I would be the Society's only female member, my brother said, and I would have the

---

Ester Gasperoni Rota
Paris, France.

[1] Some of the episodes mentioned in this memoir are recounted in my books (Gasperoni Rota, 1995 and 2006).

E. Damiani et al. (eds.), *From Combinatorics to Philosophy,*
DOI 10.1007/978-0-387-88753-1_1, © Springer Science+Business Media, LLC 2009

honor of cleaning up after the meetings. I hesitated, then accepted, simply because I was intrigued by the Society's goal: "the quest for truth". To this day, I don't know whether that quest was successful. In any case, while looking for truth, my brother found time to hunt for rare stamps and put together a remarkable stamp collection.

At about this time, father, a music lover, took us to La Scala, in Milan, for the opening of Verdi's *Aida*. I was awed by the impressive décor while Gianco found the music fascinating, particularly Aida's famous march, which he took to humming incessantly and off-key, much to everybody's annoyance. It was perhaps this memorable soirée at La Scala that awakened Gianco's interest in the theater. He would be a director, he announced. Father approved and immediately asked a carpenter to build a full fledged stage.

Gianco's first play was a modern version of *Snow White*. He easily recruited the actors among his friends, who were eager to embark on this venture, albeit as dwarfs. I was sure I would be given the part of Snow White. I had seen the Walt Disney film and had listened over and over to a Snow White record mother had bought at my insistence. I knew the part by heart. But my hopes were dashed when Gianco chose a precocious beauty named Angela who, in spite of her early age, flirted outrageously with the boys. I was fuming, particularly when Gianco asked me to play the part of the evil queen. Yet my desire to go on stage was stronger than my resentment: I played the queen.

For the premiere we had invited all our friends, who in turn had extended the invitation to *their* friends, who had spread the news among *their* friends. Half an our before the curtain was to rise, about fifty boys and girls stampeded into the room fighting for the best seats, so to speak, since there were no chairs and everybody sat on the floor. *Snow White* was a hit. Encouraged by his success, Gianco staged other fairy tales: *The Sleeping Beauty*, with (naturally) Angela playing the part of the Princess and I the part of the witch who puts the Princess under a spell. Only in *Puss in Boots* did my brother give me the main part, but only because Angela had flatly refused to be the cat.

One day, our parents took us to see a conjurer in a local theater. By the time we had returned home, my brother had decided that for the rest of his life he would perform magic tricks. I will never understand why father humoured my brother's bizarre penchant by supplying him with a magician's full equipment: wands, boxes with secret compartments, a deck of doctored cards. Again, Gianco's first performance attracted a crowd. The audience applauded heartily as my brother produced fake pearls out of the box and guessed the suits of the cards that three of the spectators had picked out of the doctored deck.

Then came the war. Actually, the war had already started, but we had not yet felt its full impact. Not that things were really normal in our lives. Father was an outspoken anti-fascist, which did not exactly make for a smooth life. At school, Gianco and I were frowned upon both by teachers and classmates because father wouldn't allow us to wear fascist uniforms. Some of our classmates openly ostracized us. I grumbled and complained, while my brother became taciturn and pored over his books and his stamp collection. He had developed a new interest in the occult and read anything he could find on the subject. I wondered whether his quest for truth had taken a new turn.

The bombings started. The target of the allied planes was the nearby city of Milan. A few bombs, however, were dropped for good measure on our train station and on the bridge that spanned the Ticino river and connected Vigevano to Milan. When the sirens howled, we took refuge in the cellar, where cots were always ready. Unaware of the danger, I looked forward to the night bombings, since they meant that there would be no school the following day, but they angered my brother. In spite of the hostile environment, he loved school.

The ominous day arrived when we had to leave our city and our house. After Mussolini's forced resignation in July 1943, father, sensing the forthcoming unrest, put us on a train and shipped us off to Varallo, a small town in the Valsesia . He decided to stay and wait for further developments. But in the fateful month of September, Mussolini, freed by the Germans from his confinement on the Gran Sasso mountain, announced the founding of the *Repubblica Sociale Italiana*, also known as Salo' Republic, from the name of the town on Lake Garda where Mussolini set up his headquarters. Italy was occupied by German troops. Father barely escaped the Nazi soldiers when they came to take over our house. A "black list" was posted all over town. It contained the names of 27 subversives who, if found, were to be immediately arrested. Father's name was on top of the list.

Our trek towards safety was long and tortuous. We moved from a small hotel in Varallo to a dank, dark apartment, which, however, was not dark enough to protect us from curious eyes. We left the apartment and found shelter in a nearby convent, where we were joined by another anti-fascist couple who had two children. Their daughter, Carla a few years older than I, became my friend. We spent days cooped up in a large room that served as communal dormitory and were only allowed to go as far as the dining room where we took our meals with the nuns. Fearing that we might lapse into chronic ignorance, father decided to teach us both math and Italian composition. Whereas it took me hours to add and divide and subtract and multiply, my brother solved all kinds of problems in a matter of minutes. As for his essays, they contained thoughts and expressions which were way above my head, but which drew approving nods from our father. We filled our long, empty hours at the convent with whatever games we could play within four walls. Armed with a sheet of paper and a pencil, we played *naval battle*, sinking imaginary battleships and submarines. We wrote coded diaries which the following day we couldn't decipher. We *smoked* in the bathroom, choking on cigarettes made out of paper and filled with camomile leaves. Our friend Carla taught us a card game which soon became our favorite: poker. Gianco learned fast and was soon playing like a pro. Even Carla was no match for him. My brother was a master at bluffing, but also at guessing whether the other players were bluffing. It wasn't difficult as far as I was concerned, for I blushed whenever I had a good hand. We used dry beans (kindly supplied by the nuns) for money. Gianco always ended the game with mounds of beans, whereas I was left "beanless" and had to borrow from my brother, who charged an interest: one bean for every ten borrowed.

A series of dramatic events put an end to our games and our lessons. A small village perched on a hill across from the convent was burned to the last house by the Salo' Fascists, who accused the villagers of giving shelter to Partisans. In Varallo, there were random roundups and shootings. It became imperative that we leave. The nuns suggested that we move to Rimella, a village on the Italian slope of Mount Rosa, which had so far been overlooked by the Fascists and the Nazis. Armed with fake identity cards, which gave us the unlikely name of Porcheddu, we piled onto a horse-drawn cart with our meagre possessions. The cart was led by a man the nuns trusted who could only take us part of the way. To reach the village, which clung to a steep slope, we had to climb the mountain, following a narrow path inaccessible to any kind of vehicle. I huffed and puffed and complained, while Gianco, his face a mask, his back bent under his rucksack, moved on with dogged determination, humming *Aida's* march. I started crying, as much for myself and my tired legs, as for my brother, whose apparent stoicism, I suspected, hid a suffering that was far greater than mine. In the village, we spread the story that we were escaping city bombings. Our parents would have liked to let a large house near the church, overlooking the valley, but the owner asked a rent that was beyond our means. We moved into a smaller house with rudimentary accommodations. Food was scarce, but we never went hungry. Fearing that we might reveal dangerous secrets, our parents refused to enrol us in the local school. Gianco and I resumed our Italian and math lessons with father, to which were added Latin lessons imparted by the parish priest who had, amazingly, become a friend of our anti-clerical father with whom he shared the same political ideas. The priest, Don Giuseppe, praised my brother's intelligence, his quick grasp of the Latin subtleties. Luckily, I was considered too young to learn Latin.

In spite of the cold we were skimpily dressed: I wore an old coat that hardly covered my knees. My brother's wardrobe consisted of two pairs of knickerbockers that awoke the curiosity of the village boys and two sweaters that were much too small for him. He hated the knickerbockers. We had plenty of time on our hands, which we tried to fill with the usual games of poker and naval battle and with long walks through the village. I would have liked to befriend some of the girls that stared at us with curiosity but without animosity, but my brother warned me against such friendships which, he said, could become the Trojan horse of our safety. I didn't know what a Trojan horse was; still, I blindly believed everything Gianco said and steered clear of the village girls. Perhaps in order to keep us from mischief, our mother assigned simple household tasks to us. One of them was setting the table at mealtimes. Gianco, who hated this chore, devised a clever way of eluding it. Whenever his turn came to set the table, he would ask me to do it for him in exchange for an IOU. In the IOU he promised to set the table twice in my stead "at a later date", which he never specified. I collected a boxful of IOU's, waiting for my brother to fulfil his promise. He never did, since a series of dramatic events made the IOU's seem irrelevant. We had deluded ourselves into thinking that we could wait out the war in that apparently safe haven. But there were no safe havens in Italy in those terrible years. The Partisans took over the village, from which they organized raids against the Nazis and Fascists that occupied the surrounding valleys.

The Nazis sent a plane to bomb the Partisans out of the village. Luckily, there were no casualties, but, by a bizarre quirk of fate, a bomb destroyed the very house that our parents would have let, had it not been for its high rent.

A heavy snowfall had made Rimella inaccessible. But, as the snow started to melt under the first rays of sunshine, the Nazi and Fascist troops moved towards the village. From the village square, we could see a long black line of soldiers climbing the slope. The Partisans had disappeared almost overnight. It was too late for us to escape; our parents decided to stay put and wait for events. At first, the Nazis and Fascists seemed to ignore us. They did, however, requisition one room in our house to lodge some of their soldiers.

Then, one day, father was arrested: his real identity had been discovered. How could this have happened, since throughout the village we were known as Porcheddu ? Later, my parents found out that the village baker had seen our real name on a tag hanging from one of our rucksacks, which, in the rush of departure, we had forgotten to remove. Hoping to ingratiate himself with the Fascists, he had informed on us.

My brother and I cuddled up to our mother who was sobbing desperately as our father was taken away by a soldier. Before leaving the house, father took his son aside and told him that, should anything happen to him, he would have to take over as head of the family. Gianco listened in silence, a grave expression on his face. He was not yet thirteen years old. I believe it was on that very day, while listening to our father's sad voice, that my brother became a man.

But, once again, fate was on our side. Because of an error made in trascribing father's name (*Rosa* instead of *Rota*) the Fascist officer who interrogated him mistook him for a Rimella born engineer who had helped the Fascists rebuild a bridge destroyed by the Partisans in one of the surrounding valleys. Don Giuseppe, who had been summoned to help establish father's identity, confirmed, while giving father a warning look, that he was Mr. *Rosa*. The outcome of this misunderstanding was near miraculous: not only did the Fascist officer release father, but, convinced that he was, indeed, the friendly engineer, he named him mayor of Rimella!

"I have been saved by one letter", said father when he returned home. He knew, however, that the mistake would soon be uncovered. We had to leave. For a brief period we hid in a nearby hamlet, Sant'Anna, higher up on the slope and accessible only through a narrow, winding path. We slept on mounds of straw in abandoned shanties until we found decent, though modest, lodgings. Here we had to combat a new enemy: lice, a "memento" of our Fascist roommates. But father was nervous. Sant'Anna was too close to Rimella for comfort. Once again, we were on the move. As our next refuge, our parents had chosen the Anzasca Valley, on the opposite side of the Rimella slope. We started out at dawn and walked for hours, without really knowing where we would settle. Gianco and I had worn out our only pair of shoes and had to wear felt slippers mother had bought in the village. Cut by the hardened crust of snow, my legs bled. Luckily, Gianco's legs were protected by the loathsome knickerbockers.

We stopped in Vanzone, a hamlet near the Swiss border that seemed relatively calm, where we rented a small apartment owned by a Swiss lady. For some time we

led an almost tranquil life. But Nazi and Fascist troops soon swarmed into the village and started rounding up people on the slightest suspicion. Father decided to flee. Led by a guide who was a friend of our landlady, he climbed the mountain that separated the village from Switzerland, hoping to find shelter in a country which had remained neutral and which had already taken in a great number of Italian antifascists. Before leaving, he urged us to go to a farm owned by an uncle in Roncallo, on the shores of Lake Orta. Once again, we packed our rucksacks and walked over twenty kilometers to Piedimulera to find a train that would take us to the safety of our uncle's farm. Hungry, exhausted and practically shoeless, we finally arrived at the farm after and adventurous journey on a jam-packed train. Our aunt's first comment was: "My poor children! We have to get you some shoes !" In Roncallo we found clothes, food and love. Our uncle had two children, a boy who was about my age, and a younger girl. Other boys and girls who had escaped bombed out cities, became our friends.

In spite of the war that was still raging, we were a happy crowd; my brother soon became its leader.

Breaking through the fascist censorship, Father had sent a postcard to our former Swiss landlady, who had forwarded it to us thanks to a trustworthy messenger. With very guarded language he informed us that he had safely reached Switzerland.

Mother made sure that we attended regular schools: I had to walk miles to reach mine. The only existing high school in the area, which both my brother and my cousin attended, was on the opposite shore of the lake. Every morning Gianco, our cousin and several other children were rowed across the lake by a boatman named Giovanni and were rowed back in the afternoon. In normal times, it would have been a fun trip. But times were anything but normal: Nazi and Fascist soldiers arrived and occupied our side of the lake. They often came to the farm, pretending to be searching for weapons or deserters. What they were looking for was food, which mother and our aunt supplied abundantly so as to keep them at bay. In an old notebook belonging to my mother, I found the following entry: "*At noon 3 very threatening Germans arrive. They are are armed and they want to search the house and arrest its owner. I take them to the kitchen, where they eat, drink and smoke. They finally leave, taking a rabbit with them*".

The opposite shore had been taken over by the Partisans who shot intermittently at their enemies across the lake. Boating to school became a nightmare, even though some mothers, including our own, had succeeded in obtaining a safeconduct from both the Nazi and Partisan commanders. Huddled in the boat, a white flag floating in the wind, Gianco and his school mates crossed the lake daily, praying that a trigger happy soldier would not shoot at them. I remember the anguished expressions on my mother's and my aunt's faces as they peered through the afternoon mist, looking for the white flag that announced the children's return from school.

Throughout these adversities, my brother maintained an olympian calm and stood out as one of the best students in his class. One of his professors predicted a brilliant literary future for Gianco, who excelled both in the field of science and the humanities. He passed his finals with a grade of *ottimo*, excellent.

The war ended, at last, and father returned from Switzerland, thin and weary, but all in one piece. We returned to Vigevano, to find that our house had been plundered by Nazi and Fascist soldiers. Gone was our beloved theater and the German-made electric train and the stamp collection my brother had patiently put together. Father's wine cellar had been emptied and the pig which we had kept in the back yard and which father, an animal lover, had refused to kill even when meat had become a rarity, had been slaughtered and eaten.

We resumed, or tried to resume, a normal routine. But food was still scarce; fuel was nowhere to be found. At school, we kept both our coats and gloves on. This didn't keep my brother from being, once again, the top student in his class as well as the founder of a school newspaper, *Sentite anche noi*, (*Listen to us*). Perhaps in order to combat the general despondency, Gianco filled the paper with humorous articles and jokes.

Father was worried. Looting and robbery were rampant. The Marshall plan had not yet been conceived and he feared that Italy would become an easy prey to the

Soviet Union. He shuddered at the idea that after having shaken the yoke of one dictatorship, his country might fall under another, equally ruthless, tyranny.

His brother, who had emigrated to South America many years before, wrote to him describing a bright new world where his competence as an engineer and architect would be greatly appreciated. Much to my, and to Gianco's, chagrin (for, in spite of the terrible living conditions, we had many friends) father decided to cross the ocean and join his brother in Ecuador.

One cold morning in December 1946, a little over a year after the end of the war, we boarded a Danish cargo boat, the *Benny Skou*, and sailed towards South America. The winter sea was rough and we had to brave storms, a tempestuous sea and devastating sea-sickness. While bravely walking on deck during a thunderstorm, Gianco slipped and sprained his elbow. The boat's captain, who was also the boat's doctor, put his arm in a sling. My brother was rather proud of his sling, which he paraded like a trophy. To while away the hours, we resumed our games of poker, using fake bills that, at our insistence, father had drawn for us. Unfortunately, instead of adding to our fun, this counterfeit money became the source of constant conflicts. Gianco and I became greedy and hoarded the bills like a miser hoarding gold. We stole from each other; we quarrelled. Had Gianco's arm not been in a sling, I think we would have resorted to fisticuffs. One day father collected those bones of contention and threw them into the sea. "I will not tolerate greed", he commened bitterly, "even in games".

The crossing to South America took almost twenty days. We arrived at last in Montevideo, the capital of Uruguay. From there a ferry-boat took us to Buenos Aires, where we roamed the city, laughing at the winter scenes which filled the shop windows, in ludicrous contrast with the sizzling heat of the Argentinian Summer. It was in Buenos Aires that Gianco and I had our first taste of a strange drink imported from the United States, called "Coca-Cola".

To reach Ecuador, we took a number of planes, with stops in Chile, Peru and Guayaquil, on the Pacific Coast. The propeller planes of the forties were not very stable. They rocked and shook and plunged and were so badly pressurized that when we flew near the Aconcagua, the highest mountain in the Andes, we had to breathe through oxygen masks.

Settling in Quito was far from easy. Both Gianco and I missed our school, our Italian friends. Father enrolled us in an Anglo-Spanish school called *Colegio Americano*. While I trundled along, trying to master Spanish and English, Gianco learned quickly and spoke both languages fluently in a matter of months. He made friends easily and became very popular, particularly among girls. When the School elected a Carnival Queen, my brother was chosen among the *Caballeros de Honor*, the Gentlemen of Honor, who were to accompany the queen to her throne. I remember our proud father taking a picture of his son wearing a tuxedo and a bow tie.

My brother also became editor of the school newspaper, *School Views*, and participated in a number of debates, including one on psychology and psychoanalysis, which was covered by a local newspaper. Young Rota, wrote the newspaper, "*displayed an amazing erudition, quite uncommon among 18 year old youths*". His social success went hand in hand with his academic triumph. He excelled both in the humanities and in science. I was both proud and jealous. He managed to outdo me even in the field of languages, the only field in which I had some success. At the end of the year, the School gave a prize to the student who had shown the highest proficiency in English. During Gianco's Senior year, my brother and I were neck and neck for the prize. Naturally, Gianco won. Only the following year, when he left for the States, did I receive the coveted prize.

Gianco's math teacher, *Señor* Acosta, who also taught my class, was one of my brother's most enthusiastic admirers. He told our parents he had never had such a brilliant student and encouraged Gianco to follow a career in mathematics. He was less enthusiastic about my mathematical talent: my grades were so low that I anticipated complete failure. A fierce defender of the family's intellectual honour, Gianco couldn't allow his sister to fail math, or any other subject for that matter. He devised a clever ruse to help me out. *Señor* Acosta had a very original grading system. He would assign a math exercise and allot a certain span of time to finish it. If successfully completed within the allotted time, the exercise would receive a number of points, which varied according to the difficulty of the exercise. When, after a brief countdown, our teacher cried: " *Terminado !*", the students lined up in front of his desk and handed in their work. At the bottom of each exercise, our teacher wrote the number of points, followed by his signature and the student's last name. The student who, at the end of the trimester, had "collected" the highest number of

Dec. 1946
On the "Benny Skou"

points received the top grade, 20/20. My brother always *outpointed* everybody else in his class, much to the dismay of his fellow students, who were far behind him. But Gianco had always had a generous disposition. He decided to withhold some of his points, so as not to hurt his classmates. Actually, he had another purpose in mind: he planned to save me from disaster by giving me his extra points. I was touched by this display of brotherly love, yet I hesistated, not out of moral scruples, I confess, but because I feared that our teacher would uncover the trick and punish us both. But my clever brother had foreseen all possible hitches. He pointed out that our teacher only wrote the student's last name on the exercise, never his first name. "Your last name is Rota and so is mine, he concluded. *Señor* Acosta will never know the difference". I needed those points badly: I accepted. At the end of the trimester, thanks to Gianco's points, I managed to bring home an honorable grade in math. But we had underestimated our teacher's shrewdness: the following trimester *Señor* Acosta replaced *Rota* with *Ester* and *Giancarlo* on our respective exercises. Had he uncovered our trick ? Gianco didn't think so. He claimed that, had that been the case, he would have meted out a harsh punishment to both of us. I disagreed. I am convinced to this day that *Señor* Acosta was not willing to sacrifice one of his most brilliant students on the altar of inflexible ethics.

In spite of our long school days, father wanted us to continue classical studies under his direction. While I painfully coped with Latin, Gianco was at ease with both Latin and Greek and learned French on the side. He was soon reading Balzac and

Proust, who became one of his favorite authors. A private tutor gave him German lessons and was astonished at the rapidity with which he picked up this difficult language. I myself did not cease to be amazed at the nimbleness of his learning process. When preparing for our finals, I would spend hours reading and memorizing with laborious concentration while my brother would quickly skim through his books, nodding every so often and mumbling: "Yes", "Right", "O.K.", "Are you sure you are ready for the exams ?" I asked him, both worried by, and envious of, his nonchalance. "Of course", he answered. "The important thing when preparing for an exam is understanding, not memorizing". Naturally, he passed all his finals with flying colours and was admitted to Princeton University on a scholarship. When he left for the United States, I felt like a ship lost at sea, especially when, as a Senior, I had to prepare for my orals, which were held in the School's library in front of an audience. It was a frightening performance that my parents and friends of the family were planning to attend.

I was particularly worried about my Math final. I was sure that the Math examiner, who had been dispatched by the Ministry of Education, would immediately detect my mathematical weaknesses. If I failed the exam, I would disgrace my family. Luckily, Gianco, returned from the States a few days before the ordeal and came to my rescue. "Don't worry, he said calmly, I will teach you a new theorem that will dumbfound everybody, including the Math examiner".

Under Gianco's guidance I not only memorized the theorem, but managed to understand it... well, almost. My brother seemed satisfied. "Whatever they ask you at the exam", he told me a few hours before the performance, "produce the theorem. You will get twenty, I promise".

I followed Gianco's instructions to the letter. My demonstration of the theorem was met by a few moments of astonished silence followed by an enthusiastic applause. The Examination Committee gave me 20/20, the only 20 I had ever had in math. "You saved my life", I said later to my brother, squeezing his hand. His lips curved in a mischievous smile. "You did all right, except for one detail: you bungled the conclusion. Luckily, nobody noticed".

A few months later, we both left for the States.

But that is another story.

# References

Gasperoni Rota, E. (1995), *Orage sur le lac*, Paris, l'École des loisirs.
Gasperoni Rota, E. (2006), *L'arbre de Capuliès*, Arles, Actes Sud junior.

UN CERTAMEN DE PSICOLOGIA ESTUDIANDO TODAS LAS
TENDENCIAS DE ELLA OFRECIO EL COLEGIO AMERICANO

Hoy tendrá lugar el certamen de taquigrafía y mecanografía

De izquierda a derecha aparecen los alumnos Glauco Rota, Fabián Ponce, Lenín Guerra y Enrique Merizac, del Colegio Americano de Quito, quienes participaron en el Certamen de Psicología, realizado como número del Programa Conmemorativo del X Aniversario del Colegio que se celebrará el 22 de Junio.— (Foto Pacheco)

# Chapter 2
# Remembering Gian-Carlo Rota
## Invited Chapter

Joseph J. Kohn

I have known Gian-Carlo Rota for about half a century. I saw him very frequently during the periods that we overlapped in the Boston area (i.e. 1958-68 and 1996-97) and also on many other occasions. I remember Gianco as one of the most stimulating, insightful and witty persons. Here I will recall some of the indelible impressions he left on me.

Both Gian-Carlo and I went to the American School in Quito, Ecuador. We were in the same class, however we did not know each other then, I left for the US in 1945 and he came to Quito in 1946. Nevertheless I knew of him through my friends and relatives, he was an outstanding student and very popular. One of our common friends and classmates from those days was Alberto Muggia. He recently wrote me about Gianco: *"I think of him often. He is the person that really made me what I am. He taught me that what we learned at school was a negligible part of our educational experience. He taught me to separate my free time by the clock, devoting an hour to French, an hour to Latin, an hour to history etc. After school, a few of us went to his house, at least four times a week, and he showed us some of his books, and told us what we learn from them. He taught us to play many games such as chess and tarocchi. He taught me to read Flaubert and Balzac, stimulated me to learn French on my own, and helped me to translate Moliere's L'Avare from French to Spanish at the age of 13. It was more fun to go to his house, to study after school rather than doing anything else. He even tried to get me interested in Mathematics and Physics, but I was not a good pupil. I preferred the poets and writers of the 19th century, though I remember learning all about Galileo"*. Thus Gian-Carlo (or Gianco as he was frequently called) was a teacher and a scholar already in high school.

As an undergraduate Gianco went to Princeton in 1950 at the same time that I went to MIT. Gianco did his graduate work in Yale and I in Princeton. We finally met in Cambridge in 1958. At the time Gianco was working on his joint book with Birkhoff (see [1]). He worked in a closet which was just about large enough to fit a small desk and chair. The desk was covered with papers and there was a pile of books on the floor. Gianco said that working in such a crammed position forced him

E. Damiani et al. (eds.), *From Combinatorics to Philosophy,*
DOI 10.1007/978-0-387-88753-1_2, © Springer Science+Business Media, LLC 2009

to concentrate and kept him away from distractions. Many years later I visited Gianco in Los Alamos. We took a small trip through the neighboring hills. Gianco was greatly inspired by the spectacular scenery. The panorama, which changed with each turn, stretched out for miles in all directions. The contrast between these two physical surroundings parallels Gianco's strengths. On the one hand his ability to zero in on the critical technical details that underlie any discipline and on the other his ability to see the "big picture" and to deal with the ideas that drive any intellectual endeavor.

For many years Gianco was editor of the Advances in Mathematics. He had a very clear view of how a journal should be run and followed it. In that way he reminded me of Lefschetz when he was editor of the Annals of Mathematics. As with Lefschetz, his top priority was excellence. Thus many authors who submitted important articles were left waiting because outstanding took precedence over important and Gianco had very strong convictions about what was truly outstanding. A case in point is the a truly remarkable paper written by Charles Fefferman (see [2]). When I heard Fefferman's lectures on these results I was very impressed, I told Gianco about them, he asked for a preprint and after receiving it he called me. He was very excited, he declared that this paper was a "tour-de-force" and that, despite its length, he would like to publish it in the Advances as soon as possible. He did publish it promptly, even though it delayed the publication of a long line of previously accepted shorter papers.

A groundbreaking feature of the Advances were the short reviews written by Gianco. These were very succinct snap judgements which were almost always on target. Here are two examples:

"J. Passmore, *Recent Philosophers*, Open Court London 1985. When pygmies cast such long shadows, it must be very late in the day".

This review illustrates two recurring themes that were part of Gianco's "Weltanschaung". One is the disproportionate influence that mediocre thinkers have. When decrying the "fashions" in mathematics his usual example of a "pygmy" is a "Professor Neanderthal". This refers to a fussy pedant who refuses to see the woods for the trees. The other theme (the lateness of the day) is Gianco's repeated insistence that we (especially mathematicians), in order to prevail over the "pygmies", must change our ways before it is too late. He was especially concerned with the neglect of good exposition. He was also preoccupied by the fact that mathematicians are unable, or unwilling, to unite to advance the influence of their profession. I was amazed how masterfully he could fight for this cause when he led the Mathematics Section of the United States National Academy of Science. The main problem for mathematicians in the Academy is their underrepresentation as compared with the physicist and chemists. Gianco mastered the Byzantine politics involved with great skill and was successful in enhancing our section.

The following review illustrates Gianco's unbounded enthusiasm when he encounters something that he really likes.

"J. H. Conway and N. J. A. Sloane, *Sphere Packings, Lattices and Groups*, Springer, New York, 1988. This is the best survey of the best work in one of the best fields of combinatorics, written by the best people. It will make the best reading by the best students interested in the best mathematics that is now going on".

Gianco had very definite views about what was best and what was worst in mathematics and he expressed his views very bluntly and eloquently, especially in informal discussions as well as in his book *IndiscreteThoughts* (see [3]).

Most of my meetings with Gianco were punctuated by a meal. In the sixties we met, usually on Saturdays, to explore the bookstores on Harvard Square and invariably ended with a late lunch. There were also innumerable Sunday brunches and dinners after colloquia. Gianco was a gourmet. In the early days he would prepare, with skill and precision, eggs Benedict for brunch using his own special recipe for hollandaise sauce. Later, when waistlines and cholesterol intake became issues, his favorite brunch consisted of a dozen oysters (usually in the Charles Hotel in Cambridge). He maintained that the zinc in the oysters was good for the brain. During the academic year 1996-97 I spent a sabbatical year at Harvard and Sunday brunch with Gianco became a ritual. However, one Sunday Gianco announced that for the next month he will not be able participate in these brunches. The reason was that he had won the MIT yearly presidential teaching award and he had to give a one hour lecture to a general audience in one month. He felt that he had to use all the time available to prepare for that lecture. I went to the lecture and I concur with Peter Lax when he said: "A lecture by Rota is like a double martini". It was packed with surprising insights, bold statements and humor, it was both informative and provocative. The preparation was unbelievable every move was accurately timed. This included his signature sip of Coca Cola which he performed with elegant deliberation, walking across the stage to pour himself a glass which gave the audience time to reflect on his last pronouncement.

Gianco was a remarkable teacher Not only in his incomparable lectures but in the rapport with students at all levels. His teaching credo is summed up in the following statement that he wrote in [3]. *"A good teacher does not teach facts, he or she teaches enthusiasm. Young people need encouragement. Left to themselves, they may not know how to decide what is worthwhile. They may drop an original idea because they think someone else must have thought of it already. Students need to be taught to believe in themselves and not to give up"*. I have met many of Gianco's students, he always left a lasting impression. My wife's first cousin, Riccardo DiCapua, was a student of Gianco's undergraduate course 18.025 "The World of Mathematics" in 1969. Riccardo was profoundly influenced by Gianco, he writes: *"...He would enter the lecture hall in 10-250 impeccably attired in a black suit, jacket buttoned, a crisp white shirt, and a bright, red tie. The blackboard had to be wiped immaculately clean of any stray chalk marks left over from the previous class. The students waited breathless, in eager anticipation of a clean slate every class. Every non-lecture task was assigned to a student. One student had been given a certain amount of cash the first day of classes, and assigned the duty of buying and bringing in a can of Coca Cola from the vending machine by Room 26-100 (the only vending machine on campus that is set to the exact temperature). It had to be placed before class started on the front, right corner of the instructors table. Another was charged with collecting homework at the beginning of class, while a third was responsible for handing back previous homework assignments and tests to the students. A fourth would take attendance ("quietly, please"). A fifth student had the most onerous responsibility: taking notes. Prof. Rota was insistent that his students pay attention to what he was saying: note-taking would distract their attention, he insisted. The designated note-taker would take copious notes, which the department secretary would photocopy after every class. Of course, there was a sixth student charged with handing out said notes the following day! The years passed. I worked for a few years and returned to Harvard University for an MBA. I went into management of diversified manufacturing companies. The years turned into decades, and thirty-five years later, I decided to "give back to the community" by teaching one or two courses every term at Broward College in Fort Lauderdale, FL, USA.... And maybe one of my students will carry that tiny spark of joyful inspiration many years hence to yet a third generation of practitioners and students of mathematics. Grazie, Gian-Carlo!"*

Years ago I dined with Gianco and a number of colleagues at an upscale San Francisco restaurant. Gianco asked for the wine list and was very pleased to find one of his favorite Barolos which he declared worth the elevated price. The waiter brought the wine in an uncorked bottle, Gianco was furious: "How is this possible? If the bottle is already opened I can't know that the wine is genuine. Take it back!" The waiter called the restaurant owner. All the people in the restaurant stopped eating and talking to watch the spectacle. The scene was theatrical: both Gianco and the restaurant owner were outraged and vented their emotions in colorful Italian-accented language which resonated in the hushed establishment. At last the restauranteur challenged Gianco to tell the difference between wine from a bottle to be opened at the table and the original opened bottle. He brought out a new bottle opened it and poured out two glasses, one from each bottle, behind Gianco's back. In the last scene Gianco tasted the wine from both glasses with a professional demeanor and declared that there was no difference between them. The restauranteur smiled but his smile soon disappeared as Gianco bluntly explained to him in great detail why serving an opened bottle is inappropriate and entirely unacceptable.

At MIT Gianco was always surrounded by undergraduate students, graduate students, postdocs, and younger colleagues. This crowd became known as the "Rota rooters". He was their leader and role model who prodded them into thinking about mathematics and philosophy. He had great influence on them both in intellectual and gastronomic endeavors. Thus I was very surprised when (sometime in the 1960's) Gianco accepted a position at Rockefeller University. Rockefeller Institute had been dedicated to biomedical research since 1901. In 1965 it expanded its mission to education and became Rockefeller University, one the first appointments was Marc Kac who was charged with attracting a topnotch mathematics faculty. Kac convinced Rota to accept a position at Rockefeller. As a new mathematics center the department at Rockeffeller was a relatively quiet place, it did not have the hustle and bustle of MIT. It was too solitary a place for Gianco, he became depressed, after two years he returned to MIT. In this period I visited New York frequently, on these occasions I often dined with Gianco. We explored Latin American cuisine which we both enjoyed, it reminded us of our childhood in Ecuador. This cuisine is very spicy and rich. After the meal we would take Alka-Seltzer to relieve of us from the discomfort caused by this exotic repast. Invariably we got into a discussion. Gianco maintained that the most efficient way to take Alka-Seltzer is to brake it into little pieces before dropping it into a glass of water. I maintained that it is more effective to drop it in whole. Gianco argued that the braking into small pieces increases the surface area in contact with the water and thus the speed of dissolution. My argument was that it is not desirable for the tablet to dissolve very quickly because that tends to create large bubbles which evaporate outside the liquid; whereas, the slower speed achieved by submerging the tablet intact results in smaller bubbles which do not escape the liquid and are ingested.

These discussions often led to larger scientific and philosophical issues. On such occasions we would catch up on some mathematical topics with which one of us was familiar and the other not. We also recounted "mathematical gossip". This ranged from rumors about exciting new developments in mathematics to amusing incidents involving some of our colorful colleagues. One point that we frequently debated was Gianco's view that a biographical sketch should include descriptions of the subject including the "warts". I did not totally agree with this view, especially when it involved my teachers and colleagues in Princeton. My conversations with Gianco were always lively and a lot of fun. He always had thought provoking views which he presented with great eloquence and often with a humorous twist. I remember Gian-Carlo with great admiration for his many talents, his enthusiasm, his self confidence, his energy, and his relentless pursuit of those intellectual endeavors which he considered of primary importance.

# References

1. Birkhoff, G., Rota, G.-C. (1959), *Ordinary Differential Equations,* New York, Ginn and Co.
2. Fefferman, C. (1979), *Parabolic invariant theory in complex analysis,* in "Adv. Math.", 31, pp. 131-262 .
3. Rota, G.-C. (1997), *Indiscrete Thoughts*, F. Palombi (ed), Boston, Basel, Berlin, Birkhäuser.

# Part II
# Gian-Carlo Rota, the Mathematician

# Chapter 3
# A Glimpse of Vector Invariant Theory. The Points of View of Weyl, Rota, De Concini and Procesi, and Grosshans
## Invited Chapter

Andrea Brini

## 3.1 Introduction

We will try to provide a brief and elementary description of some of the founding ideas of the characteristic-zero theory as well as of the characteristic-free theory of vector invariants for the classical groups $GL(d)$, $SL(d)$, $Sp_{2m}$, $O(d)$.

The present paper is almost a *verbatim* written version of the lecture delivered in Milan, on the occasion of the Gian-Carlo Rota Memorial Conference. The lecture was devoted to an audience of Mathematicians, but not to specialists of invariant theory or representation theory: therefore, our discussion is sketchy, results and methods are not developed in their full generality, and we limited our exposition to a very few aspects of the present-day sophisticated and branched theory. As a matter of fact, our exposition is not to be meant as an *introduction* to vector invariant theory, but just as a *reading invitation* towards the original works.

The characteristic zero approach to vector invariants is traditionally based on the so-called Weyl's Theorem (or *reduction principle*) (see, e.g [36], Theorem(2.5.A), p.44 ) which essentially says that a system of generators of the algebra of (absolute) vector invariants for any subgroup $G$ of $GL(d)$ can be constructed by *polarizing* a system of generators of the invariants in the first $d$ vector variables (or even, in a more refined form, a system of generators of the invariants in the first in $d-1$ vector variables, in case by adding the determinant of the first $d$ variables). Weyl's Theorem can be regarded as an immediate corollary of a beautiful combinatorial identity, the *Gordan-Capelli-Deruyts polar expansion formula* (see, e.g. Capelli [5], Capitolo I-XXII, for its original formulation, [29] for the version reported here, and [4] for a superalgebraic and *canonical* form). We briefly discuss the way to derive from Weyl's Theorem the basic invariant theory for the groups $SL(d)$, $Sp_{2m}$, $SO(d)$, $O(d)$.

Weyl's Theorem is not valid in positive characteristic (see, e.g. [12]).

Andrea Brini
Università degli Studi di Bologna, Italy

E. Damiani et al. (eds.), *From Combinatorics to Philosophy,*
DOI 10.1007/978-0-387-88753-1_3, © Springer Science+Business Media, LLC 2009

Also the Gordan-Capelli-Deruyts polar expansion formula fails to be true in arbitrary characteristic, since it involves *rational* coefficients.

In their fundamental 1976's paper, De Concini and Procesi wrote ([6], p.331): *The classical proofs of invariant theoretic results follow essentially two equivalent paths. Polarization and the Gordan-Capelli expansion (via the theorem of E. Pascal) or double centralizer theorem, owing to the linear reductivity of the group in consideration. There is, on the other hand, another line of approach based on the standard Young tableaux which should be traced at least to Hodge (1943), to Igusa (1954) (who proves the first fundamental theorem of vector invariants in a characteristic free approach) and finally to Doubilet, Rota and Stein (1974), which gives the main technical tool: the straightening formula.*

The straightening formula for bideterminants of Doubilet, Rota and Stein [12] is to be regarded as the characteristic-free version of the Gordan-Capelli-Deruyts polar expansion formula. This result allowed Rota and his collaborators to provide characteristic-free combinatorial proofs of the First and Second Fundamental Theorems for relative vector invariants of $GL(d)$, or, equivalently, absolute invariants of $SL(d)$ [12, 10].

A quite relevant step forward was made by De Concini and Procesi in 1976 [6]. They proved two variants of Rota's Straightening Formula which hold for Pfaffians and Gramians. Furthermore, they realized (in the extremely general language of *schemes* and *formal invariants*), that a common strategy in the proofs of the First Fundamental Theorems for vector invariants of classical groups could be that of proving versions of these results on the localized rings (with respect to suitable Zariski open sets) of naturally chosen affine varieties, and then to get the global result by using the *cancellation laws*, which follow, in turn, from the Straightening Formulae. At the end of page 331 of [6], they wrote: *This is the main contribution of the paper*.

In the second part of this paper, we sketchy describe some aspects of the characteristic-free approach to the invariant theory of the classical groups $GL(d)$, $Sp_{2m}$, $O(d)$, following along the lines of the pioneering work of Rota and De Concini and Procesi. The structure of this part of the exposition is the following:

- The starting point is the straightening formula for bideterminants (Doubilet, Rota, Stein, 1974, [12]) and its analogs for Pfaffians and Gramians (De Concini, Procesi 1976, [6]).
- From the straightening formulae one derives versions of the so-called E. Pascal theorems (or, as we shall see, *Second Fundamental Theorems*) for scalar products, symplectic products and inner products. These results lead to the main combinatorial tool: the *cancellation laws*.
- The first Fundamental Theorem for $G = GL(d), Sp_{2m}, O(d)$ is proved into three further steps:

  1. Consider the coordinate ring **S** of a suitable affine variety on which the group $G$ acts, and a subalgebra **B** of the invariant algebra $\mathbf{S}^G$ which we want to show equals $\mathbf{S}^G$.

2. (De Concini-Procesi, 1976) Consider the localization with respect to a suitable Zariski open set and prove the equality between the localized rings

$$S[1/\Delta]^G = B[1/\Delta],$$

$\Delta \in \mathbf{B}$.

3. Let $\psi \in \mathbf{S}^G \subseteq \mathbf{S}[1/\Delta]^G = \mathbf{B}[1/\Delta]$. Then, there exists an integer $k$ such that $\Delta^k \psi \in \mathbf{B}$; the *cancellation laws* imply that $\psi \in \mathbf{B}$.

Therefore, the First Fundamental Theorems follows:

$$\mathbf{S}^G = \mathbf{B}.$$

In 2007, Grosshans [24] proved a quite general result which provides a bridge between Weyl's approach and the approaches of Rota, De Concini and Procesi.

Grosshans's main result is a far-reaching generalization of Weyl's Theorem. Informally speaking, Grosshans's main result says that, given an algebraically closed field $\mathbb{K}$ of arbitrary characteristic, for *any* subgroup $H$ of $GL(d, \mathbb{K})$, the $H-$vector invariants in $n$ variables ($n > d$) can still be obtained from the $H-$vector invariants in $d$ variables; more precisely, the algebra of $H-$vector invariants in $n$ variables is the $p-root\ closure$ of the algebra of the "polarized" $H-$invariants in $d$ variables.

I thank Frank D. Grosshans and Francesco Regonati for their advice, encouragement, and invaluable suggestions.

## 3.2 The Characteristic Zero Approach

### 3.2.1 Preliminaries

Let $\mathbb{K}$ be a field of characteristic zero.

Let $L = \{x_1, x_2, \ldots, x_n\}$ and $P = \{1, 2, 3, \ldots, d\}$ be two alphabets.

The letterplace algebra on the pair $(L, P)$ is the associative and commutative $\mathbb{K}$-algebra

$$\mathbb{K}[(x_i|j)]_{i=1,\ldots,n; j=1,\ldots,d}$$

generated by the variables $(x_i|j)$ and, therefore, if $char(\mathbb{K}) = 0$, it can be regarded as the algebra $\mathbb{K}[V^{\oplus n}]$ of *polynomial functions over the direct sum* of $n$ copies of a vector space $V$, $dim(V) = d$. Actually, given a basis of the dual space $V^*$, we can read the letterplace variable $(x_i|j)$ as the *j-th coordinate function over the i-th copy of $V$* in the direct sum $V^{\oplus n}$, for every $i = 1, 2, \ldots, n$, $j = 1, 2, \ldots, d$.

The *contragradient action* of the general linear group $GL(d)$ on

$$\mathbb{K}[(x_i|j)] \cong \mathbb{K}[V^{\oplus n}]$$

is given in the obvious way.

The (letter) polarization operator $D_{x_h,x_k}$ is the linear operator from $\mathbb{K}[(x_i|j)]$ to itself defined as follows:

$$D_{x_h,x_k} = \sum_{j=1}^{d} (x_h|j) \frac{\partial}{\partial(x_k|j)}.$$

The operator $D_{x_h,x_k}$ is the unique derivation of the algebra $\mathbb{K}[(x_i|j)]$ such that

$$D_{x_h,x_k}((x_i|j)) = \delta_{k,i}(x_h|j), \qquad j = 1,\ldots,d.$$

More in general, a *(letter) polarization process* is an operator that can be written as a polynomial in the $D_{x_h,x_k}$, $h,k = 1,\ldots,n$.

The crucial fact is that polarization processes commute with the action of the general linear group $GL(d)$, or, equivalently, they are $GL(d)$-equivariant linear endomorphism of $\mathbb{K}[(x_i|j)]$. In the classical terminology, they are also said to be *invariantive processes*.

**Definition 3.1.** Given a subgroup $G \subset GL(d)$, an element $f \in \mathbb{K}[(x_i|j)]$ is called an absolute $G$-invariant if

$$\varrho \cdot f = f,$$

for every $\varrho \in G$.

Clearly, polarization processes map invariants to invariants.

A basic class of examples is provided by the *brackets*

$$[x_{i_1} \ldots x_{i_d}] = det\left[(x_{i_h}|j)\right]_{h,j=1,\ldots,d}.$$

Clearly, brackets are absolute $SL(d)$-invariants.

Another crucial fact is that the action of a polarization process on a product of brackets produces a polynomial in the brackets.

### 3.2.2 Weyl's Theorem (The Reduction Principle)

**Theorem 3.1.** *(Gordan-Capelli-Deruyts Polar Expansion Formula - weak form)*
*Let $f \in \mathbb{K}[(x_i|j)]$. Then $f$ can be written in the form*

$$f = \sum_{q,s} \mathcal{P}_{q,s} \mathcal{F}_{q,s}, \qquad where$$

1. *$\mathcal{P}_{q,s}$ is a polarization process, for every $q$ and $s$.*
2. *$\mathcal{F}_{q,s} = [x_1 \ldots x_d]^q \varphi_s(x_1,\ldots,x_{d-1})$, where $\varphi_s(x_1,\ldots,x_{d-1})$ is a polynomial in the variables $(x_i|j)$, $i = 1,\ldots,d-1$, $j = 1,\ldots,d$. In the classical language, one says that $\varphi_s$ is a polynomial function in the first $d-1$ vector variables.*
3. *$\mathcal{F}_{q,s} = \mathcal{P}'_{q,s}(f)$, where $\mathcal{P}'_{q,s}$ is a polarization process.*

In plain words, the invariant theoretical meaning of the preceding result is the following.

Let $f$ be an absolute $G$-invariant; $f$ can be written as a linear combination of the "polarized" polynomials $\mathcal{P}_{q,s} \mathcal{F}_{q,s}$. The crucial point 3) is that the $\mathcal{F}_{q,s}$'s can be obtained, in turn, by polarizing the original polynomial $f$, and then, since polarization processes are invariantive processes, they *must be* absolute $G$-invariants.

**Corollary 3.1.** *(The reduction principle of classical invariant theory - Weyl's Theorem)*

*Let $G$ be a subgroup of $GL(d)$, and let $f = f(x_1, x_2, \ldots, x_n)$ be a $G$-invariant polynomial function in $n$ vector variables, $n \geq d$. Then*

$$f = \sum_{q,s} \mathcal{P}_{q,s}([x_1 x_2 \ldots x_d]^q \varphi_s(x_1, x_2, \ldots, x_{d-1})),$$

*where the $[x_1 x_2 \ldots x_d]^q \varphi_s(x_1, x_2, \ldots, x_{d-1})$ are $G$-invariants, and the $\varphi_s$'s are polynomial functions involving only the first $d-1$ vector variables $x_1, x_2, \ldots, x_{d-1}$.*

### 3.2.3 Absolute Invariants for the Special Linear Group $SL(d)$

First of all, we have to determine the polynomials $[x_1 x_2 \ldots x_d]^q \varphi_s(x_1, x_2, \ldots, x_{d-1})$ that are absolute $SL(d)$-invariants. Since the bracket $[x_1 x_2 \ldots x_d]$ is clearly an absolute $SL(d)$-invariant, it follows that $\varphi_s(x_1, x_2, \ldots, x_{d-1})$ must be an absolute $SL(d)$-invariant. It is not difficult to prove that $\varphi_s(x_1, x_2, \ldots, x_{d-1})$ is forced to be a *constant*.

Then, any absolute $SL(d)$-invariant is of the form

$$f = \sum_{q,s} \mathcal{P}_{q,s}([x_1 x_2 \ldots x_d]^q);$$

since $\mathcal{P}_{q,s}([x_1 x_2 \ldots x_d]^q)$ is a polynomial in the brackets

$$[x_{i_1}, \ldots, x_{i_d}], \quad \{i_1, \ldots, i_d\} \subseteq \{1, 2, \ldots, n\},$$

the subalgebra of absolute $SL(d)$-invariants $\mathbb{K}[(x_i|j)]^{SL(d)} \cong \mathbb{K}[V^{\oplus n}]^{SL(d)}$ is precisely the subalgebra generated by the brackets.

### 3.2.4 Absolute Invariants for the Symplectic Group $Sp_{2m}$

Let $d$ be even, say $d = 2m$. Consider the *symplectic products* (in standard form)

$$[x_h, x_k] = \sum_{j=1}^{m} ((x_h|j)(x_k|m+j) - (x_h|j+m)(x_k|j)) \in \mathbb{K}[(x_i|j)].$$

The symplectic group can be defined as the subgroup $Sp_{2m} \subset GL(2m)$ of all the elements that leave the symplectic products fixed. In other words, $Sp_{2m}$ is the largest subgroup of $GL(2m)$ for which the symplectic products $[x_h, x_k]$ are absolute invariants. An element of $Sp_{2m}$ is called a symplectic transformation; we recall that the determinant of any symplectic transformation equals 1.

Again, we have to determine the polynomials $[x_1 x_2 \ldots x_{2m}]^q \varphi_s(x_1, x_2, \ldots, x_{2m-1})$ that are absolute $Sp_{2m}$-invariants. Since $Sp_{2m} \subset SL(2m)$, the bracket $[x_1 x_2 \ldots x_{2m}]$ is clearly an absolute $Sp_{2m}$-invariant, it follows that $\varphi_s(x_1, x_2, \ldots, x_{2m-1})$ must be an absolute $Sp_{2m}$-invariant.

One then proves that $\varphi_s(x_1, x_2, \ldots, x_{d-1})$ is a polynomial in the symplectic products $[x_h, x_k]$, with $h, k = 1, \ldots, 2m-1$, $h \neq k$.

Furthermore, from a well-known determinantal identity we infer:

$$[x_1 x_2 \ldots x_{2m}] = \pm Pf\,[[x_h, x_k]]_{h,k=1,\ldots,d=2m}$$

where $Pf\,[[x_h, x_k]]_{h,k=1,\ldots,d=2m}$ denotes the *Pfaffian* of the skew-symmetric matrix

$$[[x_h, x_k]]_{h,k=1,\ldots,d=2m},$$

namely, the polynomial

$$Pf\,[[x_h, x_k]]_{h,k=1,\ldots,d=2m} = \sum sgn(\sigma)\,[x_{\sigma(1)}, x_{\sigma(2)}] \cdots [x_{\sigma(2m-1)}, x_{\sigma(2m)}] \in \mathbb{K}[(x_i|j)],$$

where the sum is over all permutations of $\{1, \ldots, 2m\}$ such that

$$\sigma(1) < \sigma(2), \ldots, \sigma(2m-1) < \sigma(2m), \text{ and } \sigma(1) < \sigma(3) < \ldots < \sigma(2m-1).$$

Hence, any absolute $Sp_{2m}$-invariant

$$[x_1 x_2 \ldots x_{2m}]^q \varphi_s(x_1, x_2, \ldots, x_{2m-1})$$

turns out to be a polynomial in the symplectic products $[x_h, x_k]$, with $h, k = 1, \ldots, 2m = d$.

Since the action of a polarization process on a polynomial in the symplectic products $[x_h, x_k]$ produces a polynomial in the symplectic products $[x_h, x_k]$, it follows that the subalgebra of absolute $Sp_{2m}$-invariants

$$\mathbb{K}[(x_i|j)]^{Sp_{2m}} \cong \mathbb{K}[V^{\oplus n}]^{Sp_{2m}}$$

is precisely the subalgebra generated by the symplectic products $[x_h, x_k]$, with $h, k = 1, \ldots, n$.

### 3.2.5 Absolute Invariants for the Special Orthogonal Group $SO(d)$

Consider the *inner products* (in standard form)

$$< x_h, x_k > = \sum_{j=1}^{d} (x_h|j)(x_k|j) \in \mathbb{K}[(x_i|j)].$$

The special orthogonal group can be defined as the subgroup $SO(d) \subset SL(d)$ of all the elements that leave the inner products fixed.

In other words, $SO(d)$ is the largest subgroup of $SL(d)$ for which the inner products $< x_h, x_k >$ are absolute invariants.

An element of $SO(d)$ is called a proper orthogonal transformation.

Again, we have to determine the polynomials $[x_1 x_2 \ldots x_d]^q \varphi_s(x_1, x_2, \ldots, x_{d-1})$ that are absolute $SO(d)$−invariants. Since $SO(d) \subset SL(d)$, the bracket $[x_1 x_2 \ldots x_d]$ is clearly an absolute $SO(d)$−invariant; it follows that $\varphi_s(x_1, x_2, \ldots, x_{d-1})$ must be an absolute $SO(d)$−invariant.

One then proves that $\varphi_s(x_1, x_2, \ldots, x_{d-1})$ is a polynomial in the inner products

$$< x_h, x_k >,$$

with $h, k = 1, \ldots, d-1$.

Clearly,

$$D_{x_p, x_q}(< x_h, x_k >) = \delta_{qh} < x_p, x_k > + \delta_{qk} < x_h, x_p >.$$

Since the polarization operators $D_{x_h, x_k}$ are *derivations* of the algebra $\mathbb{K}[(x_i|j)]$, it follows that the subalgebra of absolute $SO(d)$-invariants

$$\mathbb{K}[(x_i|j)]^{SO(d)} \cong \mathbb{K}[V^{\oplus n}]^{SO(d)}$$

is precisely the subalgebra generated by the brackets $[x_{i_1}, \ldots, x_{i_d}]$ and the inner products $< x_h, x_k >$, with $h, k = 1, \ldots, n$.

### 3.2.6 Absolute Invariants for the Orthogonal Group $O(d)$

The orthogonal group can be defined as the subgroup $O(d) \subset GL(d)$ of all the elements that leave the inner products fixed.

In other words, $O(d)$ is the largest subgroup of $GL(d)$ for which the inner products $< x_h, x_k >$ are absolute invariants.

An element of $O(d)$ is called a proper transformation if its determinant equals 1 and an improper transformation if its determinant equals $-1$. The polynomials $[x_1 x_2 \ldots x_d]^q \varphi_s(x_1, x_2, \ldots, x_{d-1})$ that are absolute $O(d)$−invariants are, a fortiori, $SO(d)$−invariants.

Therefore, $\varphi_s(x_1, x_2, \ldots, x_{d-1})$ is a polynomial in the inner products

$$< x_h, x_k >,$$

with $h, k = 1, \ldots, d - 1$.

Since $\varphi_s(x_1, x_2, \ldots, x_{d-1})$ is also an absolute $O(d)$-invariant, it immediately follows that $[x_1 x_2 \ldots x_d]^q \varphi_s(x_1, x_2, \ldots, x_{d-1})$ is an absolute $O(d)$-invariant if and only if $q$ is even.

Note that

$$[x_1 x_2 \ldots x_d]^2 = det [< x_h, x_k >]_{h,k=1,\ldots,d}.$$

It follows that the subalgebra of absolute $O(d)$-invariants

$$\mathbb{K}[(x_i|j)]^{O(d)} \cong \mathbb{K}[V^{\oplus n}]^{O(d)}$$

is precisely the subalgebra generated by the inner products $< x_h, x_k >$, with $h, k = 1, \ldots, n$.

## 3.3 The Straightening Formulae

In this section, we state the *Straightening Formulae* for bideterminants, Pfaffians and Gramians (minors of a "generic" symmetric matrix).

The Straightening Formulae for bideterminants, Pfaffians and Gramians are the starting points of a quite general theory, nowadays known as *Standard Monomial Theory* (abbreviated, SMT). For a thorough introduction to SMT and to its relations with the invariant theory of the classical groups, see [34], Ch. 13. For a comprehensive treatment of the subject, we refer the reader to the recent book of Lakshmibai and Raghavan [30].

In 1987, Grosshans , Rota and Stein discovered a superalgebraic version of the Straightening Formula for bideterminants [19]; the Straightening Formulae for Pfaffians and Gramians can be derived from that for (super)bideterminants, by using the *umbral-symbolic method* [3].

In the following, for the sake of simplicity, $\mathbb{K}$ will denote a field of arbitrary characteristic.

### 3.3.1 Standard Young Tableaux

A Young tableau $Y$ is an array of positive integers

$$
\begin{array}{l}
i_{11} i_{12} \ldots\ldots\ldots i_{1\lambda_1} \\
i_{21} i_{22} \ldots\ldots i_{2\lambda_2} \\
\vdots \\
i_{p1} i_{p2} \ldots i_{p\lambda_p}
\end{array}
,
$$

associated to a partition $\lambda = (\lambda_1 \geq \lambda_2 \geq \cdots \geq \lambda_p)$.

The partition $\lambda$ is called the *shape* of the tableau $Y$, denoted by the symbol $sh(Y)$.

A Young tableau is said to bo *standard* if each of its rows is increasing and each of its columns is non decreasing.

### 3.3.2 Bideterminants

Let $L = \{x_1, x_2, \ldots, x_n\}$ and $P = \{\xi_1, \xi_2, \ldots, \xi_m\}$ be alphabets, and let

$$\mathbb{K}[(x_i|\xi_j)]_{i=1,\ldots,n,\ j=1,\ldots,m}$$

be the letterplace algebra generated over $\mathbb{K}$ by the pair $(L,P)$, that is the associative and commutative $\mathbb{K}$−algebra generated by the variables $(x_i|\xi_j)$. Let

$$(S,T) = \begin{pmatrix} a_{11}a_{12}\ldots\ldots\ldots a_{1\lambda_1} & b_{11}b_{12}\ldots\ldots\ldots b_{1\lambda_1} \\ a_{21}a_{22}\ldots\ldots a_{2\lambda_2} & b_{21}b_{22}\ldots\ldots b_{2\lambda_2} \\ \vdots & \vdots \\ a_{p1}a_{p2}\ldots a_{p\lambda_p} & b_{p1}b_{p2}\ldots b_{p\lambda_p} \end{pmatrix},$$

$$a_{hk} \in \{1,\ldots,n\}, \quad b_{hk} \in \{1,\ldots,m\}$$

be a pair of Young tableaux of the same shape $\lambda$ over the sets $\{1,\ldots,n\}$ and $\{1,\ldots,m\}$, respectively.

The *bideterminant* of $(S,T)$, usually written $(S|T)$, is the polynomial

$$(S|T) = det\left[(x_{a_{1h}}|\xi_{b_{1k}})\right]_{h,k=1,\ldots,\lambda_1} \times \cdots \times det\left[(x_{a_{ph}}|\xi_{b_{pk}})\right]_{h,k=1,\ldots,\lambda_p}.$$

**Theorem 3.2.** *(The Straightening Formula for Bideterminats [12, 10])*

*The set of bideterminants $(S|T)$, with $S$ and $T$ standard Young tableaux, is a $\mathbb{K}$−basis of the letterplace algebra $\mathbb{K}[(x_i|\xi_j)]_{i=1,\ldots,n,\ j=1,\ldots,m}$.*

### 3.3.3 Pfaffians

Let $L = \{x_1, x_2, \ldots, x_n\}$ be an alphabet, and let $\mathbb{K}[(x_i|x_j)]_{i,j=1,\ldots,n}$ be the letterplace algebra generated over $\mathbb{K}$ by the pair $(L,L)$; let

$$Skew_{\mathbb{K}}[L]$$

be its quotient with respect to the ideal generated by the elements

$$(x_i|x_j) + (x_j|x_i), i \neq j, \quad (x_i|x_i), i = 1,\ldots,n.$$

Clearly, $Skew_{\mathbb{K}}[L]$ is the same as the plethystic algebra $S(\wedge^2(W))$, where $W$ is the $n$–dimensional $\mathbb{K}$–vector space with basis $L = \{x_1, x_2, \ldots, x_n\}$.

Consider the skew-symmetric matrix

$$Z_n = \left[(x_i|x_j)\right]_{i,j=1,\ldots,n},$$

with entries in $Skew_{\mathbb{K}}[L]$. If $i_1, \ldots, i_{2v}$ are indices in $\{1, 2, \ldots, n\}$, we will denote by

$$Pf[i_1, \ldots, i_{2v}]$$

the Pfaffian of the skew-symmetric matrix obtained from $Z_n$ by taking the rows and the columns of indices $i_1, \ldots, i_{2v}$.

To a given Young tableau (with even row lengths)

$$S = \begin{matrix} i_{11} \ldots\ldots\ldots\ldots i_{12v_1} \\ i_{21} \ldots\ldots\ldots i_{22v_2} \\ \vdots \\ i_{p1} \ldots i_{p2v_p} \end{matrix}$$

with $i_{hk} \in \{1, 2, \ldots, n\}$, we associate the "polynomial"

$$Pf[S] = Pf[i_{1,1}, \ldots, i_{12v_1}] \times Pf[i_{21}, \ldots, i_{22v_2}] \times \cdots \times Pf[i_{p1}, \ldots, i_{p2v_p}].$$

**Theorem 3.3.** *(The Straightening Formula for Pfaffians [6])*
    *The set of Pfaffians $Pf[S]$, with $S$ standard Young tableaux, is a $\mathbb{K}$–basis of the algebra $Skew_{\mathbb{K}}[L]$.*


### 3.3.4 Gramians


Let $L = \{x_1, x_2, \ldots, x_n\}$ be an alphabet, and let $\mathbb{K}[(x_i|x_j)]_{i,j=1,\ldots,n}$ be the letterplace algebra generated over $\mathbb{K}$ by the pair $(L, L)$; let

$$\mathbf{S}_{\mathbb{K}}[L]$$

be its quotient with respect to the ideal generated by the elements

$$(x_i|x_j) - (x_j|x_i), \quad i, j = 1, \ldots, n.$$

Clearly, $\mathbf{S}_{\mathbb{K}}[L]$ is the same as the plethystic algebra $S(S^2(W))$, where $W$ is the $n$–dimensional $\mathbb{K}$–vector space with basis $L = \{x_1, x_2, \ldots, x_n\}$.

Let

$$(S,T) = \begin{pmatrix} a_{11}a_{12}\ldots\ldots\ldots a_{1\lambda_1} & b_{11}b_{12}\ldots\ldots\ldots b_{1\lambda_1} \\ a_{21}a_{22}\ldots\ldots a_{2\lambda_2} & b_{21}b_{22}\ldots\ldots b_{2\lambda_2} \\ \vdots & \vdots \\ a_{p1}a_{p2}\ldots a_{p\lambda_p} & b_{p1}b_{p2}\ldots b_{p\lambda_p} \end{pmatrix},$$

$$a_{hk}, b_{hk} \in \{1,\ldots,n\}.$$

be a pair of Young tableaux of the same shape $\lambda = (\lambda_1 \geq \lambda_2 \geq \ldots \geq \lambda_p)$ over the set $\{1,\ldots,n\}$.

The pair $(S,T)$ is said to be *doubly standard* (*d-standard*, for short) whenever

- $S$ and $T$ are standard tableaux;
- the "nested" tableau

$$\begin{array}{l} a_{11}a_{12}\ldots\ldots\ldots a_{1\lambda_1} \\ b_{11}b_{12}\ldots\ldots\ldots b_{1\lambda_1} \\ a_{21}a_{22}\ldots\ldots a_{2\lambda_2} \\ b_{21}b_{22}\ldots\ldots b_{2\lambda_2} \\ \vdots \\ a_{p1}a_{p2}\ldots a_{p\lambda_p} \\ b_{p1}b_{p2}\ldots b_{p\lambda_p} \end{array}$$

is a standard tableau.

The *Gramian* of $(S,T)$, written $Gr(S|T)$, is the product

$$Gr(S|T) = det\left[(x_{a_{1h}}|x_{b_{1k}})\right]_{h,k=1,\ldots,\lambda_1} \times \cdots \times det\left[(x_{a_{ph}}|x_{b_{pk}})\right]_{h,k=1,\ldots,\lambda_p} \in \mathbf{S}_{\mathbb{K}}[L].$$

**Theorem 3.4.** (*The Straightening Formula for Gramians [6]*)
*The set of Gramians $Gr(S|T)$, with $(S,T)$ doubly standard, is a $\mathbb{K}-$basis of the algebra $S_{\mathbb{K}}[L]$.*

## 3.4 The E. Pascal Theorems and the Cancellation Laws

In the sequel, we assume $\mathbb{K}$ to be an infinite field of arbitrary characteristic.

In this section, we introduce the *factorization laws* and the *cancellation laws* for vector/covector scalar products, symplectic products and inner products. These results were first explicitly stated and proved in [6]. Here, we follow a slightly different path, which is closer to the approach of Désarménien, Kung, and Rota [10]; this approach is, in turn, suggested by the pioneering work of E. Pascal [32].

The starting point is the remark that the algebras generated by scalar products, symplectic products and inner products are the images of the algebras $\mathbb{K}[(x_i|\xi_j)]$, $Skew_{\mathbb{K}}[L]$ and $\mathbf{S}_{\mathbb{K}}[L]$ (defined in the previous section) with respect to natural algebra morphisms, called *E. Pascal morphisms*. The crucial fact is that $\mathbb{K}-$linear bases of the kernels of these algebra morphisms are given by subsets of the standard bases described by the Straightening Formulae. Therefore, one immediately gets standard basis theorems for the images of the E. Pascal morphisms, and, from them, deduces the factorization laws and the cancellation laws.

As a matter of fact, as we shall see, the characterizations of the kernels of the E. Pascal morphisms, combined with the First Fundamental Theorems, immediately yield the Second Fundamental Theorems in the traditional form (see, e.g. [36, 6]).

### 3.4.1 The E. Pascal Theorem and the Cancellation Law for Scalar Products

Let $n, m, d$ be positive integers. Consider the $\mathbb{K}$−algebra

$$\mathbf{S} = \mathbb{K}[x_{is}, \xi_{jt}]_{i=1,\ldots,n \ j=1,\ldots,m, \ s,t=1,\ldots,d}.$$

Given a $\mathbb{K}$−vector space $V$ of dimension $d$, a basis $B$ of $V$ and a dual cobasis $B^*$ of $V^*$, the algebra $\mathbf{S} = \mathbb{K}[x_{is}, \xi_{jt}]$ has to be regarded as the algebra of polynomial functions in $n$ "vector variables" $x_1, \ldots, x_n$ in $V$ and $m$ "covector variables" $\xi_1, \ldots, \xi_m$ in $V^*$, in symbols,

$$\mathbf{S} = \mathbb{K}[V^{\oplus n} \oplus (V^*)^{\oplus m}].$$

Let us consider the *E.Pascal* $\mathbb{K}$−*algebra morphism*

$$\Phi : \mathbb{K}[(x_i|\xi_j)]_{i=1,\ldots,n, \ j=1,\ldots,m} \to \mathbf{S}$$

defined by setting

$$(x_i|\xi_j) \to < x_i|\xi_j > = \sum_{h=1}^{d} x_{ih}\xi_{jh}.$$

**Theorem 3.5.** *(The E. Pascal theorem for scalar products)*
*The kernel $Ker\Phi$ of the Pascal morphism is spanned by the bideterminants $(S|T)$, $S, T$ standard, of shape $sh(S) = sh(T) = \lambda$, with $\lambda_1 > d$.*

Let **B** be the image of the Pascal morphism, that is **B** is the subalgebra of $\mathbf{S} = \mathbb{K}_m[V^{\oplus n} \oplus (V^*)^{\oplus m}]$ generated by the scalar products $< x_i|\xi_j >, i = 1, \ldots, n, \ j = 1, \ldots, m$.

**Corollary 3.2.** *The set of all $\Phi[(S|T)]$, $S, T$ standard, of shape $sh(S) = sh(T) = \lambda$, with $\lambda_1 \leq d$ is a $\mathbb{K}$− linear basis of **B**.*

**Theorem 3.6.** *(The factorization law [6])*
*Let $\varphi = \sum c_i \Phi[(S_i|T_i)] \in \mathbf{B}$, $S_i, T_i$ standard with rows of length at most $d$. If $\varphi$ vanishes when the vector variables $x_1, \ldots, x_d$ or the covector variables $\xi_1, \ldots, \xi_d$ are evaluated on linearly dependent elements, then the first row of each $S_i$ and of each $T_i$ is $12 \cdots d$ exactly, or, equivalently,*

$$\varphi = \Delta \cdot \varphi', \quad \varphi' \in \mathbf{B},$$

*where $\Delta = det\left[< x_i|\xi_j >\right]_{i,j=1,\dots,d}$.*

**Corollary 3.3.** *(The cancellation law for scalar products)*
*Let $\psi \in S = \mathbb{K}[V^{\oplus n} \oplus (V^*)^{\oplus m}]$ be such that $\Delta \cdot \psi \in \mathbf{B}$. Then $\psi \in \mathbf{B}$.*

## *3.4.2 The E. Pascal Theorem and the Cancellation Law for Symplectic Products*

Let $L = \{x_1, x_2, \dots, x_n\}$ be an alphabet. We recall that the algebra

$$S\,kew_{\mathbb{K}}[L]$$

can be regarded as the algebra generated by the variables

$$(x_h|x_k), (x_k|x_h), h \neq k, \quad (x_h|x_k) = -(x_k|x_h).$$

Let $d$ an even integer, say $d = 2m$, and let

$$\mathbf{S} = \mathbb{K}\left[(x_h|j)\right]_{h=1,\dots,n,\ j=1,\dots,2m} = \mathbb{K}[V^{\oplus n}],$$

where $V$ is a vector space of dimension $d = 2m$.

Let us consider the *E.Pascal $\mathbb{K}$−algebra morphism*

$$\Phi : S\,kew_{\mathbb{K}}[L] \to \mathbf{S}$$

defined by setting

$$(x_h|x_k) \to [x_h, x_k] = \sum_{j=1}^{m} ((x_h|j)(x_k|m+j) - (x_h|j+m)(x_k|j)).$$

**Theorem 3.7.** *(The E. Pascal theorem for symplectic products)*
*The kernel $Ker\Phi$ of the Pascal morphism is spanned by the Pfaffians $Pf[S]$, $S$ standard (with even row lengths) such that the first row has length $2q$, with $q > m$.*

Let $\mathbf{B}$ be the image of the Pascal morphism, that is $\mathbf{B}$ is the subalgebra of $\mathbf{S} = \mathbb{K}[V^{\oplus n}]$ generated by the symplectic products $[x_h, x_k]$, $h, k = 1, \dots, n, \quad h \neq k$.

**Corollary 3.4.** *The set of all $\Phi[Pf[S]]$, $S$ standard (with even row lengths) such that the first row has length $2q$, with $q \leq m$ is a $\mathbb{K}-$ linear basis of $\mathbf{B}$.*

**Theorem 3.8.** *(The factorization law [6])*
*Let $\varphi = \sum c_i \Phi[Pf[S_i]] \in \mathbf{B}$, $S_i$ standard with rows of length at most $2q$, with $q \leq m$. If $\varphi$ vanishes when the vector variables $x_1, \dots, x_{2m}$ are evaluated on linearly dependent elements, then the first row of each $S_i$ is $12 \cdots 2m$ exactly, or, equivalently,*

$$\varphi = \Delta \cdot \varphi', \quad \varphi' \in \mathbf{B},$$

*where $\Delta = Pf[i_1,\ldots,i_{2m}]$.*

**Corollary 3.5.** *(The cancellation law for symplectic products)*
  *Let $\psi \in \mathbf{S} = \mathbb{K}[V^{\oplus n}]$ be such that $\Delta \cdot \psi \in \mathbf{B}$. Then $\psi \in \mathbf{B}$.*

### 3.4.3 The E. Pascal Theorem and the Cancellation Law for Inner Products

Let $L = \{x_1, x_2, \ldots, x_n\}$ be an alphabet. The algebra $\mathbf{S}_{\mathbb{K}}[L]$ can be regarded as the algebra generated by the variables

$$(x_h|x_k), (x_k|x_h), \quad (x_h|x_k) = (x_k|x_h).$$

Let $d$ an integer, and let

$$\mathbf{S} = \mathbb{K}[(x_h|j)]_{h=1,\ldots,n,\ j=1,\ldots,2m} = \mathbb{K}[V^{\oplus n}],$$

where $V$ is a vector space of dimension $d$.
  Let us consider the *E.Pascal $\mathbb{K}$-algebra morphism*

$$\Phi : \mathbf{S}_{\mathbb{K}}[L] \to \mathbf{S}$$

defined by setting

$$(x_h|x_k) \to\, <x_h, x_k> = \sum_{j=1}^{d} (x_h|j)(x_k|j).$$

**Theorem 3.9.** *(The E. Pascal theorem for inner products)*
  *The kernel $Ker\Phi$ of the Pascal morphism is spanned by the Gramians $Gr(S|T)$, with $(S,T)$ doubly standard, of shape $sh(S) = sh(T) = \lambda$, with $\lambda_1 > d$.*

Let $\mathbf{B}$ be the image of the Pascal morphism, that is $\mathbf{B}$ is the subalgebra of $\mathbf{S} = \mathbb{K}[V^{\oplus n}]$ generated by the inner products $<x_h, x_k>$, $h, k = 1, \ldots, n$.

**Corollary 3.6.** *The set of all $\Phi[Gr(S|T)]$, with $(S,T)$ doubly standard, of shape $sh(S) = sh(T) = \lambda$, with $\lambda_1 \leq d$ is a $\mathbb{K}-$ linear basis of $\mathbf{B}$.*

**Theorem 3.10.** *(The factorization law [6])*
  *Let $\varphi = \sum c_i \Phi[Gr(S_i|T_i)] \in \mathbf{B}$, $(S_i, T_i)$ doubly standard with rows of length at most $d$. If $\varphi$ vanishes when the vector variables $x_1, \ldots, x_d$ are evaluated on linearly dependent elements, then the first row of each $S_i$ and of each $T_i$ is $12\cdots d$ exactly, or, equivalently,*

$$\varphi = \Delta \cdot \varphi', \quad \varphi' \in \mathbf{B},$$

*where $\Delta = \Phi[Gr(12\ldots d|12\ldots d)] = det\left[<x_i|x_j>\right]_{i,j=1,\ldots,d}$.*

**Corollary 3.7.** *(The cancellation law for inner products)*
  *Let $\psi \in \mathbf{S} = \mathbb{K}[V^{\oplus n}]$ be such that $\Delta \cdot \psi \in \mathbf{B}$. Then $\psi \in \mathbf{B}$.*

## 3.5 The First and the Second Fundamental Theorems

We will consider three cases:

1. The absolute $GL(d)$−invariants in vector and covector variables.
2. The absolute $Sp_{2m}$−invariants in vector variables.
3. The absolute $O(d)$− invariants in vector variables, in characteristic different from 2.

One can describe the approach to the three cases in a parallel way.

### 3.5.1 The First Fundamental Theorems

The main steps of the proofs are the following.

- Consider an algebra $\mathbf{S}$ and the invariant algebra $\mathbf{S}^\mathbf{G}$, $\mathbf{G} = GL(d), Sp_{2m}, O(d)$. Look for a subalgebra $\mathbf{B}$ of $\mathbf{S}$ which is, in turn, a subalgebra of $\mathbf{S}^\mathbf{G}$.
- Consider the localization with respect to a suitable Zariski open set and prove the equality between the localized rings

$$\mathbf{S}[1/\Delta]^\mathbf{G} = \mathbf{B}[1/\Delta],$$

$\Delta \in \mathbf{B}$.
- Let $\psi \in \mathbf{S}^\mathbf{G} \subseteq \mathbf{S}[1/\Delta]^\mathbf{G} = \mathbf{B}[1/\Delta]$. Then, there exists an integer $k$ such that $\Delta^k \psi \in \mathbf{B}$; the *factorization/cancellation* laws imply that $\psi \in \mathbf{B}$.

Therefore, the First Fundamental Theorems follows:

$$\mathbf{S}^\mathbf{G} = \mathbf{B},$$

where the algebras $\mathbf{S}$, $\mathbf{B}$ and the elements $\Delta$ to be considered are:

#### $GL(d)$−invariants in vector and covector variables.

- $\mathbf{S} = \mathbb{K}[x_{is}, \xi_{jt}]_{i=1,...,n\ j=1,...,m,\ s,t=1,...,d} = \mathbb{K}[V^{\oplus n} \oplus (V^*)^{\oplus m}]$.
- $\mathbf{B}$ is the subalgebra generated by the scalar products
  $< x_i|\xi_j >, i = 1,\ldots,n,\ j = 1,\ldots,m$.
- $\Delta = det\left[< x_i|\xi_j >\right]_{i,j=1,...,d}$.

#### $Sp_{2m}$−invariants in vector variables.

- $\mathbf{S} = \mathbb{K}[(x_h|j)]_{h=1,...,n,\ j=1,...,2m} = \mathbb{K}[V^{\oplus n}]$.
- $\mathbf{B}$ is the subalgebra generated by the symplectic products
  $[x_h, x_k],\ h, k = 1,\ldots,n,\quad h \neq k$.
- $\Delta$ is the Pfaffian $Pf[[x_h, x_k]]_{h,k=1,...,d=2m}$.

$O(d)$–**invariants in vector variables in characteristic different from** 2.

- $S = \mathbb{K}[(x_h|j)]_{h=1,\dots,n,\ j=1,\dots,2m} = \mathbb{K}[V^{\oplus n}]$.
- **B** is the subalgebra generated by the inner products

$< x_h, x_k >,\ h, k = 1, \dots, n, \quad h \neq k$.

- $\Delta = det[< x_h, x_k >]_{h,k=1,\dots,d}$.

### 3.5.2 The Second Fundamental Theorems

The Second Fundamental Theorems provide (minimal) sets of generators for the relation ideals $J_d$ in the algebras $\mathbf{S^G}$.

From the results of the preceding subsection, we know that the algebras of absolute invariants for $GL(d)$, $Sp_{2m}$ and $O(d)$ are precisely the images of the E. Pascal morphism for scalar products, for symplectic products and for inner products $(char(\mathbb{K}) \neq 2)$, respectively.

Therefore, we infer:

$GL(d)$–**invariants in vector and covector variables.**
From Laplace expansion:

*The ideal of relations among the the scalar products $< x_i|\xi_j >$ is generated by the $d+1 \times d+1$ minors of the matrix $\left[< x_i|\xi_j >\right]_{i=1,\dots,n,\ j=1,\dots,m}$.*

$Sp_{2m}$–**invariants in vector variables.**
We recall a classical identity for Pfaffians. Given a square skew-symmetric matrix $A = [a_{ij}]$ of even order $m$, $Pf[A] = \sum_{i=2}^{m} (-1)^i a_{1i} Pf[A_{\hat{1}\hat{i}}]$, where $A_{\hat{1}\hat{i}}$ is the skew-symmetric matrix of order $m-2$ obtained from $A$ by deleting both the first and the i-th rows and columns. Hence, we have:

*The ideal of relations among the symplectic products $[x_i, x_j]$ is generated by the Pfaffians $Pf[i_1, i_2, \dots, i_{2d+2}]$ of the matrix $\left[[x_i, x_j]\right]_{i,j=1,\dots,n}$.*

$O(d)$–**invariants in vector variables in characteristic different from** 2.
From Laplace expansion:

*The ideal of relations among the the inner products $< x_h|x_k >$ is generated by the $d+1 \times d+1$ minors of the matrix $[< x_h, x_k >]_{h,k=1,\dots,n}$.*

## 3.6 Grosshans's Theorem

In this section, the field $\mathbb{K}$ is assumed to be an *algebraically closed* field of arbitrary characteristic.

Let $d, n$ be positive integers, $d \le n$. Let $V_d$ be a $\mathbb{K}$-vector space of dimension $d$. We consider the imbedding $j : \mathbb{K}[V_d^{\oplus d}] \hookrightarrow \mathbb{K}[V_d^{\oplus n}]$ such that $(jF)(x_1, \ldots, x_n) = F(x_1, \ldots, x_d)$, for $F \in \mathbb{K}[V_d^{\oplus d}]$.

Clearly, a (contravariant) action of the group $GL(d, \mathbb{K})$ is defined on the algebras $\mathbb{K}[V_d^{\oplus d}]$ and $\mathbb{K}[V_d^{\oplus n}]$.

Given the group $GL(n, \mathbb{K})$, let $E_{hk}(\lambda)$, $h, k = 1, \ldots, n$, $\lambda \in \mathbb{K}$, with $\lambda \ne -1$ if $h = k$, be the standard system of generators of the group $GL(n, \mathbb{K})$.

By setting

$$(E_{hk}(\lambda) \cdot F)(x_1, \ldots, x_n) = F(x_1, \ldots, x_k + \lambda x_h, \ldots, x_n),$$

for every $F \in \mathbb{K}[V_d^{\oplus n}]$, one defines a group action $GL(n, \mathbb{K}) * \mathbb{K}[V_d^{\oplus n}]$ of $GL(n, \mathbb{K})$ on $\mathbb{K}[V_d^{\oplus n}]$.

Given a subgroup $H$ of $GL(d, \mathbb{K})$, the actions of $H$ and $GL(n, \mathbb{K})$ commute, so $GL(n, \mathbb{K})$ maps the algebra of $H$-invariants $\mathbb{K}[V_d^{\oplus n}]^H$ to itself.

Let $GL(n, \mathbb{K}) * j\left[\mathbb{K}[V_d^{\oplus d}]\right]^H$ be the algebra generated by all $g * j(f)$ for $g \in GL(n, \mathbb{K})$ and $f \in \mathbb{K}[V_d^{\oplus d}]^H$. This algebra is called the *algebra of invariants of $n$ vectors obtained from those of $d$ vectors by polarization.*

A natural question arises: what is the relationship between $\mathbb{K}[V_d^{\oplus n}]^H$ and $GL(n, \mathbb{K}) * j\left[\mathbb{K}[V_d^{\oplus d}]\right]^H$?

Let $char(\mathbb{K}) = p \ge 0$ and let $R$ and $S$ be $\mathbb{K}$- algebras with $R \subset S$. We say that $S$ is contained in the *$p$-root closure* of $R$ (or *purely inseparable closure* of $R$) if for every $s \in S$, there is a non- negative integer $m$ so that $s^{p^m} \in R$. (If $char(\mathbb{K}) = 0$, $S$ is contained in the $p$-root closure of $R$ if and only if $R = S$.)

**Theorem 3.11.** *(Grosshans [24])*

*Let $\mathbb{K}$ be an algebraically closed field of characteristic $p \ge 0$. Let $H$ be any subgroup of $GL(d, \mathbb{K})$. The algebra of $H$-invariants of $n$ vectors $\mathbb{K}[V_d^{\oplus n}]^H$ is the $p$-root closure of the algebra of polarized invariants of $d$ vectors $GL(n, \mathbb{K}) * j\left[\mathbb{K}[V_d^{\oplus d}]\right]^H$.*

*Remark 3.1.* (Grosshans's Theorem and Weyl's Theorem in characteristic zero)

Let $char(\mathbb{K}) = 0$. Grosshans's Theorem implies that

$$\mathbb{K}[V_d^{\oplus n}]^H = GL(n, \mathbb{K}) * j\left[\mathbb{K}[V_d^{\oplus d}]\right]^H \qquad (\dagger).$$

Since, in characteristic zero, the algebra $GL(n, \mathbb{K}) * j\left[\mathbb{K}[V_d^{\oplus d}]\right]^H$ is the same as the algebra obtained from the algebra $\mathbb{K}[V_d^{\oplus d}]^H$ by applying polarization processes (in the elementary sense of Subsection 2.1), statement ($\dagger$) is precisely the original statement of Weyl's Theorem, when $\mathbb{K} = \mathbb{C}$.

# References

1. Brini, A., Palareti, A., Teolis, A. (1988), *Gordan–Capelli Series in Superalgebras*, in "Proc. Natl. Acad. Sci. USA", 85, pp. 1330-3.
2. Brini, A., and Teolis, A. (1989), *Young–Capelli Symmetrizers in Superalgebras*, in "Proc. Natl. Acad. Sci. USA", 86, pp. 775-8.
3. Brini, A., Huang, R.Q., Teolis, A. (1992), *The Umbral Symbolic Method for Supersymmetric Tensors*, in "Adv. Math.", 96, pp. 123-93.
4. Brini, A., Regonati, F., Teolis, A. (2005), *Combinatorics and Representations of General Linear Lie Superalgebras over Letterplace Superalgebras*, in *Computer Algebra and Clifford Algebra with Applications*, H. Li, P. Olver, G. Sommer (eds.), (Lecture Notes in Computer Sciences, 3519), Berlin, Springer, pp. 239-57.
5. Capelli, A. (1902), *Lezioni sulla teoria delle forme algebriche*, Neaples, Pellerano.
6. De Concini, C., Procesi, C. (1976), *A Characteristic Free Approach to Invariant Theory*, in "Adv. Math.", 21, pp. 330-54.
7. De Concini, C., Eisembud, D., Procesi, C. (1980), *Young Diagrams and Determinantal Varieties*, in "Invent. Math.", 56, pp. 129-65.
8. De Concini, C., Eisembud, D., Procesi, C. (1982), *Hodge Algebras*, in "Astérisque", 91.
9. Deruyts, J. (1892), *Essai d'une théorie générale des formes algébriques*, in "Mémoires. Société Royale des Sciences de Liège", 17, pp. 1-156.
10. Désarménien, J., Kung, J. P. S., Rota, G.-C. (1978), *Invariant Theory, Young Bitableaux and Combinatorics*, in "Adv. Math.", 27, pp. 63-92.
11. Dieudonné, J. A., Carrell, J. B. (1971), *Invariant Theory, Old and New*, New York, Academic Press.
12. Domokos, M. (2003), *Matrix Invariants and the Failure of Weyl's Theorem. in Polynomial Identities and Combinatorial Methods*, in "Lect. Notes Pure Appl. Math.", 235, pp. 215-36.
13. Doubilet, P., Rota, G.-C., and Stein, J. (1974), *On the Foundation of Combinatorial Theory IX. Combinatorial Methods in Invariant Theory*, in "Stud. Appl. Math.", 53, pp. 185-216.
14. Fulton, W., Harris, J. (1991), *Representation Theory. A first Course*, (Graduate Texts in Mathematics, 129), New York, Springer.
15. Goodman, R., Wallach, N. R. (1998), *Representations and Invariants of the Classical Groups*, (Encyclopedia of Mathematics and its Applications, 68), Cambridge, Cambridge University Press.
16. Grace, J. H., Young, A. (1903), *The Algebra of Invariants*, Cambridge, Cambridge University Press.
17. Green, J. A. (1980), *Polynomial Representations of GLn*, in "Lecture Notes in Math.", 830.
18. Green, J. A. (1991), *Classical Invariants and the General Linear Group*, in "Progr. Math.", 95, pp. 247-72.
19. Grosshans, F. D., Rota, G.-C., Stein, J. (1987), *Invariant Theory and Superalgebras*, in "Amer. Math. Soc.".
20. Grosshans, F. D. (1973), *Observable Groups and Hilbert's Fourteenth Problem*, in "Amer. J. Math.", 95, pp. 229-53.
21. Grosshans, F. D. (1993), *The Symbolic Method and Representation Theory*, in "Adv. Math.", 98, pp. 113-42.
22. Grosshans, F. D. (1997), *Algebraic Homogeneous Spaces and Invariant Theory*, in "Lecture Notes in Math.", 1673.
23. Grosshans, F. D. (2003), *The Work of Gian-Carlo Rota on Invariant Theory*, in "Algebra Universalis", 49, pp. 213-58.
24. Grosshans, F. D. (2007), *Vector Invariants in Arbitrary Characteristic*, in "Transform. Groups", 12, pp. 499-514.
25. Hodge, W. V. D. (1943), *Some Enumerative Results in the Theory of Forms*, in "Proc. Cambridge Philos. Soc.", 39, pp. 22-30.
26. Howe, R. (1989), *Remarks on Classical Invariant Theory*, in "Trans. Amer. Math. Soc.", 313, pp. 539-70.

27. Howe, R. (1992), *Perspective on Invariant Theory: Schur Duality, Multilpicity-free Actions and Beyond*, in "The Schur Lectures, Israel Math. Conf. Proc.", 8, pp. 1-182.
28. Igusa, J. (1954), *On the Arithmetic Normality of the Grassmann Variety*, in "Proc. Natl. Acad. Sci. USA", 40, pp. 309-13.
29. Kraft, H., Procesi, C., *Classical Invariant Theory, A Primer*, http://www. math. unibas. ch.
30. Lakshmibai, V., Raghavan, K. N. (2008), *Standard Monomial Theory. Invariant Theoretic Approach*, Berlin, Springer.
31. Losik, M., Michor, P. W., Popov, V. L. (2006), *On Polarizations in Invariant Theory*, in "J. Algebra", 301, pp. 406-24.
32. Pascal, E. (1888), *Sopra un teorema fondamentale nella teoria del calcolo simbolico delle Forme ennarie*, in "Atti Accad. Naz. Lincei", 5, pp. 119-24.
33. Procesi, C. (1982), *A Primer of Invariant Theory*, G. Boffi (notes by), in *Brandeis Lecture Notes*, 1, Waltham, MA, Brandeis University.
34. Procesi, C. (2006), *Lie Groups. An Approach Through Invariants and Representation*, New York, Springer.
35. Procesi, C., Rogora, E. (1991), *Aspetti geometrici e combinatori della teoria delle rappresentazioni del gruppo unitario*, in "Quaderni U. M. I.", 36.
36. Weyl, H. (1946), *The Classical Groups*, Princeton, NY, Princeton Univ. Press.

Informally speaking, Straightening Formulae are theorems that describe special bases of certain algebras, together with an algorithm that allow the elements of a prescribed system of generators to be expressed as linear combinations of the elements of these bases. These bases are usually called *standard bases*. For the sake of simplicity, we will identify the Straightening Formulae with the Standard basis Theorems they imply. For an "axiomatic" approach to this theme, see. eg. [34], p. 499 ff.).

# Chapter 4
# Partitions of a Finite Partially Ordered Set
## Contributed Chapter

Pietro Codara

**Abstract** In this paper, we investigate the notion of partition of a finite partially ordered set (poset, for short). We will define three different notions of partition of a poset, namely, monotone, regular, and open partition. For each of these notions we will find three equivalent definitions, that will be shown to be equivalent. We start by defining partitions of a poset in terms of fibres of some surjection having the poset as domain. We then obtain combinatorial characterisations of such notions in terms of blocks, without reference to surjection. Finally, we give a further, equivalent definition of each kind of partition by means of analogues of equivalence relations.

## 4.1 Introduction

A partition of a set $A$ is a collection of nonempty, pairwise disjoint subsets, often called *blocks*, whose union is $A$. Equivalently, partitions can be defined by means of *equivalence relations*, by saying that a partition of a set $A$ is the set of equivalence classes of an equivalence relation on $A$. A third definition of a partition can be given in terms of *fibres* of a surjection: a partition of a set $A$ is the set $\{f^{-1}(y) \mid y \in B\}$ of fibres of a surjection $f : A \to B$.

In this paper, we investigate the notion of partition of a finite poset. Our goal is to find the analogues of the three definitions of partition of a set mentioned above, in terms of blocks, relations, and fibres, respectively. For the rest of this paper, we shall omit the term finite, as we only deal with finite posets; 'poset partition' means 'partition of a poset'.

We begin our study of poset partitions with the notion of partition given in terms of fibres. Some background in category theory, and a few preliminary results on two

Pietro Codara
Università degli Studi di Milano, Italy

E. Damiani et al. (eds.), *From Combinatorics to Philosophy,*
DOI 10.1007/978-0-387-88753-1_4, © Springer Science+Business Media, LLC 2009

different categories having posets as objects (Section 4.2), will allow us to identify three kind of morphisms appropriate to introduce our first definition. In Section 4.3, we introduce monotone, regular, and open partitions, according to the morphisms we are considering.

The definitions given in Section 4.3 need to mention morphism to describe what a poset partition is. To investigate the combinatorial structure of a poset partition, it is useful to have intrinsic notions of poset partitions making no reference to morphisms. In Section 4.4 we obtain such a combinatorial characterisation of each kind of partition.

In Section 4.5, we investigate the analogues of the definition of partition of a set given by means of equivalence relations. Again, we obtain characterisations of monotone, regular, and open partitions in this terms.

This piece of work contains a revisited and extended version of some of the results obtained in [3], where the author defines and analyses the notions of monotone and regular partitions of a poset, and in [4], where open partitions are introduced.

**Acknowledgment.** The author wishes to thank Prof. O. D'Antona, and Dr. V. Marra for their helpful comments and suggestions, and for the many discussions on this topic.

## 4.2 Background, and Preliminary Results

When one defines a partition of a set in terms of fibres, one makes use of a special class of morphisms of the category[1] Set of sets and functions. In fact, such definition exploits the notion of surjection, which can be shown to coincide in Set with the notion of *epimorphism*. Recall that an epimorphism in a category is a morphism $f : A \rightarrow B$ that is right-cancellable with respect to composition: whenever $h \circ f = k \circ f$, for $h, k : B \rightarrow C$, we have $h = k$. The category-theoretic dual of the notion of epimorphism is *monomorphism*. In Set, monomorphisms coincide with injections. The well-known fact that each function between sets factorises (in an essentially unique way) as a surjection followed by an injection can be reformulated in categorical terms by saying that the classes of epimorphisms and monomorphism form a factorisation system for Set, or, equivalently, that (epi,mono) is a factorisation system for Set. Epimorphism and factorisation systems will play a key role in the following.

First, consider the category Pos of posets and *order-preserving maps* (also called *monotone maps*), i.e., functions $f : P \rightarrow Q$, with $P, Q$ posets such that $x \leqslant y$ in $P$ implies $f(x) \leqslant f(y)$ in $Q$, for each $x, y \in P$. In Pos, (epi,mono) is not a factorisation system; to obtain one we need to isolate a subclass of epimorphisms, called

---

[1] For background on category theory we refer, *e.g.*, to [1].

regular epimorphisms. A morphism $e : B \to C$ in an arbitrary category is a *regular epimorphism* if and only if there exists a pair $f, g : A \to B$ of morphisms such that

(1) $e \circ f = e \circ g$,

(2) for any morphism $e' : B \to C'$ with $e' \circ f = e' \circ g$, there exists a unique morphism $\psi : C \to C'$ such that $e' = \psi \circ e$.

Regular epimorphisms are epimorphisms, as one easily checks. While in Set regular epimorphisms and epimorphisms coincide, that is not the case in Pos. The dual notion of regular epimorphism (obtained by reversing arrows in the above statement) is *regular monomorphism*. It can be shown (see, e.g., [3, Proposition 2.5]) that (regular epi,mono) is a factorisation system for the category Pos. A second factorisation system for Pos is given by the classes of epimorphisms and regular monomorphisms. In other words, each order-preserving map between posets factorises in an essentially unique way both as a regular epimorphism followed by a monomorphism, and as an epimorphism followed by a regular monomorphism.

The existence of two distinct factorisation systems in Pos leads us to introduce two different notions of partition of a poset, one based on the use of epimorphisms, the other based on the use of regular epimorphisms.

Our next step is to characterise regular epimorphisms.

**Notation.** If $\pi$ is a partition of a set $A$, and $a \in A$, we denote by $[a]_\pi$ the block of $a$ in $\pi$. When no confusion is possible, we shall write $[a]$ instead of $[a]_\pi$. Further, let us stress our usage of different symbols for representing different types of binary relations. The symbol $\leqslant$ denotes the partial order relation between elements of a poset. A second symbol, $\lhd$, represents the associated covering relation. Finally, the symbol $\lesssim$ denotes *quasiorder relations*, sometimes called *preorders*, i.e, reflexive and transitive relations.

**Definition 4.1 (Blockwise quasiorder).** Let $(P, \leqslant)$ be a poset and let $\pi$ be a partition of the set $P$. For $x, y \in P$, $x$ is *blockwise under $y$ with respect to $\pi$*, written $x \lesssim_\pi y$, if and only if there exists a sequence $x = x_0, y_0, x_1, y_1, \ldots, x_n, y_n = y \in P$ satisfying the following conditions.

(1) For all $i \in \{0, \ldots, n\}$, $[x_i] = [y_i]$.

(2) For all $i \in \{0, \ldots, n-1\}$, $y_i \leqslant x_{i+1}$.

Observe that the relation $\lesssim_\pi$ in Definition 4.1 indeed is a quasiorder. In fact, if $x \leqslant y$ and $y \leqslant z$ for $x, y, z \in P$, then there exist two sequences $x = x_0, y_0, x_1, y_1, \ldots, x_n, y_n = y$ and $y = y_{n+1}, z_{n+1}, y_{n+2}, z_{n+2}, \ldots, y_{n+m}, z_{n+m} = z$ satisfying (1) and (2), and a sequence $x = x_0, y_0, x_1, y_1, \ldots, x_n, y_n = y_{n+1}, z_{n+1}, y_{n+2}, z_{n+2}, \ldots, y_{n+m}, z_{n+m} = z$ satisfying (1) and (2), too. Thus, $x \lesssim_\pi z$ and the relation $\lesssim_\pi$ is transitive. The reflexivity of $\lesssim_\pi$ results trivially.

The definition of blockwise quasiorder allows us to isolate a special kind of order-preserving map.

**Definition 4.2 (Fibre-coherent map).** Consider two posets $P$ and $Q$. Let $f : P \to Q$ be a function, and let $\pi_f = \{f^{-1}(q) | q \in f(P)\}$ be the set of fibres of $f$. We say $f$ is a *fibre-coherent map* whenever for any $p_1, p_2 \in P$, $f(p_1) \leqslant f(p_2)$ if and only if $p_1 \lesssim_{\pi_f} p_2$.

**Proposition 4.1.** *In* Pos, *regular epimorphisms are precisely fibre-coherent surjections.*

*Proof.* ($\Rightarrow$) Let $(P, \leqslant_P)$ and $(Q, \leqslant)$ be posets, let $e : P \to Q$ be a regular epimorphism and let $\pi_e = \{e^{-1}(q) | q \in Q\}$. Since $e$ is epi, it is an order-preserving surjection. Moreover, by the definition of regular epimorphism, there exists a pair $f, g : R \to P$ of morphisms such that $e \circ f = e \circ g$.

Suppose, by way of contradiction, that $e$ is not fibre-coherent. If $x \lesssim_{\pi_e} y$ for some $x, y \in P$, then there exists a sequence $x = x_0, y_0, x_1, y_1, \ldots, x_n, y_n = y \in P$ satisfying conditions (1) and (2) in Definition 4.1. For such a sequence, since $e$ is order-preserving, we have $e(x) = e(x_0) = e(y_0) \leqslant e(x_1) = e(y_1) \leqslant \cdots \leqslant e(x_n) = e(y_n) = e(y)$. Thus, to satisfy the *absurdum hypothesis* there must exist $p_1, p_2 \in P$, with $e(p_1) = q_1$ and $e(p_2) = q_2$, such that $q_1 \leqslant q_2$ but $p_1 \not\lesssim_{\pi_e} p_2$. Note that $p_1$ and $p_2$ must be incomparable, and that $q_1 \neq q_2$.

Case (i). Suppose $q_1 \lhd q_2$, where $\lhd$ is the covering relation induced by $\leqslant$. Consider the poset $Q'$ having $Q$ as underlying set, endowed with the relation $\leqslant'$ obtained by removing from $\leqslant$ the pair $(q_1, q_2)$. In other words, the only difference between $\leqslant'$ and $\leqslant$ is that $q_1 \leqslant q_2$, but $q_1 \not\leqslant' q_2$. Since $q_1 \lhd q_2$, removing $(q_1, q_2)$ from $\leqslant$ does not impair transitivity and $\leqslant'$ indeed is a partial order.

Now, consider the function $e' : P \to Q'$ that coincides with $e$ on the underlying sets. We want to show that $e'$ is order-preserving. For this, let $x, y \in P$. It suffices to consider two cases only: $e(x) = q_1$ and $e(y) = q_2$, and viceversa. In any other case, $e'$ preserves order just because $e$ does. Suppose, without loss of generality, $x \in e^{-1}(q_1)$ and $y \in e^{-1}(q_2)$. Then, $x \not\leqslant_P y$, for else the chain $p_1, x, y, p_2$ would satisfy conditions (1) and (2) in Definition 4.1, contradicting $p_1 \not\lesssim_{\pi_e} p_2$. Moreover, $y \not\leqslant_P x$, because $e$ is order preserving. Thus, for each $x \in e^{-1}(q_1)$ and $y \in e^{-1}(q_2)$, $x$ and $y$ are incomparable. Summing up, $e'$ is order preserving. Since $e'$ coincides with $e$ on the underlying sets, we obtain $e' \circ f = e' \circ g$.

Since $e$ is a regular epimorphism, by definition, we can find a unique morphism $\psi : Q \to Q'$ such that the diagram in Figure 4.1 commutes. Take $x, y \in P$ such

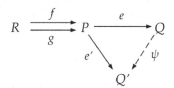

**Fig. 4.1** Proof of Proposition 4.1.

that $e(x) = e'(x) = q_1$ and $e(y) = e'(y) = q_2$. Thus, we should have $\psi(q_1) = q_1$ and $\psi(q_2) = q_2$ but, by hypothesis, $q_1 \leqslant q_2$ and $q_1 \not\leqslant' q_2$, and such a $\psi$ would not be order-preserving. Since $f$ and $g$ satisfy $e \circ f = e \circ g$ but are otherwise arbitrary, this would contradict the fact that $e$ is a regular epimorphism. Therefore, $e$ has to be fibre-coherent.

Case (ii). Suppose $q_1 \not\vartriangleleft q_2$. Then there exists a sequence $k_1, k_2, \ldots, k_u \in Q$ such that $q_1 = k_1 \vartriangleleft k_2 \vartriangleleft \cdots \vartriangleleft k_u = q_2$. Let $x_1, x_2, \ldots, x_u \in Q$ be such that $x_i \in e^{-1}(k_i)$, for each $i \in \{1, u\}$, and suppose $x_1 \lesssim_{\pi_e} x_2 \lesssim_{\pi_e} \cdots \lesssim_{\pi_e} x_u$. Since $p_1 \in e^{-1}(k_1)$ and $p_2 \in e^{-1}(k_u)$ imply $p_1 \lesssim_{\pi_e} x_1$ and $x_u \lesssim_{\pi_e} p_2$, then, by transitivity, $p_1 \lesssim_{\pi_e} p_2$, contradicting our hypothesis. Thus, there exists an index $j$ such that $k_j \vartriangleleft k_{j+1}$, but $x_j \not\lesssim_{\pi_e} x_{j+1}$. The proof follows now the same steps of *Case* (i), with $x_j$ and $x_{j+1}$ playing the role of $p_1$ and $p_2$, respectively.

($\Leftarrow$) Let $(P, \leqslant_P)$ and $(Q, \leqslant)$ be posets, and let $e : P \to Q$ be a fibre-coherent surjection. Consider the poset $R \subseteq P \times P$, having underlying set $\{(r_1, r_2) \in P \times P \mid e(r_1) = e(r_2)\}$, endowed with the order $\leqslant_R$ defined by $(r_1, r_2) \leqslant_R (s_1, s_2)$ if and only if $r_1 \leqslant_P s_1$ and $r_2 \leqslant_P s_2$. Let $f, g : R \to P$ be the projection functions of $R$, i.e., $f$ and $g$ are the order-preserving maps such that, for each $r = (r_1, r_2) \in R$, $f(r) = r_1$, $g(r) = r_2$. Clearly, $e \circ f = e \circ g$.

We need to show that $e$ is a regular epimorphism. Consider a poset $(Q', \leqslant')$ and an order-preserving map $e' : P \to Q'$ such that $e' \circ f = e' \circ g$. Note that, for each $q \in Q$, if $x, y \in e^{-1}(q)$, there exists $r \in R$ such that $f(r) = x$ and $g(r) = y$. Thus, from $e' \circ f = e' \circ g$ follows that $e'(x) = e'(y)$. Since $e$ is a surjection, we can construct a map $\psi : Q \to Q'$ by setting $\psi(q) = e'(x)$ for some $x \in e^{-1}(q)$, where $q \in Q$.

Let now $q_1, q_2 \in Q$ with $q_1 \leqslant q_2$, and let $x_1, x_2 \in P$ be such that $e(x_1) = q_1$, $e(x_2) = q_2$. By Definition 4.2, we have $x_1 \lesssim_{\pi_e} x_2$. Thus, by Definition 4.1 there exists a sequence $y_0, z_0, y_1, z_1, \ldots, y_n, z_n \in P$ with $x_1 = y_0$ and $x_2 = z_n$ such that $e(y_i) = e(z_i)$, for $i = 0, \ldots, n$, and $z_j \leqslant_P y_{j+1}$, for $j = 0, 1, \ldots, n-1$. Moreover, from $e' \circ f = e' \circ g$ it follows that $e'(y_i) = e'(z_i)$, and, since $e'$ is order-preserving, we have $e'(z_j) \leqslant' e'(y_{j+1})$. Thus, $e'(y_0) = e'(z_0) \leqslant' e'(y_1) = e'(z_1) \leqslant' \cdots \leqslant' e'(y_n) = e'(z_n)$. Therefore, we have $\psi(q_1) = e'(x_1) \leqslant \psi(q_2) = e'(x_2)$ and $\psi$ is order-preserving. The morphism $\psi$ is now well defined and, by construction, satisfies $e'(x) = \psi(e(x))$ for all $x \in P$.

Let $\psi'$ be another map from $Q$ to $Q'$, $\psi \neq \psi'$, and let $\overline{q}$ be an element of $Q$ such that $\psi(\overline{q}) \neq \psi'(\overline{q})$. Since $e$ is surjective, there exists $x \in P$ such that $e(x) = \overline{q}$. Then from $\psi'(\overline{q}) \neq \psi(\overline{q})$ and $e'(x) = \psi(\overline{q})$ we have $\psi'(\overline{q}) \neq e'(x)$ and $\psi' \circ e \neq e'$. Hence, $\psi : Q \to Q'$ is the unique function such that $\psi \circ e = e'$. Summing up, for an arbitrary morphism $e' : P \to Q'$ such that $e' \circ f = e' \circ g$, there exists a unique order preserving map $\psi : Q \to Q'$ such that $\psi \circ e = e'$, i.e., $e$ is a regular epimorphism.

The second category we are going to consider is the category OPos of posets and open maps. Such maps arise naturally in the investigation of intuitionistic logic; cf. the notion of p-morphisms of Kripke frames, e.g. in [2]. An order-preserving function $f : P \to Q$ between posets is called *open* if whenever $f(u) \geqslant v'$ for $u \in P$ and $v' \in Q$, there is $v \in P$ such that $u \geqslant v$ and $f(v) = v'$. One can check that epimorphisms in OPos are surjective open maps, and monomorphisms are injective open maps. As in Set, in OPos each epimorphism is a regular epimorphism. Further, OPos admits an (epi,mono) factorisation system. In the next section, we will see that surjective open maps in OPos induce a third kind of partition of a poset.

## 4.3 Partitions as Sets of Fibres

Poset partitions can be defined in terms of fibres. From the notions of epimorphism and regular epimorphism in Pos, we derive immediately the two following definitions.

**Definition 4.3 (Monotone partition).** A *monotone partition* of a poset $P$ is a poset $(\pi_f, \preccurlyeq)$, where $\pi_f$ is the set of fibres[2] of an order-preserving surjection $f : P \to Q$, for some poset $Q$, and $\preccurlyeq$ is the partial order on $\pi_f$ defined by

$$f^{-1}(q_1) \preccurlyeq f^{-1}(q_2) \text{ if and only if } q_1 \leqslant q_2, \tag{4.1}$$

for each $q_1, q_2 \in Q$.

**Definition 4.4 (Regular partition).** A *regular partition* of a poset $P$ is a poset $(\pi_f, \preccurlyeq)$, where $\pi_f$ is the set of fibres of a fibre-coherent surjection $f : P \to Q$, for some poset $Q$, and $\preccurlyeq$ is the partial order on $\pi_f$ defined by

$$f^{-1}(q_1) \preccurlyeq f^{-1}(q_2) \text{ if and only if } q_1 \leqslant q_2, \tag{4.2}$$

for each $q_1, q_2 \in Q$.

*Remark 4.1.* Since a fibre-coherent map is order-preserving, it follows immediately that each regular partition of a poset is a monotone partition.

Consider now the category OPos.

**Definition 4.5 (Open partition).** An *open partition* of a poset $P$ is a poset $(\pi_f, \preccurlyeq)$, where $\pi_f$ is the set of fibres of a surjective open map $f : P \to Q$, for some poset $Q$, and $\preccurlyeq$ is the partial order on $\pi_f$ defined by

$$f^{-1}(q_1) \preccurlyeq f^{-1}(q_2) \text{ if and only if } q_1 \leqslant q_2, \tag{4.3}$$

for each $q_1, q_2 \in Q$.

*Remark 4.2.* One can check that an open map is fibre-coherent. It follows that each open partition of a poset is a regular partition, whence a monotone one.

There are regular partitions that are not open, and monotone partitions that are not regular; cf. Example 4.1.

---

[2] Note that, since $f$ is surjective, $\pi_f$ is a partition of the underlying set of $P$.

## 4.4 Partitions as Partially Ordered Sets of Blocks

For each definition in the previous section, we give a new definition in terms of partially ordered blocks without mentioning morphisms. We prove, for each notion of partition, that the two kinds of definition describe exactly the same objects.

**Definition 4.6 (Monotone partition).** A *monotone partition* of a poset $P$ is a poset $(\pi, \preccurlyeq)$ where

(i) $\pi$ is a partition of the underlying set of $P$,
(ii) for each $p_1, p_2 \in P$, $p_1 \leqslant p_2$ implies $[p_1] \preccurlyeq [p_2]$.

**Theorem 4.1.** *Definitions 4.3 and 4.6 are equivalent.*

*Proof.* (Definition 4.6 $\Rightarrow$ Definition 4.3). Let $\pi$ be a partition of the underlying set of a poset $P$, and let $\preccurlyeq$ be a partial order on $\pi$ satisfying (ii). Consider the projection map $f : P \to \pi$ which sends each element of $P$ to its block in $\pi$. Since a partition does not have empty blocks, $f$ is a surjection. By (ii), $f$ is order-preserving. Since, by construction, $f^{-1}(B) = B$ for each $B \in \pi$, the partial order $\preccurlyeq$ satisfies (4.1). Thus, $(\pi, \preccurlyeq)$ is a monotone partition of $P$ according to Definition 4.3.

(Definition 4.3 $\Rightarrow$ Definition 4.6). Let $f : P \to Q$ be an order-preserving surjection, and let $(\pi_f, \preccurlyeq)$ be a monotone partition of $P$, according to Definition 4.3. Since $f$ is surjective, $\pi_f$ is a partition of the underlying set of $P$. Consider $p_1, p_2 \in P$ such that $p_1 \leqslant p_2$. Since $f$ is order-preserving, $f(p_1) \leqslant f(p_2)$ holds and, by (4.1), $[p_1] = f^{-1}(f(p_1)) \preccurlyeq [p_2] = f^{-1}(f(p_2))$. We have so proved (ii). Thus, $(\pi_f, \preccurlyeq)$ is a monotone partition of $P$ according to Definition 4.6.

**Corollary 4.1.** *Let $(\pi, \preccurlyeq)$ be a monotone partition of a poset $P$. Then, for each $p_1, p_2 \in P$,*

$$[p_1] = [p_2] \text{ if and only if } (p_1 \lesssim_\pi p_2 \text{ and } p_2 \lesssim_\pi p_1). \tag{4.4}$$

*Proof.* ($\Rightarrow$) Directly from Definition 4.1.

($\Leftarrow$) Let $p_1, p_2 \in P$ be such that $p_1 \lesssim_\pi p_2$ and $p_2 \lesssim_\pi p_1$. Then, there exist two sequences $p_1 = x_0, y_0, x_1, y_1, \ldots, x_n, y_n = p_2$ and $p_2 = z_0, w_0, \ldots, z_m, w_m = p_1$ of elements of $P$ satisfying Conditions (1) and (2) in Definition 4.1, with respect to $\pi$. By Condition (ii) in Definition 4.6, we have $[p_1] = [x_0] = [y_0] \preccurlyeq [x_1] = [y_1] \preccurlyeq \cdots \preccurlyeq [x_n] = [y_n] = [p_2]$, and $[p_2] = [z_0] = [w_0] \preccurlyeq [z_1] = [w_1] \preccurlyeq \cdots \preccurlyeq [z_m] = [w_m] = [p_1]$. Thus, $[p_1] = [p_2]$.

**Definition 4.7 (Regular partition).** A *regular partition* of a poset $P$ is a poset $(\pi, \preccurlyeq)$ where

(i) $\pi$ is a partition of the underlying set of $P$,
(ii) for each $p_1, p_2 \in P$, $p_1 \lesssim_\pi p_2$ if and only if $[p_1] \preccurlyeq [p_2]$.

**Theorem 4.2.** *Definitions 4.4 and 4.7 are equivalent.*

*Proof.* (Definition 4.7 ⇒ Definition 4.4). Let $\pi$ be a partition of the underlying set of a poset $P$ and let $\preccurlyeq$ be a partial order on $\pi$ satisfying (ii). Consider the projection map $f : P \to \pi$ which sends each element of $P$ to its block in $\pi$. Clearly, $f$ is a surjection. Since, for all $p \in P$, $f(p) = [p]$, Condition (ii) is equivalent to the fibre-coherent condition in Definition 4.2. Thus, $f$ is a fibre-coherent surjection, having $\pi$ as its set of fibres. Moreover, since $f^{-1}(B) = B$ for all $B \in \pi$, the partial order $\preccurlyeq$ satisfies (4.2). Thus, $(\pi, \preccurlyeq)$ is a regular partition of $P$ according to Definition 4.4.

(Definition 4.4 ⇒ Definition 4.7). Let $f : P \to Q$ be a fibre-coherent surjection, and let $(\pi_f, \preccurlyeq)$ be a regular partition of $P$, according to Definition 4.4. Since $f$ is surjective, $\pi_f$ is a partition of the underlying set of $P$. Consider $p_1, p_2 \in P$. By Definition 4.2, $p_1 \preccurlyeq_{\pi_f} p_2$ if and only if $f(p_1) \preccurlyeq f(p_2)$. By Definition 4.4, $f(p_1) \preccurlyeq f(p_2)$ if and only if $[p_1] = f^{-1}(f(p_1)) \preccurlyeq [p_2] = f^{-1}(f(p_2))$. Thus, $p_1 \preccurlyeq_{\pi_f} p_2$ if and only if $[p_1] \preccurlyeq [p_2]$, and (ii) is proved. Thus, $(\pi_f, \preccurlyeq)$ is a regular partition of $P$ according to Definition 4.7.

**Corollary 4.2.** *Let $(\pi, \preccurlyeq)$ be a partition of the underlying set of a poset $P$ satisfying (4.4). Then, there exists exactly one regular partition of $P$ having $\pi$ as underlying set.*

*Proof.* Let $\pi$ be a partition of the underlying set of a poset $P$ satisfying (4.4). Define the binary relation $\preccurlyeq$ on $\pi$ by prescribing that, for all $p_1, p_2 \in P$, $p_1 \preccurlyeq_\pi p_2$ if and only if $[p_1] \preccurlyeq [p_2]$. It is straightforward to check that $\preccurlyeq$ is a partial order on $P$. By Definition 4.7, $(\pi, \preccurlyeq)$ is a regular partition of $P$. To see that it is the unique regular partition of $P$ having $\pi$ as underlying set, just observe that $\preccurlyeq$ must be a partial order satisfying Condition (ii) in Definition 4.7.

The uniqueness property of regular partitions proved in the above corollary does not hold, in general, for monotone partitions; cf. Figure 4.2, which shows three distinct monotone partitions of a given poset $P$ having the same underlying set.

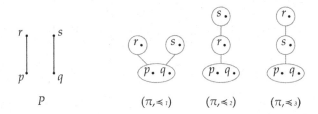

**Fig. 4.2** Distinct monotone partitions with the same support $\pi$.

Given a poset $P$ and a subset $S \subseteq P$, the *lower set* generated by $S$ is

$$\downarrow S = \{p \in P \mid p \preccurlyeq s, \text{ for some } s \in S\}.$$

Analogously, the *upper set* generated by $S$ is $\uparrow S = \{p \in P \mid p \geqslant s$, for some $s \in S\}$. When $S$ is a singleton $\{s\}$, we write $\downarrow s$ for $\downarrow \{s\}$, and $\uparrow s$ for $\uparrow \{s\}$.

**Definition 4.8 (Open partition).** An *open partition* of a poset $P$ is a poset $(\pi, \preccurlyeq)$ where

(i) $\pi$ is a partition of the underlying set of $P$ such that for each $B \in \pi$ there exist $B_1, \ldots, B_k \in \pi$ satisfying [3]

$$\uparrow B = B_1 \cup \cdots \cup B_k . \tag{4.5}$$

(ii) for each $B, C \in \pi$, $B \preccurlyeq C$ if and only if there are $p_1 \in B$, $p_2 \in C$ such that $p_1 \leqslant p_2$.

**Theorem 4.3.** *Definitions 4.5 and 4.8 are equivalent.*

*Proof.* (Claim). If $\pi$ satisfies (i), then, for each $B, C \in \pi$, $C \subseteq \uparrow B$ if and only if there are $p_1 \in B$, $p_2 \in C$ such that $p_1 \leqslant p_2$.

Whenever $p_1 \leqslant p_2$, the block $C$ intersects the upper set of the block $B$. By (4.5), $C$ must be entirely contained in $\uparrow B$. The converse is trivial.

(Definition 4.8 $\Rightarrow$ Definition 4.5). Let $\pi$ be a partition of the underlying set of a poset $P$ satisfying (i) and let $\preccurlyeq$ be a partial order on $\pi$ satisfying (ii).

Let us consider the projection map $f : P \to \pi$ which sends each element of $P$ to its block. Let $p_1, p_2 \in P$. If $p_1 \leqslant p_2$ then, by (ii), $f(p_1) = [p_1] \preccurlyeq f(p_2) = [p_2]$ and $f$ is order-preserving. Since $\pi$ does not have empty blocks, $f$ is surjective. To show $f$ is open, we consider $p_1 \in P$, and $B \in \pi$, such that $B \preccurlyeq [p_1]$. By (ii) and (Claim), $[p_1] \subseteq \uparrow B$. Thus, there exists $p_2 \in B$ such that $p_2 \leqslant p_1$. Since $f(p_2) = [p_2] = B$, $f$ is open. Therefore, $(\pi, \preccurlyeq)$ is a regular partition of $P$ according to Definition 4.5.

(Definition 4.5 $\Rightarrow$ Definition 4.8). Let $f : P \to Q$ be a surjective open map, and let $(\pi_f, \preccurlyeq)$ be an open partition of $P$, according to Definition 4.5. Suppose, by way of contradiction, that (4.5) does not hold. Thus, there exist $p_1, p_2 \in C$ such that $p_1 \in \uparrow B$, but $p_2 \notin \uparrow B$, for some $B, C \in \pi_f$. Let $f(B) = q$. Since $f$ is order-preserving, $q \in \downarrow f(p_1)$. Since $f$ is open, $q \notin \downarrow f(p_2)$, for else we would find $p \in B$ with $p \leqslant p_2$, against the hypothesis $p_2 \notin \uparrow B$. Since $f(p_1) = f(p_2)$, $q \in \downarrow f(p_1)$ and $q \notin \downarrow f(p_2)$ is a contradiction. Thus, (i) holds.

To prove (ii) consider $p_1, p_2 \in P$, and suppose $p_1 \leqslant p_2$. Since $f$ is order-preserving, $f(p_1) \leqslant f(p_2)$. By Condition (4.3) in Definition 4.5, $[p_1] = f^{-1}(f(p_1)) \preccurlyeq [p_2] = f^{-1}(f(p_2))$, and one side of (ii) is proved.

Suppose now that for some $B, C \in \pi_f$, $B \preccurlyeq C$. Let $q_1 = f(B)$ and $q_2 = f(C)$. By Condition (4.3), $q_1 \leqslant q_2$ in $Q$. Since $f$ is surjective, fibres are not empty, and there exists $p_1 \in B$. Moreover, since $f$ is open, there exists $p_2 \in f^{-1}(q_1) = C$ such that $q_2 \leqslant q_1$. We have so proved (ii), and the proof is complete. Thus, $(\pi_f, \preccurlyeq)$ is an open partition of $P$ according to Definition 4.8.

---

[3] Here and in the following lower and upper sets are always relative to the order $\leqslant$ of the poset.

Next we prove that each open partition is solely determined by its underlying set.

**Corollary 4.3.** *Let $P$ be a poset, and let $\pi$ be a partition of the underlying set of $P$ satisfying* (i) *in Definition 4.8. Then, there exists a unique partial order $\leqslant$ on $P$ such that $(\pi, \leqslant)$ is an open partition of $\pi$.*

*Proof.* (Claim). For each $B, C \in \pi$, $C \subseteq \uparrow B$ if and only if there are $p_1 \in B$, $p_2 \in C$ such that $p_1 \leqslant p_2$.

Whenever $p_1 \leqslant p_2$, the block $C$ intersects the upper set of the block $B$. By (4.5), $C$ must be entirely contained in $\uparrow B$. The converse is trivial.

Let $\leqslant$ be the binary relation on $\pi$ satisfying (ii) in Definition 4.8. Clearly, $\leqslant$ is uniquely determined. We need to show that $\leqslant$ is a partial order on $\pi$. Reflexivity and transitivity of $\leqslant$ hold trivially. To show antisymmetry, let $B, C \in \pi$ be such that $B \leqslant C$ and $C \leqslant B$. Let $b_1 \in B$. Since $C \leqslant B$, by (Claim), there exists $c_1 \in C$ such that $c_1 \leqslant b_1$. Since $B \leqslant C$ there exists $b_2 \in B$ such that $b_2 \leqslant c_1 \leqslant b_1$. We can construct in this way an infinite chain $\cdots \leqslant c_i \leqslant b_i \leqslant \cdots \leqslant b_2 \leqslant c_1 \leqslant b_1$. Since $P$ is finite, we can find an element $b_s \in B$ which occurs in the chain more then once, and, by construction, an element $c_t \in C$ such that $b_s \leqslant c_t \leqslant b_s$. Therefore, $b_s = c_t$. Since $\pi$ is a set partition, and its blocks are pairwise disjoints, we obtain $B = C$. Thus, the relation $\leqslant$ is antisymmetric; therefore it is a partial order on $\pi$.

We close this section with an example.

*Example 4.1.* We refer to Figure 4.3, and consider the poset $P$.

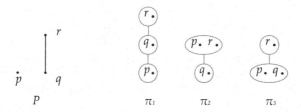

**Fig. 4.3** Example 4.1.

One can easily check, using the characterisations of poset partitions provided in Definitions 4.6, 4.7, and 4.8, that the following hold.

- $\pi_1$ is a monotone partition of $P$, but it is not regular.
- $\pi_2$ is a regular partition of $P$, but it is not open.
- $\pi_3$ is an open partition of $P$.

## 4.5 Partitions Induced by Quasiorders

A quasiorder relation $\lesssim$ on a set $A$ induces on $A$ an equivalence relation $\equiv$ defined as

$$x \equiv y \text{ if and only if } x \lesssim y \text{ and } y \lesssim x, \text{ for any } x, y \in A. \qquad (4.6)$$

The set $\pi$ of equivalence classes of $\equiv$ is a partition of $A$.

**Notation.** In the following we denote by $[x]_\lesssim$ the equivalence class (the block) of the element $x$ induced by the quasiorder $\lesssim$ via the equivalence relation defined in (4.6).

Further, the quasiorder $\lesssim$ induces on $\pi$ a partial order $\preccurlyeq$ defined by

$$x \lesssim y \text{ if and only if } [x]_\lesssim \preccurlyeq [y]_\lesssim, \text{ for any } x, y \in A. \qquad (4.7)$$

We call $(\pi, \preccurlyeq)$ *the poset of equivalence classes induced by* $\lesssim$.

This correspondence allows us to give a further definition of monotone, regular, and open partition of a poset by means of quasiorders.

**Definition 4.9 (Monotone partition).** A *monotone partition* of a poset $(P, \leqslant)$ is the poset of equivalence classes induced by a quasiorder $\lesssim$ on $P$ extending $\leqslant$, *i.e.* such that $\leqslant \subseteq \lesssim$.

**Theorem 4.4.** *Definitions 4.6 and 4.9 are equivalent.*

*Proof.* (Definition 4.6 $\Rightarrow$ Definition 4.9). Let $\pi$ be a partition of the underlying set of a poset $P$, and let $\preccurlyeq$ be a partial order on $\pi$ satisfying Condition (ii) in Definition 4.6. Consider the relation $\lesssim$ on $P$ defined by

$$p_1 \lesssim p_2 \text{ if and only if } [p_1]_\pi \preccurlyeq [p_2]_\pi, \text{ for each } p_1, p_2 \in P. \qquad (4.8)$$

One can easily check that $\lesssim$ is reflexive and transitive, and thus it is a quasiorder on $P$. Moreover, $[p_1]_\pi = [p_2]_\pi$ if and only if $[p_1]_\pi \preccurlyeq [p_2]_\pi$ and $[p_2]_\pi \preccurlyeq [p_1]_\pi$. Thus, by (4.8), $[p_1]_\pi = [p_2]_\pi$ if and only if $p_1 \lesssim p_2$ and $p_2 \lesssim p_1$. By (4.6), $[p_1]_\pi = [p_2]_\pi$ if and only if $[p_1]_\lesssim = [p_2]_\lesssim$. Therefore, $\pi$ coincide with the set of equivalence classes induced by $\lesssim$. Moreover, by (4.8) and (4.7), $\preccurlyeq$ coincide with the partial order on $\pi$ induced by $\lesssim$. Thus, $(\pi, \preccurlyeq)$ is the poset of equivalence classes of $P$ induced by $\lesssim$.

Suppose now that $p_1 \leqslant p_2$, for some $p_1, p_2 \in P$. By (ii) in Definition 4.6 we have $[p_1]_\pi \preccurlyeq [p_2]_\pi$. By (4.8), $p_1 \lesssim p_2$. Thus, $\lesssim$ extends $\leqslant$, and one side of the statement is proved.

(Definition 4.9 $\Rightarrow$ Definition 4.6). Let $(P, \leqslant)$ be a poset, let $\lesssim$ be a quasiorder on $P$ extending $\leqslant$, and let $(\pi, \preccurlyeq)$ be the poset of equivalence classes induced by $\lesssim$. Since $\lesssim$ extends $\leqslant$, for each $p_1, p_2 \in P$, $p_1 \leqslant p_2$ implies $p_1 \lesssim p_2$. Moreover, by (4.7), $p_1 \lesssim p_2$ implies $[p_1] \preccurlyeq [p_2]$. Thus, Condition (ii) in Definition 4.6 hold. We obtain that $(\pi, \preccurlyeq)$ is a monotone partition of $P$ according to Definition 4.6.

For what regular partitions are concerned, the definition in terms of quasiorders can be formulated as follows.

**Definition 4.10 (Regular partition).** A *regular partition* of a poset $(P, \leqslant)$ is the poset of equivalence classes induced by a quasiorder $\lesssim$ on $P$ extending $\leqslant$ and satisfying

$$\lesssim \; = \; \mathrm{tr}(\lesssim \setminus \rho), \tag{4.9}$$

where $\mathrm{tr}(R)$ denotes the transitive closure of the relation $R$, and $\rho$ is a binary relation defined by

$$\rho = \{(x, y) \in P \times P \mid x \lesssim y, \, x \not\leqslant y, \, y \not\leqslant x\}.$$

**Theorem 4.5.** *Definitions 4.7 and 4.10 are equivalent.*

*Proof.* (Definition 4.7 $\Rightarrow$ Definition 4.10). Let $\pi$ be a partition of the underlying set of a poset $P$, and let $\preccurlyeq$ be a partial order on $\pi$ satisfying Condition (ii) in Definition 4.7. Consider the relation $\lesssim$ on $P$ defined by

$$p_1 \lesssim p_2 \text{ if and only if } [p_1]_\pi \preccurlyeq [p_2]_\pi, \text{ for each } p_1, p_2 \in P. \tag{4.10}$$

Since $(\pi, \preccurlyeq)$ is a monotone partition of $P$, proceeding as in the first part of the proof of Theorem 4.4, we obtain that $(\pi, \preccurlyeq)$ is the poset of equivalence classes induced by $\lesssim$, and that $\lesssim$ extends $\leqslant$. Moreover, by Condition (ii) in Definition 4.7, $\lesssim$ coincides with $\lesssim_\pi$.

It remains to prove that $\lesssim$ satisfies (4.9). Let $p_1, p_2 \in P$ be such that $p_1 \lesssim p_2$. If $(p_1, p_2) \notin \rho$, then $(p_1, p_2) \in \mathrm{tr}(\lesssim \setminus \rho)$, trivially. Let then $(p_1, p_2) \in \rho$, that is $p_1 \lesssim p_2$, $p_2 \not\lesssim p_1$, $p_1 \not\leqslant p_2$. Since $\lesssim$ coincides with $\lesssim_\pi$, there exists a sequence $p_1 = x_0, y_0, x_1, y_1, \ldots, x_n, y_n = p_2$ satisfying Conditions (1) and (2) in Definition 4.1. By Condition (1) in Definition 4.1, and by (4.6), for each $i \in \{0, \ldots, n\}$, $x_i \lesssim y_i$ and $y_i \lesssim x_i$ hold. Thus, $(x_i, y_i) \notin \rho$. By Condition (2) in Definition 4.1, and since $\lesssim$ extends $\leqslant$, for each $i \in \{0, \ldots, n-1\}$, we have that $y_i \lesssim x_{i+1}$. Thus, $(y_i, x_{i+1}) \notin \rho$. Summing up, no pair of adjacent element in the sequence $x_0, y_0, x_1, y_1, \ldots, x_n, y_n$ belongs to $\rho$. Since each of these pairs belong to $\lesssim$, $(p_1, p_2) \in \mathrm{tr}(\lesssim \setminus \rho)$. Thus, (4.9) holds, and the first part of the statement is settled.

(Definition 4.10 $\Rightarrow$ Definition 4.7). Let $(P, \leqslant)$ be a poset, let $\lesssim$ be a quasiorder on $P$ extending $\leqslant$ and satisfying (4.9) and let $(\pi, \preccurlyeq)$ be the poset of equivalence classes induced by $\lesssim$.

(Claim 1). $p_1 \lesssim_\pi p_2$ implies $p_1 \lesssim p_2$, for each $p_1, p_2 \in P$.

Suppose that $p_1 \lesssim_\pi p_2$, for some $p_1, p_2 \in P$. Thus, there exists a sequence $p_1 = x_0, y_0, x_1, y_1, \ldots, x_n, y_n = p_2 \in P$ satisfying Conditions (1) and (2) in Definition 4.1. By Condition (1), for each $i \in \{0, \ldots, n\}$, $[x_i]_\pi = [y_i]_\pi$. Thus, by (4.7), $x_i \lesssim y_i$. By Condition (2), for each $i \in \{0, \ldots, n-1\}$, $y_i \preccurlyeq x_{i+1}$. Since, by hypothesis, $\lesssim$ extends $\leqslant$, we have $y_i \lesssim x_{i+1}$. Therefore, whenever $p_1 \lesssim_\pi p_2$, we have $p_1 = x_0 \lesssim y_0 \lesssim \cdots \lesssim x_n \lesssim y_n = p_2$. By transitivity $p_1 \lesssim p_2$.

(Claim 2). $p_1 \lesssim p_2$ implies $p_1 \lesssim_\pi p_2$, for each $p_1, p_2 \in P$.

Let $p_1 \lesssim p_2$, for some $p_1, p_2 \in P$. We shall analyze three different cases, covering all the possibilities for $p_1$ and $p_2$.

Case (a). $p_2 \lesssim p_1$. Since $p_1 \lesssim p_2$ and $p_2 \lesssim p_1$, by (4.6), we have $[p_1]_\pi = [p_2]_\pi$. Thus, the sequence $p_1, p_2$ satisfies Conditions (1) and (2) in Definition 4.1, with respect to $\pi$. We obtain $p_1 \lesssim_\pi p_2$.

Case (b). $p_2 \not\lesssim p_1$, and $p_1 \leqslant p_2$. Since the sequence $p_1, p_1, p_2, p_2$ satisfies Conditions (1) and (2) in Definition 4.1, we have $p_1 \lesssim_\pi p_2$.

Case (c). $p_2 \not\lesssim p_1$, $p_1 \not\leqslant p_2$, that is, $(p_1, p_2) \in \rho$. Let $R = \lesssim \backslash \rho$. Since $p_1 \lesssim p_2$, the pair $(p_1, p_2)$ arises in $\lesssim$ from the transitive closure of the binary relation $R$. Thus, there exists a sequence $p_1 = z_0 \lesssim z_1 \lesssim \cdots \lesssim z_r = p_2$ of elements of $P$ such that, for all $i \in \{0, \ldots, r-1\}$, $z_i \, R \, z_{i+1}$. For each of this pair of elements, either Case (a) or Case (b) apply. Thus, $z_i \lesssim_\pi z_{i+1}$. Since $\lesssim_\pi$ is transitive, we obtain immediately $p_1 \lesssim_\pi p_2$.

In any case, whenever $p_1 \lesssim p_2$, we have $p \lesssim_\pi q$, and the claim is settled.

By (Claim 1) and (Claim 2), the quasiorders $\lesssim$ and $\lesssim_\pi$ coincide, and Condition (ii) in Definition 4.7 is trivially verified. We obtain that $(\pi, \leqslant)$ is a regular partition of $P$ according to Definition 4.7.

*Remark 4.3.* Let $(\pi, \leqslant)$ be the monotone partition induced by a quasiorder $\lesssim$. By the construction given in the proof of Theorem 4.5, we infer that $(\pi, \leqslant)$ is a regular partition if and only if $\lesssim$ coincides with $\lesssim_\pi$.

In the case of open partitions, the definition in terms of quasiorders is as follows.

**Definition 4.11 (Open partition).** An *open partition* of a poset $(P, \leqslant)$ is the poset of equivalence classes induced by a quasiorder $\lesssim$ on $P$ extending $\leqslant$ and such that, for every $p, q \in P$,

$$\text{if } p \lesssim q, \text{ then there exists } p' \in P \text{ such that } p \lesssim p', \; p' \lesssim p, \text{ and } p' \leqslant q. \quad (4.11)$$

**Theorem 4.6.** *Definitions 4.8 and 4.11 are equivalent.*

*Proof.* (Definition 4.8 $\Rightarrow$ Definition 4.11). Let $(\pi, \leqslant)$ be an open partition of a poset $P$, satisfying thus Conditions (i) and (ii) in Definition 4.8. Consider the relation $\lesssim$ on $P$ defined by

$$p_1 \lesssim p_2 \text{ if and only if } [p_1]_\pi \leqslant [p_2]_\pi, \text{ for each } p_1, p_2 \in P. \quad (4.12)$$

Since $(\pi, \leqslant)$ is a monotone partition of $P$, proceeding as in the first part of the proof of Theorem 4.4, we obtain that $(\pi, \leqslant)$ is the poset of equivalence classes induced by $\lesssim$, and that $\lesssim$ extends $\leqslant$. Let $p \lesssim q$ for some $p, q \in P$. By (4.12), $[p]_\pi \leqslant [q]_\pi$. By Conditions (i) and (ii) in Definition 4.8, $[p]_\pi \leqslant [q]_\pi$ implies $[q]_\pi \subseteq\uparrow [p]_\pi$. Thus, there exists $p' \in [p]_\pi$ such that $p' \leqslant q$, and (4.11) is satisfied. Therefore, $(\pi, \leqslant)$ is an open partition of $P$ according to Definition 4.11.

(Definition 4.11 $\Rightarrow$ Definition 4.8). Let $(\pi, \leqslant)$ be the poset of equivalence classes induced by $\lesssim$.

(Claim). Let $B, C \in \pi$. Then, $B \lesssim C$ if and only if $C \subseteq\uparrow B$.

($\Leftarrow$) Suppose $C \subseteq\uparrow B$. Since blocks are nonempty, there exist $p \in B$ and $q \in C$ such that $p \leqslant q$. Since $\lesssim$ extends $\leqslant$, we have $p \lesssim q$. Hence, by (4.7), $B \lesssim C$.

($\Rightarrow$) Suppose $B \lesssim C$. By (4.7), for each $q \in C$, $p \in B$, we have $p \lesssim q$. Moreover, by (4.11), there exists $p' \in B$ such that $p' \leqslant q$. Thus, for each $q \in C$ there exists $p' \in B$ such that $p' \leqslant q$. In other words, $C \subseteq\uparrow B$.

Conditions (ii) in Definition 4.8 follows immediately from (Claim). Let $B, C \in \pi$, and suppose that $p \leqslant q$, for some $p \in B$, $q \in C$. Since $\lesssim$ extends $\leqslant$, we have $p \lesssim q$. Moreover, by (4.7), $B \lesssim C$. By (Claim), $C$ is entirely contained in $\uparrow B$. We have so proved that whenever a block $C$ intersects $\uparrow B$, the block $C$ is entirely contained in $\uparrow B$. Thus, $\uparrow B$ can be written as a union of blocks of $\pi$, as in Conditions (i) in Definition 4.8, and the theorem is proved.

## 4.6 Further Work

The further development of a theory of partitions of finite posets can follow several research direction. One of these concerns enumeration problems. It is easy to realize that enumerating poset partitions is never a simple problem, except in trivial cases – *e.g.* counting monotone, regular, and open partitions of chains, or of trivially ordered posets. Indeed, as we have shown, the enumeration of poset partitions is tightly related to the problem of counting the number of quasiorders on a finite set of points. Special cases may be tractable; we plan to address the issue in future work, following the relatively simple cases investigated in [3].

Another direction the we plan to follow in our future research involves the investigation of the ordered structure of all poset partitions of a poset. In [3], we proved that the classes of all monotone and regular partitions of a poset form a lattice, and we analysed its first properties.

Applications of the notions of monotone, regular, and open partitions can also be found in the study of distributive lattices, and Heyting algebras. Indeed, we recall here the well-known fact that the category Pos is dually equivalent to the category of finite bounded distributive lattices with their {0, 1}-preserving lattice homomorphisms (for details see, *e.g.*, [5]). In the dual category of finite distributive lattices our notions of monotone and regular partition correspond precisely to the notions of sublattice and regular sublattice (see [3, Chapter 5]), respectively, of a distributive lattice.

Concerning open partitions, the category OPos can be proved to be dually equivalent to the category of finite Heyting algebras with their homomorphisms (for details on Heyting algebras, see, *e.g.*, [6]). In the dual category of finite Heyting algebras our notion of open partition corresponds precisely to the notion of subalgebra.

The above duality allows, for instance, the application of our results on open partitions to the study of the notion of probability distribution in Gödel logic (an extension of the intuitionistic propositional calculus); please see [4].

# References

1. Adámek, J., Herrlich, H., Strecker, G. E. (2004), *Abstract and Concrete Categories - The Joy of Cats*, electronic edition.
2. Chagrov, A., Zakharyaschev, M. (1997), *Modal logic*, (Oxford Logic Guides, 35), Oxford, Clarendon.
3. Codara, P. (2008), *A theory of partitions of partially ordered sets*. Ph.D. Diss., Milan, Università degli Studi di Milano.
4. Codara, P., D'Antona, O., Marra, V. (2009), *Open Partitions and Probability Assignments in Gödel Logic*, in *ECSQARU 2009*, LNCS (LNAI), in press.
5. Davey, B. A., Priestley, H. A. (2002), *Introduction to Lattices and Order*, 2nd ed., Cambridge, Cambridge University Press.
6. Johnstone, P. T. (1982), *Stone Spaces*, (Cambridge Studies in Advanced Mathematics, 3), Cambridge, Cambridge University Press.

# Chapter 5
# An Algebra of Pieces of Space — Hermann Grassmann to Gian Carlo Rota
## Invited Chapter

Henry Crapo

**Abstract** We sketch the outlines of Gian Carlo Rota's interaction with the ideas that Hermann Grassmann developed in his *Ausdehnungslehre*[13, 15] of 1844 and 1862, as adapted and explained by Giuseppe Peano in 1888. This leads us past what Gian Carlo variously called *Grassmann-Cayley algebra* and *Peano spaces* to the *Whitney algebra* of a matroid, and finally to a resolution of the question "What, really, was Grassmann's *regressive product*?". This final question is the subject of ongoing joint work with Andrea Brini, Francesco Regonati, and William Schmitt.

## 5.1 Almost Ten Years Later

We are gathered today in order to renew and deepen our recollection of the ways in which our paths intersected that of Gian Carlo Rota. We do this in poignant sadness, but with a bitter-sweet touch: we are pleased to have this opportunity to meet and to discuss his life and work, since we know how Gian Carlo transformed us through his friendship and his love of mathematics.

We will deal only with the most elementary of geometric questions; how to represent pieces of space of various dimensions, in their relation to one another. It's a simple story, but one that extends over a period of some 160 years. We'll start and finish with Hermann Grassmann's project, but the trail will lead us by Giuseppe Peano, Hassler Whitney, to Gian Carlo Rota and his colleagues.

Before I start, let me pause for a moment to recall a late afternoon at the Accademia Nazionale dei Lincei, in 1973, on the eve of another talk I was petrified to give, when Gian Carlo decided to teach me how to *talk*, so I wouldn't make a fool of myself the following day. The procedure was for me to start my talk, with an audience of one, and he would interrupt whenever there was a problem. We were in that otherwise empty conference hall for over two hours, and I never got past my first

Henry Crapo
Centre de Recherche "Les Moutons matheux", France

E. Damiani et al. (eds.), *From Combinatorics to Philosophy,*
DOI 10.1007/978-0-387-88753-1_5, © Springer Science+Business Media, LLC 2009

paragraph. It was terrifying, but it at least got me through the first battle with my fears and apprehensions, disguised as they usually are by timidity, self-effacement, and other forms of apologetic behavior.

## 5.2 Synthetic Projective Geometry

Grassmann's plan was to develop a purely formal algebra to model natural (synthetic) operations on geometric objects: *flat*, or *linear* pieces of space of all possible dimensions. His approach was to be *synthetic*, so that the symbols in his algebra would denote geometric objects themselves, not just numbers (typically, coordinates) that could be derived from those objects by measurement. His was not to be an algebra of numerical quantities, but an algebra of pieces of space.

In the *analytic* approach, so typical in the teaching of Euclidean geometry, we are encouraged to assign "unknown" variables to the coordinates of variable points, to express our hypotheses as equations in those coordinates, and to derive equations that will express our desired conclusions.

The main advantage of a synthetic approach is that the logic of geometric thought and the logic of algebraic manipulations may conceivably remain parallel, and may continue to cast light upon one another. Grassmann expressed this clearly in his introduction to the *Ausdehnungslehre*[13, 14]:

Grassmann (1844): *"Each step from one formula to another appears at once as just the symbolic expression of a parallel act of abstract reasoning. The methods formerly used require the introduction of arbitrary coordinates that have nothing to do with the problem and completely obscure the basic idea, leaving the calculation as a mechanical generation of formulas, inaccessible and thus deadening to the intellect. Here, however, where the idea is no longer strangely obscured but radiates through the formulas in complete clarity, the intellect grasps the progressive development of the idea with each formal development of the mathematics"*.

In our contemporary setting, a synthetic approach to geometry yields additional benefits. At the completion of a synthetic calculation, there is no need to *climb back up* from scalars (real numbers, necessarily subject to round-off errors, often rendered totally useless by division by zero) or from drawings, fraught with their own approximations of incidence, to statements of geometric incidence. In the synthetic approach, one even receives precise warnings as to particular positions of degeneracy. The synthetic approach is thus tailor-made for machine computation.

Gian Carlo was a stalwart proponent of the synthetic approach to geometry during the decade of the 1960's, when he studied the combinatorics of ordered sets and lattices, and in particular, matroid theory. But this attitude did not withstand his encounter with invariant theory, beginning with his lectures on the invariant theory of the symmetric group at the A.M.S. summer school at Bowdoin College in 1971.

As Gian Carlo later said, with his admirable fluency of expression in his adopted tongue,

*"Synthetic projective geometry in the plane held great sway between 1850 and 1940. It is an instance of a theory whose beauty was largely in the eyes of its beholders. Numerous expositions were written of this theory by English and Italian mathematicians (the definitive one being the one given by the American mathematicians Veblen and Young). These expositions vied with one another in elegance of presentation and in cleverness of proof; the subject became required by universities in several countries. In retrospect, one wonders what all the fuss was about".*

*"Nowadays, synthetic geometry is largely cultivated by historians, and the average mathematician ignores the main results of this once flourishing branch of mathematics. The claim that has been raised by defenders of synthetic geometry, that synthetic proofs are more beautiful than analytic proofs, is demonstrably false. Even in the nineteenth century, invariant-theoretic techniques were available that could have provided elegant, coordinate-free analytic proofs of geometric facts without resorting to the gymnastics of synthetic reasoning and without having to stoop to using coordinates".*

Once one adopts an invariant-theoretic approach, much attention must be paid to the reduction of expressions to standard form, where one can recognize whether a polynomial in determinants is equal to zero. The process is called *straightening*, and it was mainly to the algorithmic process of straightening in a succession of algebraic contexts that Gian Carlo devoted his creative talents during three decades. We filled pages with calculations such as the following, for the bracket algebra of six points in a projective plane:

$$\begin{array}{ccc} b\,c\,d \\ a\,e\,f \end{array} - \begin{array}{ccc} a\,c\,d \\ b\,e\,f \end{array} + \begin{array}{ccc} a\,b\,d \\ c\,e\,f \end{array} - \begin{array}{ccc} a\,b\,c \\ d\,e\,f \end{array} = 0$$

the *straightening* of the two-rowed tableau on the left. This expression we would write in *dotted* form

$$\begin{array}{ccc} \dot b & \dot c & \dot d \\ \dot a & e & f \end{array}$$

that would be *expanded* to the above expression by summing, with alternating sign, over all permutations of the dotted letters. The basic principle is that *dotting a dependent set of points yields zero.*

To make a long and fascinating story short, Gian Carlo finally settled upon a most satisfactory compromise, a formal super-algebraic calculus, developed with Joel Stein, Andrea Brini, Marilena Barnabei [2, 3, 4, 5] and a generation of graduate students, that managed to stay reasonably close to its synthetic geometric roots. His brilliant students Rosa Huang and Wendy Chan [7, 8, 4, 18] carried the torch of synthetic reasoning across this new territory. They rendered feasible a unification of super-algebraic and synthetic geometry, but, as we soon realize, the process is far from complete. First, we should take a closer look at Grassmann's program for synthetic geometry.

## 5.3 Hermann Grassmann's Algebra

Grassmann emphasizes that he is building an *abstract* theory that can then be *applied to real physical space*. He starts not with geometric axioms, but simply with the notion of a skew-symmetric product of letters, which is assumed to be distributive over addition and modifiable by scalar multiplication. Thus if $a$ and $b$ are letters, and $A$ is any product of letters, then $Aab + Aba = 0$. This is the skew-symmetry. It follows that any product with a repeated letter is equal to zero.

He also develops a notion of *dependency*, saying that a letter $e$ is *dependent* upon a set $\{a, b, c, d\}$ if and only if there are scalars $\alpha, \beta, \gamma, \delta$ such that

$$\alpha a + \beta b + \gamma c + \delta d = e.$$

Grassmann realizes that such an expression is possible if and only if the point $e$ lies in the *common system*, or projective subspace spanned by the points $a, b, c, d$. He uses an axiom of distributivity of product over sums to prove that the product of letters forming a dependent set is equal to zero. With $a, b, c, d, e$ as above:

$$abcde = \alpha\,abcda + \beta\,abcdb + \gamma\,abcdc + \delta\,abcdd = 0$$

the terms on the right being zero as products because each has a repeated letter.

As far as I can see, he establishes no formal axiomatization of the relation of linear dependence, and in particular, no statement of the exchange property. For that, we must wait until 1935, and the matroid theory of Whitney, MacLane and Birkhoff.

The application to geometry is proposed via *an interpretation* of this abstract algebra. The individual letters may be understood as *points*,

*The center of gravity of several points can be interpreted as their sum, the displacement between two points as their product, the surface area lying between three points as their product, and the region (a pyramid) between four points as their product.*

Grassmann is delighted to find that, in contrast to earlier formalizations of geometry, there need be no a priori maximum to the rank of the overall space being described:

*The origins of this science are as immediate as those of arithmetic; and in regard to content, the limitation to three dimensions is absent. Only thus do the laws come to light in their full clarity, and their essential interrelationships are revealed.*

Giuseppe Peano, in rewriting Grassmann's theory, chose to assume a few basic principles of *comparison of signed areas and volumes*, and to base all proofs on those principles. For instance, given a line segment $ab$ between distinct points $a, b$ in the plane, and points $c, d$ off that line, then the signed areas $abc$ and $abd$ will be equal (equal in magnitude, and equally oriented CW or CCW) if and only if $c$ and $d$ lie on a line *parallel* to $ab$. See Figure 5.1.

The corresponding statement for three-dimensional space is that signed volumes $abcd$ and $abce$, for non-collinear points $a, b, c$, are equal (equal volume, and with the same chirality, or handedness) if and only if $d$ and $e$ lie on a line *parallel* to the plane $abc$. See Figure 5.2. Since Peano restricts his attention to three-dimensional space, this principle is his main axiom, with a notion of parallelism taken to be understood. This means that even the simplest geometric properties are ultimately rephrased as equations among measured volumes. For instance, Peano wishes to show that three points $a, b, c$ are *collinear* if and only if the linear combination $bc - ac + ab$ of products of points is equal to zero. He shows that for every pair $p, q$ of points, if the points $a, b, c$ are in that order on a line, the tetrahedron $acpq$ is the disjoint union of the tetrahedra $abpq$ and $bcpq$, so their volumes add:

$$abpq + bcpq = acpq.$$

A further argument about symmetry shows that $ab + bc = ac$ holds independent of the order of $a, b, c$ on the line. The statement $abpq + bcpq = acpq$, quantified over all choices of $p$ and $q$, is Peano's *definition* of the equality $ab + bc = ac$. So his proof is complete. Perhaps we can agree that this is putting the cart before the horse. (Grassmann took expressions of the form $ab + bc = ac$, for three collinear points $a, b, c$, to be axiomatic in his algebra.)

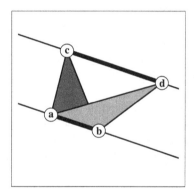

**Fig. 5.1** $abc = abd$ *iff* line $ab$ parallel to line $cd$.

I mention this strange feature of Peano's version because it became something of an *idée fixe* for Gian Carlo's work on Peano spaces. It gave rise to the technique of *filling brackets* in order to verify equations involving flats of low rank, a technique that unnecessarily relies on information concerning the rank of the overall space.

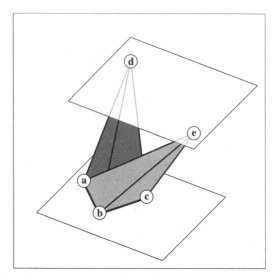

**Fig. 5.2** *abcd* = *abce* *iff* line *de* parallel to plane *abc*.

## 5.4 Extensors and Vectors

We are all familiar with the formation of linear combinations of points in space, having studied linear algebra and having learned to replace points by their *position vectors*, which can then be multiplied by scalars and added. The origin serves as reference point, and that is all we need.

Addition of points is not well-defined in real projective geometry, because although points in a space of rank $n$ (projective dimension $n-1$) may be represented as $n$-tuples of real numbers, the $n$-tuples are only determined up to an overall non-zero scalar multiple, and addition of these vectors will not produce a well-defined result. The usual approach is to consider *weighted points*, consisting of a position, given by *standard homogeneous coordinates*, of the form $(a_1, \ldots, a_{n-1}, 1)$, and a *weight* $\mu$, to form a *point* $(\mu a_1, \ldots, \mu a_{n-1}, \mu)$ *of weight* $\mu$. This is what worked for Möbius in his *barycentric calculus*. And it is the crucial step used by Peano to clarify the presentation of Grassmann's algebra.

This amounts to fixing a choice of *hyperplane at infinity* with equation $x_n = 0$. The *finite points* are represented as above, with weight 1. A linear combination $\lambda a + \mu c$, for scalars $\lambda$ and $\mu$ positive, is a point $b$ situated between $a$ and $c$, such that the ratio of the distance $a \to b$ to the distance $b \to c$ is in the inverse proportion $\mu/\lambda$, and the resulting weighted point has weight equal to $\lambda + \mu$, as illustrated in Figure 5.3.

In particular, for points $a$ and $b$ of weight 1, $a + b$ is located the midpoint of the interval $ab$, and has weight 2, while $2a + b$ is located twice as far from $b$ as from $a$, and has weight 3.

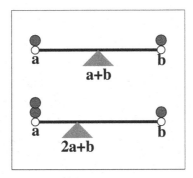

**Fig. 5.3**  Addition of weighted points.

Both Grassmann and Peano are careful to distinguish between products $ab$ of points, which have come to be called 2-*extensors*, and differences $b - a$ of points, which Grassmann calls *displacements* and Peano calls 1-*vectors*. The distinction between such types of objects is easily explained in terms of modern notation, in homogeneous coordinates.

In Figure 5.4 we show two equal 2-extensors, $ab = cd$, in red, and their difference vector, $v = b - a = d - c = f - e$ in blue, which represents a projective point, not a line, namely, the point at infinity on the line $ab$, with weight equal to the length from $a$ to $b$ and sign indicating the orientation from $a$ to $b$. Check that $ab = av = bv$, so multiplication of a point $a$ on the right by a vector $v$ creates a line segment of length and direction $v$ starting at $a$.

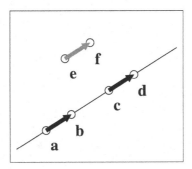

**Fig. 5.4**  $ab = cd$, while $b - a = d - c = f - e$ is a point at infinity.

To avoid cumbersome notation, we will henceforth follow Peano's example, and give all examples with reference to a real projective 3-space, rank 4.

The homogeneous coordinates of any product of $k \leq 4$ weighted points (called a $k$-*extensor*), are the $k \times k$ minors of the matrix whose rows are the coordinate vectors of the $k$ points in question. So for any four weighted points $a, b, c, d$ in a space of rank 4,

$$a = \begin{pmatrix} 1 & 2 & 3 & 4 \\ a_1 & a_2 & a_3 & a_4 \end{pmatrix}$$

$$ab = \begin{pmatrix} \mathbf{12} & \mathbf{13} & \mathbf{23} & \mathbf{14} & \mathbf{24} & \mathbf{34} \\ \begin{vmatrix} a_1 & a_2 \\ b_1 & b_2 \end{vmatrix} & \begin{vmatrix} a_1 & a_3 \\ b_1 & b_3 \end{vmatrix} & \begin{vmatrix} a_2 & a_3 \\ b_2 & b_3 \end{vmatrix} & \begin{vmatrix} a_1 & a_4 \\ b_1 & b_4 \end{vmatrix} & \begin{vmatrix} a_2 & a_4 \\ b_2 & b_4 \end{vmatrix} & \begin{vmatrix} a_3 & a_4 \\ b_3 & b_4 \end{vmatrix} \end{pmatrix}$$

$$abc = \begin{pmatrix} \mathbf{123} & \mathbf{124} & \mathbf{134} & \mathbf{234} \\ \begin{vmatrix} a_1 & a_2 & a_3 \\ b_1 & b_2 & b_3 \\ c_1 & c_2 & c_3 \end{vmatrix} & \begin{vmatrix} a_1 & a_2 & a_4 \\ b_1 & b_2 & b_4 \\ c_1 & c_2 & c_4 \end{vmatrix} & \begin{vmatrix} a_1 & a_3 & a_4 \\ b_1 & b_3 & b_4 \\ c_1 & c_3 & c_4 \end{vmatrix} & \begin{vmatrix} a_2 & a_3 & a_4 \\ b_2 & b_3 & b_4 \\ c_2 & c_3 & c_4 \end{vmatrix} \end{pmatrix}$$

$$abcd = \begin{pmatrix} \mathbf{1234} \\ \begin{vmatrix} a_1 & a_2 & a_3 & a_4 \\ b_1 & b_2 & b_3 & b_4 \\ c_1 & c_2 & c_3 & c_4 \\ d_1 & d_2 & d_3 & d_4 \end{vmatrix} \end{pmatrix}$$

If $a,b,c,d$ are points of weight 1, the 2-extensors $ab$ and $cd$ are equal if and only if the line segments from $a$ to $b$ and from $c$ to $d$ are *collinear*, of equal length, and similarly oriented. More generally, Grassmann showed that two $k$-extensors $a\ldots b$ and $c\ldots d$ differ only by a non-zero scalar multiple, $a\ldots b = \sigma\, a\ldots d$, if and only if the sets of points obtainable as linear combinations of $a,\ldots,b$ and those from $c,\ldots,d$ form what we would these days call *the same projective subspace*. Such a subspace, considered as a set of projective points, we call a *projective flat*.

Coplanar 2-extensors add the way coplanar forces do in physical systems. Say you are forming the sum $ab + cd$ as in Figure 5.5. You slide the line segments representing the forces $ab$ and $cd$ along their lines of action until the ends $a$ and $c$ coincide at the point $e$ of incidence of those two lines. The sum is then represented as the diagonal line segment of the parallelogram they generate.

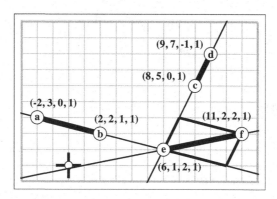

**Fig. 5.5** $ab + cd = ef$ for these six points of weight 1.

Let's carry out the explicit extensor calculations, so you know I'm not bluffing: let $a, b, c, d$ be four points coplanar in 3-space (rank 4),

|   | 1 | 2 | 3 | 4 |
|---|---|---|---|---|
| **a** | −2 | 3 | 0 | 1 |
| **b** | 2 | 2 | 1 | 1 |
| **c** | 8 | 5 | 0 | 1 |
| **d** | 9 | 7 | −1 | 1 |

That the four points are coplanar is clear from the fact that the four vectors are dependent: $-a + 2b - 3c + 2d = 0$. The 2-extensors $ab$ and $cd$ are, with their sum:

|   | 12 | 13 | 23 | 14 | 24 | 34 |
|---|---|---|---|---|---|---|
| **ab** | −10 | −2 | 3 | −4 | 1 | −1 |
| **cd** | 11 | −8 | −5 | −1 | −2 | 1 |
| **ab + cd** | 1 | −10 | −2 | −5 | −1 | 0 |

The point $e$ of intersection of lines $ab$ and $cd$, together with the point $f$ situated at the end of the diagonal of the parallelogram formed by the translates of the two line segments to $e$, have homogeneous coordinates

|   | 1 | 2 | 3 | 4 |
|---|---|---|---|---|
| **e** | 6 | 1 | 2 | 1 |
| **f** | 11 | 2 | 2 | 1 |

and exterior product

|   | 12 | 13 | 23 | 14 | 24 | 34 |
|---|---|---|---|---|---|---|
| **ef** | 1 | −10 | −2 | −5 | −1 | 0 |

equal to the sum $ab + cd$.

Cospatial planes add in a similar fashion. For any 3-extensor $abc$ spanning a projective plane $Q$ and for any 2-extensor $pq$ in the plane $Q$, there are points $r$ in $Q$ such that $abc = pqr$. The required procedure is illustrated in Figure 5.6. We slide the point $c$ parallel to the line $ab$ until it reaches the line $pq$, shown in blue, at $c'$. Then slide the point $a$ parallel to the line $bc'$ until it reaches the line $pq$ at $a'$. The oriented plane areas $abc$, $abc'$, and $a'bc'$ are all equal, and the final triangle has an edge on the line $pq$.

So, given any 3-extensors $abc, def$, and for any pair $p, q$ of points on the line of intersection of the planes $abc$ and $def$, there exist points $r, s$ such that $abc = pqr, def = pqs$, so $abc + def$ can be expressed in the form $pqr + pqs = pq(r + s)$, and the problem of adding planes in 3-space is reduced to the problem of adding points on a line. The result is shown in Figure 5.7, where $t = r + s$.

In dimensions $\geq 3$, it is necessary to *label* the individual coordinates with the set of columns used to calculate that minor. This practice becomes even more systematic in the subsequent super-algebraic approach of Rota, Stein, Brini and colleagues,

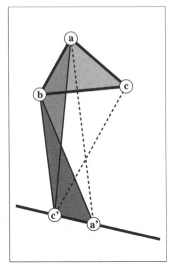

**Fig. 5.6** The 3-extensors $abc$ and $a'bc'$ are equal.

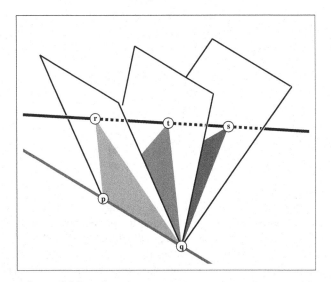

**Fig. 5.7** A sum of cospatial 3-extensors, $pqr + pqs = pqt$.

where a coordinate $abc_{ijk}$ will be denoted $(abc|ijk)$, with *negative letters* $a, b, c$ to denote vectors and *negative places* $i, j, k$ to denote coordinate positions, and the minor is calculated by the *Laplace expansion*:

$$(abc|ijk) = (a|i)(b|j)(c|k) - (a|i)(b|k)(c|j) + \cdots - (a|k)(b|j)(c|i),$$
$$= (a|i)(b|j)(c|k) - (a|i)(c|j)(b|k) + \cdots - (c|i)(b|j)(a|k)$$

as the sum of products of individual *letter-place* elements. We have chosen to list the coordinates of a line segment $ab$ in 3-space in the order $12, 13, 23, 14, 24, 34$ because that makes the negative of the difference vector $b - a$ visible in the last three coordinate places $14, 24, 34$, and the *moments* about the $3^{rd}$, $2^{nd}$, and $1^{st}$ coordinate axes, respectively, visible in the first three coordinate places.

A set of $k$ 1-extensors $a, b, \ldots d$ has a non-zero product if and only if the $k$ points are independent, and thus span a projective subspace of rank $k$ (dimension $k - 1$). This integer $k$ is called the *step* of the extensor.

We have seen that differences of points $a, b$ of weight 1 are vectors, which means simply that they are 1-extensors $e$ *at infinity*, with coordinate $e_4 = 0$.

In Grassmann's theory, there exist $k$-vectors of all steps $k$ for which $k$ is less than the rank of the entire space. In terms of standard homogeneous coordinates in a space of rank 4, they are those extensors for which all coordinates are zero whose labels involve the place 4. $k$-vectors are also definable as "boundaries" of $(k + 1)$-extensors, or as products of 1-vectors, as we shall see.

Each $k$-extensor has an associated $(k - 1)$-vector, which I shall refer to as its *boundary*,

$$\partial ab = b - a$$
$$\partial abc = bc - ac + ab$$
$$\partial abcd = bcd - acd + abd - abc$$

If the points $a, b, c, d$ are of weight 1, then the $4^{th}$ coordinate of $\partial ab$, the $14, 24, 34$ coordinates of $\partial abc$, and the $124, 134, 234$ coordinates of $\partial abcd$ are all equal to zero. Such extensors are, as elements of our affine version of projective space, pieces of space in the hyperplane at infinity. As algebraic objects they are vectors, so $\partial ab = \partial cd$, or $b - a = d - c$, for points $a, b, c, d$ of weight 1, if and only if the line segments from $a$ to $b$ and from $c$ to $d$ are parallel, of equal length, and similarly oriented. That is, they are equal as difference vectors of position.

The subspace of $k$-vectors consists exactly of those $k$-tensors obtained by taking boundaries of $k + 1$-extensors. The $k$-vectors are also expressible as exterior products of $k$ 1-vectors, since

$$\partial ab = (b - a)$$
$$\partial abc = (b - a)(c - a),$$
$$\partial abcd = (b - a)(c - a)(d - a),$$

$$\cdots$$

Remarkably, $p \partial abc = abc$ for any point $p$ in the plane spanned by $a, b, c$. In particular, $a \partial abc = b \partial abc = c \partial abc$.

So $\partial abc = bc - ac + ab$ is a 2-vector, and it can only have non-zero coordinates with labels $12, 13, 23$. Geometrically, it can be considered to be a directed line segment in the line at infinity in the plane $abc$. It is also equal to a *couple* in the plane of the points $a, b, c$, that is, the sum of two 2-extensors that are parallel to one another, of equal length and opposite orientation. Couples of forces occur in statics, and cause rotation when applied to a rigid body.

In Figure 5.8 we show two couples that are equal, though expressed as sums of quite different 2-extensors. The coordinate expressions are as follows.

$$
\begin{array}{ccc}
 & \mathbf{1\ \ 2\ 3} & \\
\mathbf{a} & 1\ \ 2\ 1 & \\
\mathbf{b} & -3\ \ 1\ 1 & \\
\mathbf{c} & -1\ \ 4\ 1 & \\
\mathbf{d} & 3\ \ 5\ 1 & \\
\end{array}
\qquad
\begin{array}{ccc}
 & \mathbf{1\ \ 2\ \ 3} & \\
\mathbf{e} & 3\ -2\ 1 & \\
\mathbf{f} & 2\ \ 0\ 1 & \\
\mathbf{g} & 8\ -2\ 1 & \\
\mathbf{h} & 9\ -4\ 1 & \\
\end{array}
$$

$$
ab + cd = \begin{pmatrix} \mathbf{12}\ \mathbf{13}\ \mathbf{23} \\ 7\ \ \ 4\ \ \ 1 \end{pmatrix} + \begin{pmatrix} \mathbf{12}\ \ \mathbf{13}\ \ \mathbf{23} \\ -17\ -4\ -1 \end{pmatrix} = \begin{pmatrix} \mathbf{12}\ \ \mathbf{13}\ \mathbf{23} \\ -10\ \ \ 0\ \ \ 0 \end{pmatrix}
$$

$$
ef + gh = \begin{pmatrix} \mathbf{12}\ \mathbf{13}\ \mathbf{23} \\ 4\ \ \ 1\ \ -2 \end{pmatrix} + \begin{pmatrix} \mathbf{12}\ \ \mathbf{13}\ \ \mathbf{23} \\ -14\ -1\ \ \ 2 \end{pmatrix} = \begin{pmatrix} \mathbf{12}\ \ \mathbf{13}\ \mathbf{23} \\ -10\ \ \ 0\ \ \ 0 \end{pmatrix}
$$

$$
ab + cd = (7, 4, 1) + (-17, -4, -1) = (-10, 0, 0)
$$
$$
ef + gh = (4, 1, -2) + (-14, -1, 2) = (-10, 0, 0)
$$

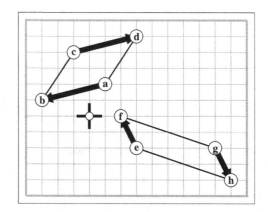

**Fig. 5.8** The couples $ab + cd$ and $ef + gh$ are equal.

## 5.5 Reduced Forms

Grassmann did much more than create a new algebra of *pieces of space*. First he showed that every sum of 1-extensors (weighted points), with sum of weights not equal to zero, is equal to a single weighted point (with weight equal to the sum of the weights of the individual points). If the sum of the weights is zero, the resulting 1-extensor is a 1-vector (which may itself be zero). He then went on to show that every sum of coplanar 2-extensors is equal to a single 2-extensor, or to the sum of two 2-extensors on parallel lines, with equal length and opposite orientation, forming a couple, or is simply zero.

The situation for linear combinations of 2-extensors in 3-space is a bit more complicated. The sum of 2-extensors that are not coplanar is not expressible as a product of points, and so is not itself an extensor. It is simply an antisymmetric tensor of step 2. Grassmann showed that such a linear combination of non-coplanar extensors can be reduced to the sum of a 2-extensor and a 2-vector, or couple.

Not bad at all, for the mid $19^{th}$ century.

At the beginning of the next millenium, in 1900, Sir Robert Ball, in his *Theory of Screws* will use a bit of Euclidean geometry to show that a sum of 2-extensors in 3-space can be expressed as the sum of a force along a line plus a moment in a plane perpendicular to that line. He called these general antisymmetric tensors *screws*. Such a combination of forces, also called a *wrench*, when applied to a rigid body produces a *screw motion*.

Much of the study of screws, with applications to statics, is to be found at the end of Chapter 2 in Grassmann. He discusses coordinate notation, change of basis, and even shows that an anti-symmetric tensor $S$ of step 2 (a screw) is an extensor if and only if the exterior product $SS$ is equal to zero. This is the first and most basic invariant property of anti-symmetric tensors.

Any linear combination of 3-extensors in rank 4 (3-space) is equal to a single 3-extensor. This is because we are getting close to the rank of the entire space. The simplicity of calculations with linear combinations of points is carried over by duality to calculations with linear combinations of copoints, here, with planes.

The extreme case of this duality becomes visible in that $k$-extensors in $k$-space add and multiply just like scalars. For this reason they are called *pseudo-scalars*.

## 5.6 Grassmann-Cayley Algebra, Peano Spaces

Gian Carlo chose to convert pseudo-scalars to ordinary scalars by taking a determinant, or *bracket* of the product[12]. These brackets provide the scalar coefficients of any invariant linear combination. Thus, for any three points $a, b, c$ on a projective line, there is a linear relation, unique up to an overall scalar multiple

$$[ab]c - [ac]b + [bc]a = 0$$

and for any four points $a, b, c, d$ on a projective plane

$$[abc]d - [abd]c + [acd]b - [bcd]a = 0$$

So for any four coplanar points $a, b, c, d$,

$$[abc]d - [abd]c = [bcd]a - [acd]b,$$

and we have two distinct expressions for a projective point that is clearly on both of the lines $cd$ and $ab$.

This is the key point in the development of the Grassmann-Cayley algebra [12, 17, 22], as introduced by Gian Carlo with his coauthors Peter Doubilet and Joel Stein. If $A$ is an $r$-extensor and $B$ is a $s$-extensor in a space of overall rank $n$, then the *meet* of A and B is defined by the two equivalent formulae

$$A \wedge B = \sum_{(A)_{r-k,k}} [A_{(1)}B]A_{(2)} = \sum_{(B)_{k,s-k}} [AB_{(2)}]B_{(1)}$$

This is the *Sweedler notation* (see below, Section 5.7) from Hopf algebra, where, for instance, in the projective 3-space, rank 4, a line *ab* and a plane *cde* as in Figure 5.9 will have meet

$$ab \wedge cde \;=\; [acde]b - [bcde]a \;=\; [deab]c - [ceab]d + [cdab]e.$$

This meet is equal to zero unless *ab* and *cde* together span the whole space, so the line meets the plane in a single point. You can check that the second equality checks with the generic relation

$$[abcd]e - [abce]d + [abde]c - [acde]b + [bcde]a \;=\; 0.$$

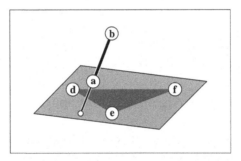

**Fig. 5.9** The meet of *ab* with *cde*.

The serious part of Grassmann-Cayley algebra starts when you try to use these simple relations to detect properties of geometric configuration in *special position*, or to classify possible interrelations between subspaces, that is, when you start work in *invariant theory*.

Gian Carlo and his colleagues made considerable progress in expressing the better-known invariants in terms of this new double algebra, and in reproving a certain number of theorems of projective geometry. Matters got a bit complicated, however, when they got to what they called the *alternating laws*, which "alternated" the operations of join (exterior product) and meet in a single equation. This required maintaining a strict accounting of the ranks of joins and meets, in order to avoid unexpected zeros along the way. They employed Peano's old technique of *filling brackets* whenever the situation got delicate. It took some real gymnastics to construct propositions of general validity when mixing joins and meets.

## 5.7 Whitney Algebra

The idea of the Whitney algebra of a matroid starts with the simple observation that the relation

$$a \otimes bc - b \otimes ac + c \otimes ab = 0$$

among tensor products of extensors holds if and only if the three points $a, b, c$ are collinear, and this *in a projective space of arbitrary rank*.

We need the notion of a *coproduct slice* of a word. For any non-negative integers $i, j$ with sum $n$, the $(i, j)$-slice of the coproduct of an $n$-letter word $W$, written $\partial_{i,j} W$, is the Sweedler sum

$$\sum_{(W)_{i,j}} W_{(1)} \otimes W_{(2)},$$

that is, it is the sum of tensor products of terms obtained by decomposing the word $W$ into two subwords of indicated lengths, both in the linear order induced from that on $W$, with a sign equal to that of the permutation obtained by passing from $W$ to its reordering as the concatenation of the two subwords. For instance,

$$\partial_{2,2}\, abcd = ab \otimes cd - ac \otimes bd + ad \otimes bc + bc \otimes ad - bd \otimes ac + cd \otimes ab.$$

We define the *Whitney algebra of a matroid* $M(S)$ as follows. Let $E$ be the exterior algebra freely generated by the underlying set $S$ of points of the matroid $M$, $T$ the direct sum of tensor powers of $E$, and $W$ the quotient of $T$ by the homogeneous ideal generated by coproduct slices of words formed from *dependent* sets of points.

Note that this is a straight-forward analogue of the principle applied in the bracket algebra of a Peano space, that *dotting a dependent set of points yields zero*.

This construction of a Whitney algebra is reasonable because *these very identities hold in the tensor algebra of any vector space*. Consider, for instance, a set of four coplanar points $a, b, c, d$ in a space of rank $n$, say for large $n$. Since $a, b, c, d$ form a dependent set, the coproduct slice $\partial_{2,2}\, abcd$ displayed above will have $ij \otimes kl$ - coordinate equal to the determinant of the matrix whose rows are the coordinate representations of the four vectors, calculated by Laplace expansion with respect to columns $i, j$ versus columns $k, l$. This determinant is zero because the vectors in question form a dependent set. Compare reference [19], where this investigation is carried to its logical conclusion.

The Whitney algebra of a matroid $M$ has a geometric product reminiscent of the meet operation in the Grassmann-Cayley algebra, but the product of two extensors is not equal to zero when the extensors fail to span the entire space. The definition is as follows, where $A$ and $B$ have ranks $r$ and $s$, the union $A \cup B$ has rank $t$, and $k = r + s - t$, which would be the "normal" rank of the intersection of the flats spanned by $A$ and $B$ if they were to form a modular pair. The *geometric product* is defined as either of the Sweedler sums

$$A \diamond B = \sum_{(A)_{r-k,k}} A_{(1)} B \otimes A_{(2)} = \sum_{(B)_{k,s-k}} A B_{(2)} \otimes B_{(1)}$$

This product of extensors is always non-zero. The terms in the first tensor position are individually either equal to zero or they span the flat obtained as the span (closure) $E$ of the union $A \cup B$. For a represented matroid, the Grassmann coordinates of the left-hand terms in the tensor products are equal up to an overall scalar multiple, because $A_{(1)}B$ and $AB_{(2)}$ are non-zero if and only if $A_{(1)} \cup B$ and $A \cup B_{(2)}$, respectively, are spanning sets for the flat $E$. These extensors with span $E$ now act like scalars for a linear combination of $k$-extensors representing the meet of $A$ and $B$, equal to that meet whenever $A$ and $B$ form a modular pair, and equal to the vector space meet in any representation of the matroid.

The development of the Whitney algebra began with an exciting exchange of email with Gian Carlo in the winter of 1995-6. On 18/11/95 he called to say that he agreed that the "tensor product" approach to non-spanning syzygies is correct, so that

$$a \otimes bc - b \otimes ac + c \otimes ab$$

is the zero tensor whenever $a, b, c$ are collinear points in any space, and gives a Hopf-algebra structure on an arbitrary matroid, potentially replacing the "bracket ring"[21], which had the disadvantage of being commutative.

Four days later he wrote: *"I just read your fax, it is exactly what I was thinking. I have gone a little further in the formalization of the Hopf algebra of a matroid, so far everything checks beautifully. The philosophical meaning of all this is that every matroid has a natural coordinatization ring, which is the infinite product of copies of a certain quotient of the free exterior algebra generated by the points of the matroid (loops and links allowed, of course). This infinite product is endowed with a coproduct which is not quite a Hopf algebra, but a new object closely related to it. Roughly, it is what one obtains when one mods out all coproducts of minimal dependent sets, and this, remarkably, gives all the exchange identities. I now believe that everything that can be done with the Grassmann-Cayley algebra can also be done with this structure, especially meets".*

On 29/11/95: *"I will try to write down something tonight and send it to you by latex. I still think this is the best idea we have been working on in years, and all your past work on syzygies will fit in beautifully".*

On 20/12/95: *"I am working on your ideas, trying to recast them in letterplace language. I tried to write down something last night, but I was too tired. Things are getting quite rough around here".*

Then, fortunately for this subject, the weather turned bad. On 9/1/96: *"Thanks for the message. I am snowbound in Cambridge, and won't be leaving for Washington until Friday, at least, so hope to redraft the remarks on Whitney algebra I have been collecting. ... "*

*"Here are some philosophical remarks. First, all of linear algebra should be done with the Whitney algebra, so no scalars are ever mentioned. Second, there is a new theorem to be stated and proved preliminarily, which seems to be a vast generalization of the second fundamental theorem of invariant theory (Why, Oh why, did I not see this before?!)"*

Here, Gian Carlo suggests a comparison between the Whitney algebra of a vector space $V$, when viewed as a matroid, and the exterior algebra of $V$.

" *I think this is the first step towards proving the big theorem. It is already difficult, and I would appreciate your help. The point is to prove classical determinant identities, such as Jacobi's identity, using only Whitney algebra methods (with an eye toward their quantum generalizations!) Only by going through the Whitney algebra proofs will we see how to carry out a quantum generalization of all this stuff*".

"*It is of the utmost importance that you familiarize yourself with the letterplace representation of the Whitney algebra, through the Feynman operators, and I will write this stuff first and send it to you*".

On 11/1/96, still snowbound in Cambridge, Gian Carlo composed a long text proposing two projects:

1. the description of a module derived from a Whitney algebra $W(M)$,
2. a faithful representation of a Whitney algebra as a quotient of a supersymmetric letter-place algebra.

This supersymmetric algebra representation is as follows. He uses the supersymmetric letter-place algebra Super[$S^-|P^+$] in negative letters and positive places. . A tensor product $W_1 \otimes W_2 \otimes \ldots \otimes W_k$ is sent to the product

$$(W_1|p_1^{|W_1|})(W_2|p_2^{|W_2|})\ldots(W_k|p_k^{|W_k|})$$

where the words $p_i^{(|w_i|)}$ are divided powers of positive letters, representing the different possible positions in the tensor product. The letter-place pairs are thus anticommutative. The linear extension of this map on $W(M)$ he termed the *Feynman entangling operator*.

Bill Schmitt joined the project in the autumn of '96. The three us met together only once. Bill managed to solve the basic problem, showing that the Whitney algebra is precisely a *lax Hopf algebra*, the quotient of the tensor algebra of a free exterior algebra by an ideal that is not a co-ideal. The main body of this work Bill and I finished [11] in 2000, too late to share the news with Gian Carlo. It was not until eight years later that Andrea Brini, Francesco Regonati, Bill Schmitt and I finally established that Gian Carlo had been completely correct about the super-symmetric representation of the Whitney algebra.

For a quick taste of the sort of calculations one does in the Whitney algebra of a matroid, and of the relevant geometric signification, consider a simple matroid $L$ on five points: two coplanar lines $bc$ and $de$, meeting at a point $a$. In Figure 5.10 we exhibit, for the matroid $L$, the geometric reasoning behind the equation $ab \otimes cde = ac \otimes bde$.

First of all,

$$ab \otimes c - ac \otimes b + bc \otimes a = 0$$

because $abc$ is dependent. Multiplying by $de$ in the second tensor position,

$$ab \otimes cde - ac \otimes bde + bc \otimes ade = 0,$$

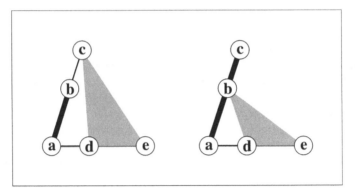

**Fig. 5.10** $ab \otimes cde = ac \otimes bde$.

but the third term is zero because *ade* is dependent in *L*, so we have the required equality $ab \otimes cde = ac \otimes bde$.

This equation of tensor products expresses the simple fact that the ratio *r* of the lengths of the oriented line segments *ac* and *ab* is equal to the ratio of the oriented areas of the triangular regions *cde* and *bde*, so, in the passage from the product on the left to that on the right, there is merely a shift of a scalar factor *r* from the second term of the tensor product to the first. *As tensor products they are equal.*

The same fact is verified algebraically as follows. Since *b* is collinear with *a* and *c*, and *d* is collinear with *a* and *e*, we may write $b = (1 - \alpha)a + \alpha c$ and $d = \beta a + (1 - \beta)e$ for some choice of non-zero scalars $\alpha, \beta$. Then

$$ab \otimes cde = (a \vee ((1 - \alpha)a + \alpha c)) \otimes (c \vee (\beta a + (1 - \beta)e) \vee e)$$
$$ac \otimes bde = (a \vee c) \otimes (((1 - \alpha)a + \alpha c) \vee (\beta a + (1 - \beta)e) \vee e)$$

both of which simplify to $-\alpha\beta(ac \otimes ace)$

Note that this equation is independent of the dimension of the overall space within which the triangular region *ace* is to be found.

## 5.8 Geometric Product

For words $u, v \in W^1$, with $|u| = r, |v| = s$, let $k = r + s - \rho(uv)$. The *geometric product* of *u* and *v* in *W*, written $u \diamond v$, is given by the expression

$$u \diamond v = \sum_{(u)_{r-k,k}} u_{(1)}v \circ u_{((2)}$$

For words $A, B$ and integers $r, s, k$ as above,

$$\sum_{(A)_{r-k,k}} A_{(1)} B \circ A_{((2)} = \sum_{(B)_{k,s-k}} A_{(1)} B \circ A_{((2)}.$$

So the geometric product is commutative:

$$A \diamond B = (-1)^{(r-k)(s-k)} B \diamond A$$

In Figure 5.11, we see how the intersection point (at $b$) of the line $ef$ with the plane $acd$ can be computed as a linear combination of points $e$ and $f$, or alternately as a linear combination of points $a$, $c$ and $d$, using the two formulations of the geometric product of $ef$ and $acd$.

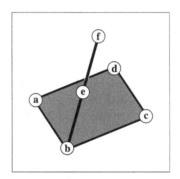

**Fig. 5.11**  The geometric product of a line with a plane.

The calculation is as follows:

$$\begin{aligned} acd \diamond ef &= acef \circ d - adef \circ c + cdef \circ a \\ &= \quad acdf \circ e - acde \circ f \end{aligned}$$

Figure 5.12 shows how the line of intersection (incident with points $a, c, f$) of planes $abc$ and $def$ can be computed as a linear combination of lines $bc$ and $ac$, or can be obtained as a single term by the alternate form of the geometric product.

The calculation is

$$\begin{aligned} abc \diamond def &= adef \circ bc - bdef \circ ac \\ &= \quad abcd \circ ef \end{aligned}$$

Compare section 56-57, pages 88-89 in Peano[20].

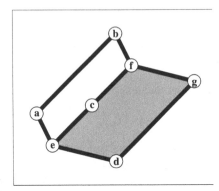

**Fig. 5.12** The geometric product of two planes.

## 5.9 Regressive Product

When I first lectured on Gian Carlo's work on the Grassmann-Cayley algebra to a seminar at McMaster University run by two eminent algebraists, Evelyn Nelson and Bernard Banaschewski, they insisted that the meet operation was *just* the dual of Grassmann's exterior product, and referred me to Bourbaki. I was never comfortable with this view, since I felt that a veritable geometric product would not be restricted to the two *extremities* of the lattice of subspaces, yielding non-trivial results only for independent joins and co-independent meets.

Lloyd Kannenberg performed an outstanding service to mathematics when he published his English translations [14, 16, 20] of these seldom-consulted works by Grassmann and Peano, the two *Ausdehnungslehre*, published under the titles *A New Branch of Mathematics* (1844) and *Extension Theory* (1862) and the *Geometric Calculus* of Peano, which, as Kannenberg says, "was published in a small print run in 1888, and has never been reissued in its entirety". Lloyd tells me that his translations of these classics were undertaken with Gian Carlo's active encouragement.

The *Ausdehnungslehre* of 1844 has an entire chapter on what Grassmann calls the regressive product. At the outset (¶125) Grassmann explains that he wants a multiplication that will produce a non-zero value for the product of magnitudes $A, B$ that are dependent upon one another. *"In order to discover this new definition we must investigate the different degrees of dependence, since according to this new definition the product of two dependent magnitudes can also have a nontrivial value"*. (We will put all direct quotations from the English translation in *italics*. We also write ∘ for the regressive product, this being somewhat more visible than Grassmann's period "." notation).

To measure the different degrees of dependence, Grassmann argues that the set of points dependent upon both $A$ and $B$ forms a *common system*, what we now call a projective subspace, the intersection of the spaces spanned by $A$ and by $B$. *"To each degree of dependence corresponds a type of multiplication: we include all these types of multiplication under the name regressive multiplication"*. The *order* of the multiplication is the value chosen for the rank of the common system.

In ¶126 Grassmann studies the modular law for ranks:

$$\rho(A) + \rho(B) = \rho(C) + \rho(D)$$

where $\rho()$ is the rank function, $C$ is the common system (the lattice-theoretic meet) and $D$ is the nearest covering system (the lattice-theoretic join). In ¶129 he explains the *meaning* of a geometric product. *" In order to bring the actual value of a real regressive product under a simple concept we must seek, for a given product whose value is sought, all forms into which it may be cast, without changing its value, as a consequence of the formal multiplication principles determined by the definition. Whatever all these forms have in common will then represent the product under a simple concept"*. So the meaning of the regressive product is synonymous with the equivalence relation *"have the same geometric product"*.

He sees that the simplest form of a product is one in *subordinate form*, that is when it is a *flag* of extensors. He thus takes the *value*, or meaning, of the product to be the *"combined observation"* of the flag of flats *"together with the (scalar) quantity to be distributed on the factors"*. A scalar multiple can be transferred from term to term in a product without changing the value of the product, that is, he is introducing a tensor product.

As a formal principle he permits the dropping of an additive term in a factor if that term has a higher degree of dependence on the other factors of the product. For instance, in the figure of three collinear points $a, b, c$ together with a line $de$ not coplanar with $a, b, c$, $ab \circ (ce + de) = ab \circ de$. We will subsequently show that this product is also equal to $1 \circ abde$.

¶130 gives the key definition. *"If A and B are the two factors of a regressive product and the magnitude C represents the common system of the two factors, then if B is set equal to CD, AD represents the nearest covering system and thus the relative system as well if the product is not zero"*. That is, we represent one of the factors, $B$, as a product $CD$ of an extensor $C$ spanning the intersection of the subspaces spanned by $A$ and $B$, times an extensor $D$ that is complementary to $C$ in the subspace spanned by $B$. We then transfer the factor $D$ to the multiplicative term involving $A$. The result is a flag of extensors, which Grassmann decides to write in decreasing order. He concludes that this flag expression is unique (as tensor product). In ¶131: *"The value of a regressive product consists in the common and nearest covering system of the two factors, if the order of the factors is given, apart from a quantity to be distributed multiplicatively on the two systems"*.

Also in ¶131, he states that the regressive product of two extensors is *equal* to its associated flag representation. With $A, B$ and $B = CD$ as above, he writes

$$A \circ B = A \circ CD = AD \circ C$$

Perhaps even more clearly (¶132), he states that *"we require that two regressive products of nonzero values $A \circ CD$ and $A' \circ C'D'$ are equal so long as generally the product of the outermost factors with the middle one is equal in both expressions,* *or if they stand in reciprocal proportion, whether the values of the orders of the corresponding factors agree or not. In particular, with this definition we can bring that regressive product into subordinate form"*. That is, any regressive product of extensors $A \circ B$ is equal to a flag product, say $E \circ C$, where $E = AD$ in the previous calculation.

He also figures out the sign law for commutativity of the regressive product. If $A \circ B = E \circ C$, and the right hand side is a flag, where the ranks of the extensors $A, B, C, E$ are $a, b, c, e$, respectively, then the sign for the exchange of order of the product is the parity of the product $(a - c)(b - c)$, the product of the *supplementary numbers* (See Figure 5.13). It's fascinating that he managed to get this right!

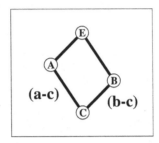

**Fig. 5.13** Values of rank in a modular lattice.

So we have (skew) commutativity for the regressive product. *How about associativity?* Grassmann calls associativity *the law of combination*. This is the extraordinary part of the story.

He has a well defined product (in fact, one for every degree of dependence), so, in principle, any multiple product is well defined (and even the correct degrees for each successive product of two factors), but he soon recognizes, to his dismay, that the regressive product is **not** generally associative.

The best way to see the non-associativity is by reference to the free modular lattice on three generators, as first found by Dedekind, in Figure 5.14. (The elements $a, b, c$ are indicated with lower-case letters, but they could be extensors of any rank. I'll also write flags in increasing order.)

In any lattice, for any pair of elements in the order $x < z$, and for any element $y$,

$$x \vee (y \wedge z) \leq (x \vee y) \wedge z$$

A lattice is *modular* if and only if, under these same conditions, equality holds:

$$x \vee (y \wedge z) = (x \vee y) \wedge z$$

The lattice of subspaces of a vector space is modular (and complemented), so any calculus of linear pieces of space will use the logic of modular lattices.

On the left of Figure 5.15, we indicate the passage from $b \circ c$ to $b \wedge c \circ b \vee c$. On the right of Figure 5.15, we show the passage from $b \wedge c \circ b \vee c$ to

$$a \circ ((b \wedge c) \circ (b \vee c)) = (a \wedge b \wedge c) \circ ((a \vee (b \wedge c)) \wedge (b \vee c)) \circ (a \vee b \vee c)$$

A value of the triple product will always land at one of the three central elements of the inner gray sublattice, but the exact position will depend on which factor entered last into the combined product. *The result carries a trace of the order in which the factors were combined!*

Try this with the simple case of three collinear points $a, b, c$. Then

$$a \circ (b \circ c) = a \circ (1 \circ bc) = 1 \circ a \circ bc,$$

but

$$(a \circ b) \circ c = (1 \circ ab) \circ c = 1 \circ c \circ ab.$$

Parentheses are not necessary in the final flags, because flag products are associative. All products in which $a$ enters last will be the same, up to sign: $a \circ (b \circ c) = (b \circ c) \circ a = -(c \circ b) \circ a$.

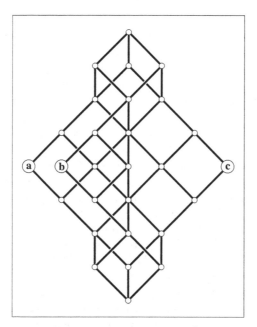

**Fig. 5.14** The free modular lattice on 3 generators.

However, the regressive product of two flag products is well defined. Grassmann showed this to be true, *an incredible feat, given the tools he had at hand!* In fact,

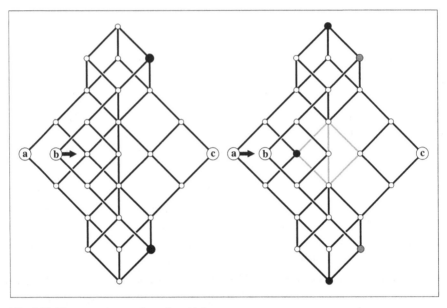

**Fig. 5.15** The regressive product of $a$ with $b \diamond c$.

he proved that the extensors of one flag can progressively multiplied into another flag, and that the result is independent not only of the order of these individual multiplications, but also independent on which flag is multiplied into which! In ¶136 Grassmann says *"Instead of multiplying by a product of mutually incident factors one can progressively multiply by the individual factors, and indeed in any order "*.

Garrett Birkhoff, in the first edition of his *Lattice Theory*, proved that the free modular lattice generated by two finite chains is a finite *distributive* lattice. This changes the whole game.

In Figure 5.16 we show the free modular lattice generated by two 2-chains. In the center of Figure 5.17 we show the result of multiplying the extensor $a$ into the flag $c \circ d$, then, on the right, the result of multiplying $b$ into that result. We end up on what might well be termed the *backbone* of the distributive lattice. In Figure 5.18 we show what happens when the flag $c \circ d$ is multiplied into the flag $a \circ b$, but starting with the top element $d$, instead. The result is the same.

By the end of his chapter on the regressive product, Grassmann seems rather disheartened. He admits clearly in ¶139 that *"the multiplicative law of combination ... is not generally valid for the mixed product of three factors"*.

He includes a footnote to say that *cases can be found in which our law still finds its application* via the results available for the product of an extensor by a flag of extensors, but concludes, with a certain degree of disillusion, that *"these cases*

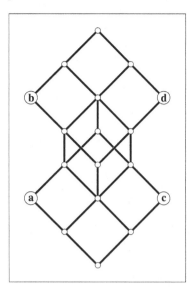

**Fig. 5.16**  The free modular lattice generated by two 2-chains.

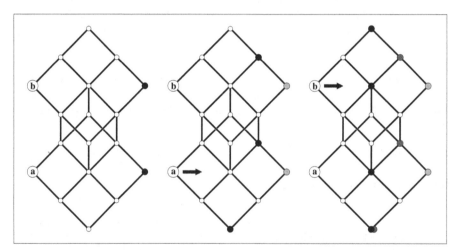

**Fig. 5.17**  The product $b \to (a \to (c < d))$

*are so isolated, and the constraints under which they occur so contrived, that no scientific advantage results from their enumeration".*

Then, having investigated duality and having proven associativity for joins of independent extensors and meets of co-independent extensors, he concludes with the note: *"the theoretical presentation of this part of extension theory now appears as completed, excepting consideration of the types of multiplication for which the law of combination is no longer valid".* He adds the footnote: *"How to treat such*

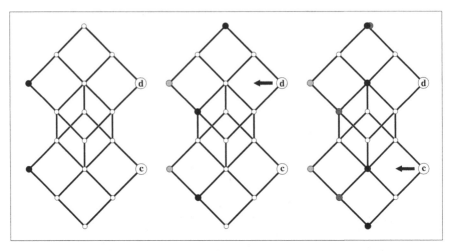

**Fig. 5.18** The product $c \to (d \to (a < b))$.

*products, which to be sure have manifold applications, I have sought to indicate at the conclusion of this work".*

Clarification of these questions involving the regressive product of two flags has been joint work with Andrea Brini, Francesco Regonati, and Bill Schmitt, last year in Bologna.

When this work is combined with the extraordinary synthesis [6] of Clifford algebra and Grassmann-Cayley algebra, all made super-symmetric, already achieved by Andrea Brini, Paolo Bravi, and Francesco Regonati, here present, you have finally the makings of a banquet that can truly be termed *geometric algebra*.

## 5.10 Higher Order Syzygies

Before closing, we should take a quick look at the *higher order syzygies* that Gian Carlo mentioned in his email messages concerning the new Whitney algebra.

Given a configuration $C$ of $n$ points in projective space of rank $k$, certain subsets $A \subseteq C$ will be dependent. The minimal dependent sets (the *circuits* of the corresponding matroid $M(C)$) will have dependencies that are uniquely defined up to an overall scalar multiple. They thus form, in themselves, a configuration of rank $n - k$ of projective points in a space of rank $n$. We call this the first *derived configuration*, $C^{(1)}$, and denote the associated matroid $M^{(1)}(C)$, the *derived matroid*.

In the same way, the circuits of $C^{(1)}$ form a new projective configuration, which we denote $C^{(2)}$, with matroid $M^{(2)}(C)$, and so on. Thus, any matroid represented as a configuration in projective space automatically acquires an infinite sequence of derived matroids. In classical terminology, $C^{(k)}$ is the configuration of $k^{th}$-*order syzygies* of $C$.

This derived information is *not*, however, fully determined by the matroid itself. The simplest example of interest is given by the uniform matroid $U_{3,6}$ of six points $\{a, b, \ldots, f\}$ in general position in the projective plane. In Figure 5.19, we show two representations of the matroid $U_{3,6}$ in the plane. In the example on the left, the three lines $ab, cd, ef$ do not meet, and the circuits $abcd, abef, cdef$ are independent, spanning the space of first order syzygies among the six points. On the right, those three lines meet, the circuits $abcd, abef, cdef$ are dependent (rank 2 in the derived matroid). Those three circuits act as linear constraints on lifting of the figure of six points into 3-space. A height vector $(h(a), h(b), \ldots, h(f))$ is orthogonal to the vector $([bcd], [acd], [abd], [bcd], 0, 0)$ if and only if the four lifted points $a' = (a_1, a_2, a_3, h(a)), \ldots, d' = (d_1, d_2, d_3, h(d))$ are coplanar, if and only if *the dependency is preserved in the lifting*. If the three circuits are of rank 3, there will be $6 - 3$ choices for the lift of the six points, and they must remain coplanar. The three circuits are of rank 2 if and only if the plane figure has a polyhedral lifting, to form a true three-dimensional triangular pyramid. And this happens if and only if the three lines $ab, cd, ef$ meet at a point, namely, the projection of the point of intersection of the three distinct planes $a'b'c'd', a'b'e'f', c'd'e'f'$.

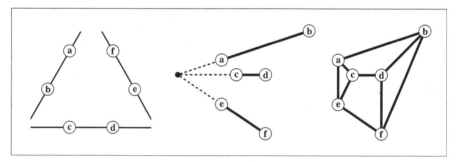

**Fig. 5.19** Representations of the uniform matroid $U_{3,6}$.

For the figure on the left, the derived configuration of first order syzygies is of rank 3 and consists simply of the intersection of 6 lines in general position in the plane. The five circuits formed from any five of the six points have rank 2, and are thus collinear, as in Figure 5.20. In the special position, when the lines $ab, cd, ef$ are concurrent, the circuits $abcd, abef, cdef$ become collinear, as on the right in Figure 5.20.

The geometric algebra can also be put to service to provide coefficients for higher-order syzygies, in much the same way that generic first-order syzygies are obtained as coproducts of dependent sets. To see how this is done, it will suffice to take another look at the present example. Check the matrix of coefficients of these first-order syzygies:

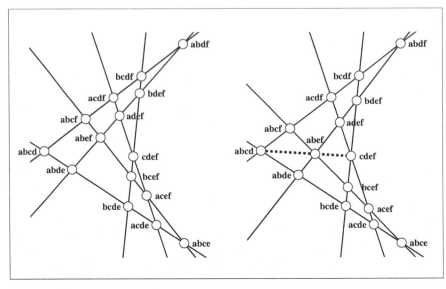

**Fig. 5.20** First order syzygies for the uniform matroid $U_{3,6}$.

$$
\begin{array}{ccccccc}
 & a & b & c & d & e & f \\
(abcd) & abc & -acd & abd & -abc & 0 & 0 \\
(abef) & bef & -aef & 0 & 0 & abf & -abe \\
(cdef) & 0 & 0 & def & -cef & cdf & -cde
\end{array}
$$

This row space $S$ is orthogonal to the space of *linear functions* on the set $S = \{a,b,c,d,e,f\}$, a space $T$ spanned by the coordinate projection functions

$$
\begin{array}{cccccc}
 & a & b & c & d & e & f \\
(1) & a_1 & b_1 & c_1 & d_1 & e_1 & f_1 \\
(2) & a_2 & b_2 & c_2 & d_2 & e_2 & f_2 \\
(3) & a_3 & b_3 & c_3 & d_3 & e_3 & f_3
\end{array}
$$

The Grassmann coordinates of these two vector subspaces differ from one another (up to an overall "scalar" multiple) by *complementation of places* and a sign of that complementation. This algebraic operation is called the *Hodge star* operator. Concretely, where $\sigma$ is the sign of the permutation merging two words into one, in a given linear order,

$$
S_{abc} = \sigma(def,abc)T_{def} = -T_{def}
$$
$$
S_{abd} = \sigma(cef,abd)T_{cef} = T_{cef}
$$
$$
\cdots
$$
$$
S_{def} = \sigma(abc,def)T_{abc} = T_{abc}
$$

The Grassmann coordinate $T_{xyz}$ for any letters $x,y,z$ is just the 3-extensor $xyz$. The "scalar" in question, which Gian Carlo and I called *the resolving bracket* [9] is obtainable by calculating any $3 \times 3$ minor of the matrix for $S$, and dividing by the 3-

extensor obtained from the complementary set of columns in the matrix for $T$. If we do this in columns $abc$ for $S$, we find the determinant $(-bcd \otimes aef + acd \otimes bef)def$, which, when divided by $-def$ yields $bcd \otimes aef - acd \otimes bef$. It helps to recognize that this expression can be obtained by joining the meet of $ab \wedge cd = a \otimes bcd - b \otimes acd$ with $ef$, so this resolving bracket is equal to zero if and only if the meet of $ab$ and $cd$ is on the line $ef$, that is, if and only if the three lines $ab, cd, ef$ are concurrent. This is the explicit synthetic condition under which the three first-order syzygies $abcd, abef, cdef$ form a dependent set in the first derived configuration.

Much work remains in order to develop an adequate set of tools for dealing with higher order syzygies in general. The concept of resolving bracket is but a first step. Gian Carlo and I spent many hours discussing these higher order syzygies, usually on the white-boards in his apartment in Boston, later in Cambridge, in his office, or in more exotic places such as Strasbourg or Firenze, during mathematical gatherings. I think he enjoyed these discussions, in the period 1985-95, difficult as it was for him to force me to express my ideas clearly. The only major breakthrough was in Gian Carlo's very fruitful collaboration with David Anick [1, 10], where they found a resolution of the bracket ring of a free exterior algebra, bases for syzygies of all orders being represented by families of totally non-standard tableaux. In this way, you have only to deal with syzygies having *single bracket coefficients*.

## 5.11 Balls in Boxes

As a closing thought, I would like to express my conviction that Gian Carlo was rightfully fascinated by probabilistic questions arising from quantum theory, but somehow never really got a proper hold on the basic issues, despite having approached them from all quarters: via general combinatorial theory, *espèces de structure*, supersymmetric algebra, umbral calculus, probability theory, and ...philosophy.

Let me suggest that it is high time we reread what he has written here and there on *balls and boxes*, as in title of today's memorial assembly, for hints he may generously have left us.

As he wrote in the introduction to *The Power of Positive Thinking* (with Wendy Chan),

> *The realization that the classical system of Cartesian coordinates can and should be enriched by the simultaneous use of two kinds of coordinates, some of which commute and some of which anticommute, has been slow in coming; its roots, like the roots of other overhaulings of our prejudices about space, go back to physics, to the mysterious duality that is found in particle physics between identical particles that obey or do not obey the Pauli exclusion principle.*

# References

1.  Anick, D., Rota, G.-C. (1991), *Higher-order Syzygies for the Bracket Ring and for the Ring of Coordinates of the Grassmannian*, in "Proc. Nat. Acad. of Sci.", 88, pp. 8087-90.
2.  Barnabei, M., Brini, A., Rota, G.-C. (1985), *On the Exterior Calculus of Invariant Theory*, in "J. of Algebra", 96, pp. 120-60.
3.  Bravi, P., Brini, A. (2001), *Remarks on Invariant Geometric Calculus, Cayley-Grassmann Algebras and Geometric Clifford Algebras*, in H. Crapo, D. Senato (eds.), *Algebraic Combinatorics and Computer Science, A Tribute to Gian-Carlo Rota*, Milan, Springer Italia, pp. 129-50.
4.  Brini, A., Huang, R. Q., Teolis, A. G. B. (1992), *The Umbral Symbolic Method for Supersymmetric Tensors*, in "Adv. Math.", 96, pp. 123-93.
5.  Brini, A., Regonati, F., Teolis, A. G. B. (2001), *Grassmann Geometric Calculus, Invariant Theory and Superalgebras*, in H. Crapo, D. Senato (eds.), *Algebraic Combinatorics and Computer Science, A Tribute to Gian-Carlo Rota*, Milan, Springer Italia, pp. 151-96.
6.  Brini, A., Teolis, A. (1996), *Grassmann's Progressive and Regressive Products and GC Coalgebras*, in G. Schubring (ed.), *Hermann Gnther Grassmann (1809-1877), Visionary Mathematician, Scientist and Neohumanist Scholar*, Dordrecht, Kluwer, pp. 231-42.
7.  Chan, W. (1998), *Classification of Trivectors in 6 – D Space*, in B. E. Sagan and R. P. Stanley (eds.), *Mathematical Essays in Honor of Gian-Carlo Rota*, Boston, Basel, Berlin, Birkhäuser, pp. 63-110.
8.  Chan, W., Rota, G.-C., Stein, J. (1995), *The Power of Positive Thinking*, in *Proceedings of the Curaçao Conference: Invariant Theory in Discrete and Computational Geometry 1994*, Dordrecht, Kluwer.
9.  Crapo, H., Rota, G.-C. (1995), *The Resolving Bracket*, in *Proceedings of the Curaçao Conference: Invariant Theory in Discrete and Computational Geometry 1994*, Dordrecht, Kluwer.
10. Crapo, H. (1993), *On the Anick-Rota Representation of the Bracket Ring of the Grassmannian*, in "Adv. Math.", 99, pp. 97-123.
11. Crapo, H., Schmitt, W. (2000), *The Whitney Algebra of a Matroid*, in "J. Combin. Theory Ser. A", 91, pp. 215-63.
12. Doubilet, P., Rota, G.-C., Stein, J. (1974), *On the Foundations of Combinatorial Geometry: IX, Combinatorial Methods in Invariant Theory*, in "Stud. Appl. Math.", 3, LIII, pp. 185-215.
13. Grassmann, H. (1844), *Die lineale Ausdehnungslehre, ein neuer Zweig der Mathematik*, Leipzig, Otto Wigand.
14. Grassmann, H. (1995), *A New Branch of Mathematics: The Ausdehnungslehre of 1844, and Other Works*, translated by Lloyd C. Kannenberg, Chicago, Open Court.
15. Grassmann, H. (1862), *Die Ausdehnungslehre, Vollständig und in strenger Form*, Berlin, Th. Cgr. Fr. Enslin (Adolph Enslin).
16. Grassmann, H. (2000), *Extension Theory*, translated by Lloyd C. Kannenberg, Providence, RI, American Mathematical Society.
17. Grosshans, F., Rota, G.-C., Stein, J., *Invariant Theory and Supersymmetric Algebras*, in "Conference Board of the Mathematical Sciences", 69.
18. Huang, R., Rota, G.-C., Stein, J. (1990), *Supersymmetric Bracket Algebra and Invariant Theory*, Rome, Centro Matematico V. Volterra, Università degli Studi di Roma II.
19. Leclerc, B. (1993), *On Identities Satisfied by Minors of a Matrix*, in "Adv. Math.", 100, pp. 101-32.
20. Peano, G. (2000), *Geometric Calculus, according to the Ausdehnungslehre of H. Grassmann*, translated by Lloyd C. Kannenberg, Boston, Basel, Berlin, Birkhäuser.
21. White, N. (1975a-b), *The Bracket Ring of a Combinatorial Geometry, I and II*, in "Trans. Amer. Math. Soc.", 202, pp. 79-95, 214, pp. 233-248.
22. White, N. (1995), *A Tutorial on Grassmann-Cayley Algebra*, in *Proceedings of the Curaçao Conference: Invariant Theory in Discrete and Computational Geometry 1994*, Dordrecht, Kluwer.

# Chapter 6
# The Eleventh and Twelveth Problems of Rota's Fubini Lectures: from Cumulants to Free Probability Theory
## Invited Chapter

Elvira Di Nardo and Domenico Senato

**Abstract** The title of the paper refers to the Fubini Lecture "Twelve problems in probability no one likes to bring up", delivered by Gian-Carlo Rota at the Institute for Scientific Interchange (Turin - Italy) in 1998. The eleventh and the twelfth problems of that lecture deal with cumulants and free probability theory respectively. Our aim is to describe the development of the classical umbral calculus, as featured by Rota and Taylor in 1994, thanks to which a unifying theory of classical, Boolean and free cumulants has been carried out. This unifying theory relies on an updating of Sheffer sequences via the classical umbral calculus. Some in-between steps lead us to develop some topics of computational statistics, which benefit of this symbolic approach in efficiency and speed-up. We resume also these applications.

## 6.1 Prologue

Gian-Carlo Rota was a real innovator. He shuns extreme technicalities to advantage of key foundational questions and concepts. Umbral calculus can be considered an instance in point. The rules of the classical umbral calculus are very simple, but it is far less elementary to deeply understand its potential and how to apply it fruitfully.

This symbolic method is no doubt a creature of Rota, involving him during most of his scientific life. It is certainly thank to Rota's efforts that the heuristic method, invented and largely employed by Rev. John Blissard between 1861 and 1868 (see references in [13]) has a solid foundation and has been spread in the mathematical community. Despite the impressive amount of the scientific literature followed to this foundation, Rota had in 1994 the bravery to turn his umbral calculus upside down, outlining what he considered the right syntax for this matter. As Rota has written in [55]: "*In the present paper we have finally decided to bite the bullet and give rigorous, simple presentation of the umbral calculus as it was meant by the*

Elvira Di Nardo · Domenico Senato
Università degli Studi della Basilicata, Italy

E. Damiani et al. (eds.), *From Combinatorics to Philosophy,*
DOI 10.1007/978-0-387-88753-1_6, © Springer Science+Business Media, LLC 2009

*founders. We have kept new notation both minimal and indispensable to avoid the misunderstanding of the past. Although the notation of Hopf algebra satisfied the most ardent advocate of spic-and-span rigor, the translation of "classical" umbral calculus into the newly found rigorous language made the method altogether unwieldy and unmanageable. Not only was the eerie feeling of witchcraft lost in the translation, but, after such a translation, the use of calculus to simplify computation and sharpen our intuition was lost by the wayside..."*

Our interest in umbrae was born by chance. In the spring of 1997, Rota was visiting at the University of Basilicata, and during one of our last conversations before his return in Cambridge, he shared with us some topics on foundational sides of probability theory. The following year Rota held his last course in Cortona and we did not miss the opportunity to spend some time with him. As it happens sometimes in the practice of mathematical investigation, the subject we deal with does not develop the original idea from which we have started. In a sultry afternoon, working on matters connected with functional composition, the Bell umbra appeared to us. After his untimely death, we have taken by heritage this new born theory and, aware of its potential, we have resumed [13] where he stopped.

Ten years later, we present a survey of results reached by the umbral syntax within cumulants and free probability theory. The topics we deal with are strictly related to the two last problems of the Fubini Lecture "Twelve problems in probability no one likes to bring up"[8], delivered by Rota at the Institute for Scientific Interchange (Turin - Italy) in 1998.

## 6.2 The Classical Umbral Calculus

The basic idea of the classical umbral calculus is simply to associate a sequence of numbers to an indeterminate, called *umbra*, which is said to represent the sequence. Such correspondence is obtained by applying a functional, called *evaluation*, to powers of the umbra. As Ira Gessel has observed [30], at first glance, the classical umbral calculus seems no more than a notation for dealing with exponential generating functions. Nevertheless, this new syntax has given rise to noteworthy computational simplifications and conceptual clarifications in different contexts. The derangement numbers have the well know exponential generating function $\sum_{n=0}^{\infty} D_n x^n / n! = e^{-x}/(1-x)$; umbral calculus give us the more interesting but considerable more recondite formula $\sum_{n=0}^{\infty} D_n^2 x^n / n! = e^x \sum_{k=0}^{\infty} k! x^k / (1+x)^{2k+2}$, see [30].

Applications of this syntax are given by Zeilberger [86], where generating functions are computed for many difficult problems dealing with counting combinatorial objects. Connections with wavelet theory have been investigated by Shen [59] and by Saliani and Senato [58].

In formal terms, an umbral calculus consists of the following data:

*i)*   a set $A = \{\alpha, \beta, \gamma, \ldots\}$, called the alphabet, whose elements are named umbrae;

*ii)* a linear functional $E$, called the evaluation, defined on a suitable polynomial ring $R[A]$ and taking value in $R$, such that

- $E[1] = 1$,
- $E[\alpha^i \beta^j \cdots \gamma^k] = E[\alpha^i] E[\beta^j] \cdots E[\gamma^k]$   (uncorrelation property);

*iii)* two special umbrae $\epsilon$ (augmentation) and $u$ (unity) such that $E[\epsilon^i] = \delta_{0,i}$ and $E[u^i] = 1$, for $i = 0, 1, \ldots$.

The idea to associate a number sequence to an indeterminate is familiar in probability theory, where the $n$-th term of a number sequence can be considered as $n$-th moment of a random variable, under suitable hypothesis. From this perspective, an umbra looks like the framework of a random variable, witout reference to a probability space. The functional $E$ can be viewed as the expectation of a random variable. The first glimmering of this approach is due to Fréchet [29] as mentioned in [10].

Let us quote again Rota ([53], problem 1: the algebra of probability): *"It has been argued that the notions of sample space and event are redundant, and that all of probability should be done in terms of random variables alone .... How would one introduce probability in terms of random variables alone? This is a subject that has been thoroughly studied and a brief mention will suffice. One takes an ordered commutative algebra over the reals, and endows it with a positive linear functional E[X]. The elements of the algebra will be the random variables and the linear functional is the expectation of a random variable".*

By analogy with the theory of random variables, the elements $a_i = E[\alpha^i]$ are called *moments* of the umbra $\alpha$. The umbra $\epsilon$ can be viewed as the random variable which takes the value 0 with probability 1 and the umbra $u$ as the random variable which takes the value 1 with probability 1.

The *Bell umbra* $\beta$ is the umbra such that $E[(\beta)_n] = 1$ for $n = 0, 1, 2, \ldots$, where $E[(\beta)_0] = 1$ and $(\beta)_n = \beta(\beta - 1) \cdots (\beta - n + 1)$ is the lower factorial. The moments of the Bell umbra are $E[\beta^n] = B_n$, where $B_n$ is the $n$-th Bell number [14]. If we call *factorial moments* of the umbra $\alpha$ the sequence $a_{(0)} = 1, a_{(n)} = E[(\alpha)_n]$ for $n > 0$, the definition of the Bell umbra $\beta$ can be reformulated as follows: the Bell umbra is the umbra whose factorial moments equal 1 for any nonnegative integer $n$. In a short while, we will show the crucial role played by this umbra within this symbolic method.

Following the analogy between umbrae and random variables, we have called *similar* the umbrae having the same sequence of moments. Actually, the notion of similarity among umbrae comes in handy in order to manipulate sequences such

$$\sum_{i=0}^{n} \binom{n}{i} a_i a_{n-i}, \quad n = 0, 1, 2, \ldots, \tag{6.1}$$

as moments of umbrae. The sequence (6.1) can not be represented by using only the umbra $\alpha$ with moments $a_0 = 1, a_1, a_2, \ldots$. Indeed, being $\alpha$ correlated to itself, the product $a_i a_{n-i}$ cannot be written as $E[\alpha^i \alpha^{n-i}]$. This is similar to what happens for random variables, where the $(n+k)$-th moment of a random variable can not be split in the product of the $n$-th moment and of the $k$-th moment of the same random variable. Instead, if $X$ and $Y$ are uncorrelated random variables, the expectation of their product can be split in the product of their expectations. So in order to umbrally represent the product $a_i a_{n-i}$, we need two distinct (and so uncorrelated) umbrae, having the same sequence of moments. Therefore, if we choose an umbra $\alpha'$ uncorrelated with $\alpha$, but with the same sequence of moments, the sequence (6.1) is umbrally represented by the umbra $(\alpha + \alpha')$

$$\sum_{i=0}^{n} \binom{n}{i} a_i a_{n-i} = E\left[\sum_{i=0}^{n} \binom{n}{i} \alpha^i (\alpha')^{n-i}\right] = E[(\alpha + \alpha')^n]. \tag{6.2}$$

A way to formalize this matter is to define two equivalence relations among umbrae. Two umbrae $\alpha$ and $\gamma$ are *umbrally equivalent* when $E[\alpha] = E[\gamma]$, in symbols $\alpha \simeq \gamma$. They are said to be *similar* when $\alpha^n \simeq \gamma^n$ for $n = 0, 1, 2, \ldots$, in symbols $\alpha \equiv \gamma$. So in (6.2), we have $\alpha \equiv \alpha'$. Outwardly, the first equivalence relation may be redundant. Instead it turns out to be essential in handling umbral polynomials $p \in R[A]$, as it will be clarified in the following.

By introducing these two equivalence relations, the drawback of Blissard's method is overcame because equations involving similar umbrae could be managed by symbolic operations. For example, the symbol $n.\alpha$ denotes the *dot-product* of $n$ and $\alpha$, that is an auxiliary umbra [55] similar to the sum $\alpha' + \alpha'' + \cdots + \alpha'''$, where $\alpha', \alpha'', \ldots, \alpha'''$ are $n$ distinct umbrae similar to the umbra $\alpha$. So the sequence (6.1) is umbrally represented by the umbra $2.\alpha$. We assume $0.\alpha \equiv \epsilon$. If the alphabet $A$ is saturated with sufficiently many umbrae similar to any expression, the dot-product of $n$ and $\alpha$ can be viewed as an umbra. It can be shown that every umbral calculus can be embedded in a saturated umbral calculus [76].

### 6.2.1 Generating Functions

The formal power series

$$u + \sum_{n \geq 1} \alpha^n \frac{t^n}{n!} \tag{6.3}$$

is the *generating function* of the umbra $\alpha$, and is denoted by $e^{\alpha t}$. The notion of umbral equivalence and similarity can be extended coefficientwise to formal power series [76], so that $\alpha \equiv \beta \Leftrightarrow e^{\alpha t} \simeq e^{\beta t}$. Moreover, any exponential formal power series[1] $f(t) = 1 + \sum_{n \geq 1} a_n t^n / n!$ can be umbrally represented by a formal power series (6.3). In fact, if the sequence $1, a_1, a_2, \ldots$ is umbrally represented by $\alpha$ then $f(t) = E[e^{\alpha t}]$, that is $f(t) \simeq e^{\alpha t}$, assuming that we naturally extend $E$ to be linear. We say that $f(t)$ is umbrally represented by $\alpha$, or, when no confusion is possible, that $f(t)$ is the generating function of $\alpha$, writing $f(\alpha, t)$. For example, the generating function of the

---

[1] Observe that with this approach we disregard of questions of whether any series converges.

augmentation umbra $\epsilon$ is $f(\epsilon,t) = 1$, and the generating function of the unity umbra $u$ is $f(u,t) = e^t$.

Let us get back to a random variable $X$. Recall that when $E[\exp(tX)]$ is a convergent function $f(t)$, it admits an exponential expansion in terms of moments which are completely determined by the corresponding distribution function (and viceversa). In this case, the moment generating function encodes all the information of $X$ and the notion of "identically distributed" random variables corresponds to the one of similar umbrae.

The first advantage of the umbral notation introduced for generating functions is the representation of operations among generating functions by means of operations among umbrae. For example, the multiplication among exponential generating functions is umbrally represented by a summation of the corresponding umbrae:

$$f(\alpha,t)f(\gamma,t) \simeq e^{(\alpha+\gamma)t} \quad \text{with} \quad f(\alpha,t) \simeq e^{\alpha t}, f(\gamma,t) \simeq e^{\gamma t}. \tag{6.4}$$

Due to (6.4), the generating function of $n.\alpha$ is $f(n.\alpha,t) = f(\alpha,t)^n$. If $\alpha$ is an umbra with generating function $f(\alpha,t)$, the umbra whose generating function is $1/f(\alpha,t)$ is denoted by $-1.\alpha$. The umbra $-1.\alpha$ is such that $-1.\alpha + \alpha \equiv \epsilon$ and is called the *inverse* of the umbra $\alpha$ [55]. The summation among exponential generating functions is umbrally represented by a disjoint sum of umbrae. The *disjoint sum* (respectively *disjoint difference*) of $\alpha$ and $\gamma$ is the umbra $\eta$ (respectively $\iota$) with moments

$$\eta^n \simeq \begin{cases} u, & n = 0 \\ \alpha^n + \gamma^n, & n > 0 \end{cases} \quad \left( \text{respectively} \quad \iota^n \simeq \begin{cases} u, & n = 0 \\ \alpha^n - \gamma^n, & n > 0 \end{cases} \right),$$

in symbols $\eta \equiv \alpha \dot{+} \gamma$ (respectively $\iota \equiv \alpha \dot{-} \gamma$) with generating functions $f(\alpha \dot{\pm} \gamma, t) = f(\alpha,t) \pm [f(\gamma,t) - 1]$.

The introduction of the generating function device leads us to introduce new auxiliary umbrae, increasing the computational potential of the umbral notation. For this purpose, we should replace $R$ by a polynomial ring having coefficients in $R$ and any desired number of indeterminates. Then we call *scalar* the umbrae whose moments are elements of $R$, whereas we call *polynomial* the umbrae whose moments are polynomials with coefficients in $R$. This replacement allows us to define the dot-product of $x$ and $\alpha$ via generating functions, that is the auxiliary umbra $x.\alpha$ has generating function $e^{(x.\alpha)} \simeq f(\alpha,t)^x$. The next step is to consider the dot-product $\gamma.\alpha$ of two umbrae. The auxiliary umbra $\gamma.\alpha$ has generating function $e^{(\gamma.\alpha)t} \simeq [f(\alpha,t)]^\gamma \simeq e^{\gamma \log f(\alpha,t)} \simeq f[\gamma, \log f(\alpha,t)]$ and moments

$$E[(\gamma.\alpha)^n] = \sum_{i=1}^n g_{(i)} B_{n,i}(a_1, a_2, \ldots, a_{n-i+1}), \quad n = 1, 2, \ldots \tag{6.5}$$

where $g_{(i)}$ are the factorial moments of the umbra $\gamma$, $B_{n,i}$ are the (partial) Bell exponential polynomials and $a_i$ are the moments of the umbra $\alpha$, see [13]. Observe that $E[\gamma.\alpha] = g_1 a_1 = E[\gamma] E[\alpha.]$

The auxiliary umbra $\gamma.\alpha$ is the umbral version of a random sum. Indeed the moment generating function $f[\gamma, \log f(\alpha,t)]$ corresponds to the random variable

$S_N = X_1 + X_2 + \cdots + X_N$, where $N$ is a discrete random variable having moment generating function $f(\gamma, t)$ and $\{X_i\}$ are independent and identically distributed random variables, having moment generating function $f(\alpha, t)$. The right-distributive property of the dot-product $\gamma.\alpha$ runs in parallel with the same for random sums. Indeed, $S_{N+M}$ is identically distributed to $S_N + S_M$, where $N$ and $M$ are independent discrete random variables. The left-distributive property of the dot-product $\gamma.\alpha$ does not hold, similarly to what happens with random sums. In fact, suppose $Z = X + Y$ with $X$ and $Y$ independent random variables and assume $\{Z_i\}$ independent random variables identically distributed to $Z$. Then $S_N = Z_1 + Z_2 + \cdots + Z_N$ is not identically distributed to $S_N^X + S_N^Y$, with $S_N^X = X_1 + X_2 + \cdots + X_N$ and $S_N^Y = Y_1 + Y_2 + \cdots + Y_N$, where $\{X_i\}$ and $\{Y_i\}$ are independent random variables, identically distributed to $X$ and $Y$ respectively.

## 6.3 Sequences of Binomial Type, Bell Umbrae and Poisson Processes

In 1973, Rota [54] claimed that compound Poisson processes are related to polynomial sequences of binomial type, that are sequences $\{p_n(x)\}$ of polynomials with degree $n$ such that $p_n(x+y) = \sum_{i=0}^{n} \binom{n}{i} p_i(x) p_{n-i}(y)$ for any $n$. This connection has ramifications into several other areas of analysis and probability: Lagrange expansions, renewal theory, exponential family distributions and infinite divisibility. Following the suggestion of Rota, Stam [71] studied polynomial sequences of binomial type in terms of an integer-valued compound Poisson process. He is especially interested in the asymptotic behavior of the probability generating function $p_n(x)/p_n(1)$, for $n \to \infty$. Partial results are obtained under conditions on the radius of convergence of a power series related to $\{p_n(x)\}$. The resulting theory involves a complicated system of notations. Due to its relevance, different authors tried to formalize this connection later. Constantine and Savit [6] derived a generalization of Dobinsky's formula by means of compound Poisson processes. Pitman [44] investigated some probabilistic aspects of Dobinsky's formula.

The theory of Bell umbrae gives a natural way to relate compound Poisson processes to polynomial sequences of binomial type.

So let us come back to the Bell umbra $\beta$. Such an umbra can be viewed as a Poisson random variable with parameter $\lambda = 1$. Indeed, the moment generating function of the Bell umbra is equal to the moment generating function of the Poisson random variable with parameter 1, that is $\exp(e^t - 1)$. The *Bell polynomial umbra $x.\beta$* can be viewed as a Poisson random variable with parameter $\lambda = x$. Indeed, the moment generating function of the Bell polynomial umbra is equal to the moment generating function of the Poisson random variable with parameter $x$, that is $\exp[x(e^t - 1)]$. The moments of this Poisson random variable, and so of $x.\beta$, are

$$\Phi_n(x) = \sum_{k=0}^{n} S(n,k) x^k. \tag{6.6}$$

The polynomials $\Phi_n(x)$ have a statistical origin and are known in the literature as exponential polynomials. Indeed, they were first introduced by Steffensen [73] and studied further by Touchard [79] and others. Rota, Kahaner and Odlyzko [54] state their basic properties via umbral operators. Their factorial moments are $(x.\beta)_n \simeq x^n$. When we replace $x$ by $n$, the resulting umbra $n.\beta$ is the sum of $n$ similar uncorrelated Bell scalar umbrae, likewise in probability theory where a Poisson random variable of parameter $n$ can be viewed as a sum of $n$ independent and identically distributed Poisson random variables with parameter 1. More in general, the closure under convolution of Poisson probability distributions, i.e. $F_s \star F_t = F_{s+t}$, where $F_t$ is a Poisson probability distribution depending on the parameter $t$, is umbrally translated by the equivalence $x.\beta + y.\beta \equiv (x+y).\beta$.

If we replace $x$ by a generic umbra $\alpha$, the auxiliary umbra $\alpha.\beta$ has moments $(\alpha.\beta)^n \simeq \Phi_n(\alpha)$, and factorial moments $(\alpha.\beta)_n \simeq \alpha^n$, for nonnegative integers $n$. The umbra $\alpha.\beta$ represents a random sum of independent Poisson random variables with parameter 1 indexed by an integer random variable $Y$, that is a randomized Poisson random variable with parameter $Y$.

If we swap the umbrae $\alpha$ and $\beta$, the result is the $\alpha$-*partition* umbra $\beta.\alpha$, that encodes the connection between compound Poisson processes and polynomial sequences of binomial type. Indeed the generating function of the $\alpha$-*partition* umbra is $\exp[f(\alpha,t)-1]$ and this suggests to interpret a partition umbra as a compound Poisson random variable with parameter 1. As well-known, a compound Poisson random variable with parameter 1 is introduced as a random sum $S_N = X_1 + X_2 + \cdots + X_N$, where $N$ has a Poisson distribution with parameter 1. The umbra $\beta.\alpha$ fits perfectly this probabilistic notion, taking into account that the Bell scalar umbra $\beta$ plays the role of a Poisson random variable with parameter 1. What is more, since the Poisson random variable with parameter $x$ is umbrally represented by the Bell polynomial umbra $x.\beta$, a compound Poisson random variable with parameter $x$ is represented by the *polynomial $\alpha$-partition* umbra $x.\beta.\alpha$, whose generating function is $\exp[x(f(\alpha,t)-1)]$. The name partition umbra has a probabilistic ground. Indeed the parameter of a Poisson random variable is usually denoted by $x = \lambda t$, with $t$ representing a time interval, so that when this interval is partitioned into non-overlapping ones, their contributions are stochastic independent and added to $S_N$. This last circumstance is umbrally expressed by the relation

$$(x+y).\beta.\alpha \equiv x.\beta.\alpha + y.\beta.\alpha, \qquad (6.7)$$

giving the binomial property for the polynomial sequence umbrally represented by $x.\beta.\alpha$. In terms of generating functions, the formula (6.7) means $h_{x+y}(t) = h_x(t)h_y(t)$, where $h_x(t)$ corresponds to the generating function of $x.\beta.\alpha$. Viceversa, every generating function $h_x(t)$, satisfying this last equality, is the generating function of a polynomial $\alpha$-partition umbra. The $\alpha$-partition umbra umbrally represents the sequence of partition polynomials $\mathcal{Y}_n = \mathcal{Y}_n(a_1, a_2, \ldots, a_n)$ (or complete Bell exponential polynomials [49]), i.e.

$$E[(\beta.\alpha)^n] = \sum_{i=1}^{n} B_{n,i}(a_1, a_2, \ldots, a_{n-i+1}) = \mathcal{Y}_n(a_1, a_2, \ldots, a_n), \qquad (6.8)$$

where $a_i$ are the moments of the umbra $\alpha$. Moreover every $\alpha$-partition umbra satisfies the relation

$$(\beta.\alpha)^n \simeq \alpha(\beta.\alpha + \alpha)^{n-1}, \qquad n = 1, 2, \ldots \qquad (6.9)$$

and conversely. The following example shows how the partition umbra allow us to umbrally represent random variables with distributions different from the Poisson law.

*Example 6.1.* The moment generating function of a Gaussian random variable with real parameter $\mu$ and $\sigma > 0$ is $\exp\left(\mu t + \sigma^2 \frac{t^2}{2}\right)$. This power series is the generating function of the umbra $\mu + \beta.(\sigma\delta)$, where $\delta$ is an umbra such that $E[\delta^2] = 1$, whereas $E[\delta^i] = 0$ for positive integers $i \neq 2$. The expression of the $n$-th moment of the umbra representing the Gaussian random variable is

$$a_n = E[(\mu + \beta.(\sigma\delta))^n] = \sum_{k=0}^{\lfloor n/2 \rfloor} \left(\frac{\sigma^2}{2}\right)^k \frac{(n)_{2k}}{k!} \mu^{n-2k}.$$

So moments of a Gaussian random variable can be expressed by using a particular sequence of orthogonal polynomials [49], the *Hermite polynomials* $H_n^{(v)}(x)$, with $\mu = x$ and $\sigma^2 = -v$.

The umbra $\beta.\alpha$ plays a central role also in the umbral representation of composition of exponential generating functions. Indeed, the *composition umbra* of $\alpha$ and $\gamma$ is the umbra $\gamma.\beta.\alpha$. This umbra has generating function $f[\gamma, f(\alpha, t) - 1]$ and moments

$$E[(\gamma.\beta.\alpha)^n] = \sum_{i=1}^{n} g_i B_{n,i}(a_1, a_2, \ldots, a_{n-i+1}), \qquad (6.10)$$

with $g_i$ and $a_i$ moments of the umbra $\gamma$ and $\alpha$ respectively. We denote by $\alpha^{<-1>}$ the *compositional inverse* of $\alpha$, i.e. the umbra having generating function $f(\alpha^{<-1>}, t)$ such that $f[\alpha^{<-1>}, f(\alpha, t) - 1] = f[\alpha, f(\alpha^{<-1>}, t) - 1] = 1 + t$. The compositional inverse of an umbra is strictly related to the Lagrange inversion formula. We will come back to this topic in Section 7. Since the umbra $\gamma.\beta$ represents a randomized Poisson random variable, it is natural to view the composition umbra as a compound randomized Poisson random variable, i.e. a random sum indexed by a randomized Poisson random variable. Moreover, since $(\gamma.\beta).\alpha \equiv \gamma.(\beta.\alpha)$, the previous relation allows us to see this random variable from another side: the umbra $\gamma.(\beta.\alpha)$ generalizes the concept of a random sum of independent and identically distributed compound Poisson random variables with parameter 1, indexed by an integer random variable $X$, i.e. a randomized compound Poisson random variable with random parameter $X$.

In the next section, we introduce the singleton umbra, which plays a complementary role to the Bell umbra. This umbra is the keystone in dealing with cumulants. Actually, we will show that any umbra is the partition umbra of its cumulant umbra, formalizing what Rota claimed in 1978 [51] *"The umbral calculus could be interpreted as a calculus of measures on Poisson algebras, generalizing Poisson processes"*.


## 6.4 Cumulants


In the Fubini Lectures ([53], problem 11: cumulants) Rota wrote:
*"Sometime in the last century, the Danish statistician Thiele observed that the variance of a random variable, namely, $Var(X) = E(X^2) - E(X)^2$, possesses notable properties, to wit:*

*1. it is invariant under translation: $Var(X + c) = Var(X)$ for any constant c.*
*2. If X and Y are independent random variables, then $Var(X + Y) = Var(X) + Var(Y)$.*
*3. $Var(X)$ is a polynomial in the moments of the random variable X.*

*He then proceeded to determine all nonlinear functionals of a random variable which have the same properties. The result of his investigations is striking. The identity of formal power series*

$$\sum_{n=0}^{\infty} \frac{E(X^n)t^n}{n!} = \exp\left(\sum_{j=1}^{\infty} \frac{\kappa_j(X)t^j}{j!}\right) \tag{6.11}$$

*determines uniquely the nonlinear functionals $\kappa_j(X)$ as polynomials in the moments, and it can be shown that every nonlinear functional of a random variable which enjoys the three above properties is a polynomial in the $\kappa_j(X)$. The nonlinear functionals $\kappa_j(X)$ are the cumulants of the random variable X. The second cumulant $\kappa_2(X)$ is the variance, the third is called the coefficient of skewness, and the fourth is called kurtosis. It is all very British.*

*Just as happened with the maximum entropy principle, it turns out that all the random variables occurring in the classical stochastic processes can be characterized by properties of their cumulants. For example, a normally distributed random variable is uniquely characterized by the fact that all its cumulants vanish beyond the second. Similarly, a random variable X which has the Poisson distribution with $E(X) = \alpha$ is the unique probability distribution for which all cumulants equal $\alpha$. It appears therefore that a random variable is better described by its cumulants than by its moments or its characteristic function"*.

Therefore, among the number sequences related to random variables, cumulants play a central role characterizing all random variables occurring in classical stochastic processes. Moreover, due to their properties of additivity and invariance under translation, cumulants are not necessarily connected with moments of probability distributions.

We can define cumulants of any sequence in disregard of convergences of involved series. By this approach, many difficulties connected to the so-called problem of cumulants smooth out, where with the problem of cumulants we refer to characterizations of sequences that are cumulants of some probability distributions. The simplest example is that the second cumulant of a probability distribution must always be nonnegative, and is zero only if all of the higher cumulants are zero. Cumulants are not subject to such constraints when they are analyzed by a symbolic point of view. Recall that there are two more families of cumulants: the Boolean and the free cumulants. In order to handle these two number sequences, Sheffer sequences will be introduced in Section 6.

A first approach to the theory of cumulants via the umbral syntax was given by Rota and Shen in [56]. In the next paragraph, we resume the umbral theory of cumulants stated in [14] by using the singleton umbra.

### 6.4.1 Singleton Umbra

We denote by $\chi$ an umbra such that $\chi^n \simeq \delta_{1,n}$ for $n = 1, 2, \ldots$. The umbra $\chi$ is called *singleton* umbra. Its generating function is $f(\chi, t) = 1 + t$.

**Proposition 6.1.** *If $\chi$ is the singleton umbra, $\beta$ the Bell umbra and $u^{<-1>}$ the compositional inverse of the unity umbra $u$, then*

$$\chi \equiv u^{<-1>}.\beta \equiv \beta.u^{<-1>} \quad and \quad \beta.\chi \equiv u \equiv \chi.\beta. \tag{6.12}$$

For the singleton umbra, the following equivalences hold:

$$\chi.(\alpha + \gamma) \equiv \chi.\alpha + \chi.\gamma \quad and \quad (\alpha + \gamma).\chi \equiv \alpha.\chi + \gamma.\chi. \tag{6.13}$$

Even if the singleton umbra has not a probabilistic counterpart, it turns out to be an effective symbolic tool in order to represent some well-known random variables. For example, the singleton umbra is involved in the umbral counterpart of mixtures of random variables. Indeed, let $\{\alpha_1, \ldots, \alpha_n\}$ be $n$ umbrae and $\{p_1, \ldots, p_n\}$ be $n$ weights such that $\sum_{i=1}^{n} p_i = 1$. The mixture of the umbrae $\{\alpha_i\}_{i=1}^{n}$ is the umbra $\gamma$ such that $\gamma \equiv \chi.p_1.\beta.\alpha_1 + \cdots + \chi.p_n.\beta.\alpha_n$, where $\beta$ denotes the Bell umbra and $\chi$ the singleton umbra. Indeed, from the first equivalence (6.13), this disjoint sum can be rewritten as $\gamma \equiv \chi.(p_1.\beta.\alpha_1 + \cdots + p_n.\beta.\alpha_n)$. Since the generating function of $\sum_{i=1}^{n} p_i.\beta.\alpha_i$ is $\exp(\sum_{i=1}^{n} p_i[f(\alpha_i, t) - 1])$, we have $e^{\gamma t} \simeq \sum_{i=1}^{n} p_i f(\alpha_i, t)$.

*Example 6.2. Bernoulli random variable.* The Bernoulli random variable $X$, with parameter $p \in [0, 1]$, has moment generating function $q + p e^t$, with $q = 1 - p$. So the Bernoulli random variable is represented by the mixture of the augmentation umbra $\epsilon$ and the unity umbra $u$, that is $\xi \equiv \chi.q.\beta.\epsilon + \chi.p.\beta.u$.

*Example 6.3. Binomial random variable.* It is well-known that a binomial random variable $Y$, with parameters $n \in \mathbb{N}$ and $p \in [0, 1]$, is the sum of $n$ independent and identically distributed Bernoulli random variables of parameter $p$. So the umbral counterpart of the binomial random variable is represented by $n.\xi \equiv n.\chi.p.\beta$. Indeed

$f(n.\xi, t) = (q + p e^t)^n$, that is the moment generating function of the binomial random variable $Y$.

## 6.4.2 Cumulant Umbra

By using the singleton umbra, the formulae commonly used to express moments of a random variable in terms of cumulants, and viceversa, can be encoded in very simple closed forms. The starting point is a suitable definition of the umbra whose moments are the cumulants of a given number sequence.

Suppose the sequence $1, a_1, a_2, \ldots$ umbrally represented by the umbra $\alpha$. Let $\kappa_1, \kappa_2, \ldots$ be the corresponding sequence of cumulants. The umbra whose moments are given by the sequence $1, \kappa_1, \kappa_2, \ldots$ is the umbra $\kappa_\alpha$ such that $\kappa_\alpha \equiv \chi.\alpha$, where $\chi$ is the singleton umbra. The proof is straightforward by means of generating functions. Indeed we have $f(\chi.\alpha, t) = 1 + \log[f(\alpha, t)]$ which gives equality (6.11). The umbra $\chi.\alpha$ is called $\alpha$-cumulant umbra. Some special cases are:

i) $\kappa_\varepsilon \equiv \varepsilon$, being $\varepsilon \equiv \chi.\varepsilon$;
ii) $\kappa_u \equiv \chi$, being $\chi \equiv \chi.u$;
iii) $\kappa_\beta \equiv u$, being $u \equiv \chi.\beta$, see the second equivalence (6.12); this parallels the same result for the Poisson random variable of parameter 1 whose cumulants equal 1.

By using generating functions, we have

iv) $\kappa_\chi \equiv \chi.\chi \equiv u^{<-1>}$, being $f(^{<-1>}, t) = 1 + \log(1 + t)$;
v) $\chi.(\chi.p.\beta) \equiv \kappa_\chi.p.\beta \equiv u^{<-1>}.p.\beta$, that is the cumulant umbra of the Bernoulli random variable, see Example 6.2;
vi) $\chi.n.\chi.p.\beta \equiv +_n u^{<-1>}.p.\beta$, that is the cumulant umbra of the binomial random variable, see Example 6.3.

On the other hand the umbra $\alpha.\chi$ has moments umbrally equivalent to the factorial moments of $\alpha$, that is $(\alpha.\chi)^n \simeq (\alpha)_n$. So we call the umbra $\alpha.\chi$ the $\alpha$-factorial umbra [14]. Factorial moments provide very concise expressions for moments of some discrete distributions, like the binomial distribution. Here we recall that $\kappa_\chi \equiv \chi.\chi$ is not only the $\chi$-cumulant umbra, but also the $\chi$-factorial umbra and in particular $(\chi.\chi)^n \simeq (\chi)_n \simeq (u^{<-1>})^n$. Since the moments of $u^{<-1>}$ are the coefficients of the exponential expansion $1 + \log(1 + t) = 1 + \sum_{i=1}^\infty (-1)^{i-1}(i-1)! t^i / i!$, we have $E[(\chi.\chi)^n] = x_{(n)} = (-1)^{n-1}(n-1)!$. Then from (6.5) the moments of $\chi.\alpha$ are

$$\kappa_n = E[(\chi.\alpha)^n] = \sum_{i=1}^n (-1)^{i-1}(i-1)! B_{n,i}(a_1, a_2, \ldots, a_{n-i+1}), \quad n = 1, 2, \ldots \quad (6.14)$$

where $a_i$ are the moments of $\alpha$. Equality (6.14) can be rewritten as the usual expression giving cumulants of a random variable in terms of its moments (cf. for example [75])

$$\kappa_n = \sum_\pi c_\pi a_\pi, \quad (6.15)$$

where the sum ranges over all partitions $\pi = [j_1^{m_1}, j_2^{m_2}, \ldots, j_k^{m_k}]$ of the integer $n, a_\pi = \prod_{j \in \pi} a_j$, and $c_\pi = d_\pi(-1)^{v_\pi - 1}(v_\pi - 1)!$ with $v_\pi = m_1 + m_2 + \cdots + m_k$, and

$$d_\pi = \frac{n!}{(j_1!)^{m_1}(j_2!)^{m_2} \cdots (j_k!)^{m_k}} \frac{1}{m_1! m_2! \cdots m_k!}. \tag{6.16}$$

So the umbral equivalence $\kappa_\alpha \equiv \chi.\alpha$ represents a closed form of equalities (6.14) and (6.15).

The three main algebraic properties of cumulants can be translated in umbral terms as follows

a)   (the additivity property) $\chi.(\alpha + \gamma) \equiv \chi.\alpha \dot{+} \chi.\gamma$, i.e. the cumulant umbra of a sum of two umbrae is equal to the disjoint sum of the corresponding cumulant umbrae;
b)   (the semi-invariance under traslation property) $\chi.(\alpha + c.u) \equiv \chi.\alpha \dot{+} \chi.c$, for any $c \in R$;
c)   (the homogeneity property) $\chi.(c\alpha) \equiv c(\chi.\alpha)$, for any $c \in R$.

The formulae to express moments of a random variable in terms of cumulants are given by

$$a_n = \sum_\pi d_\pi \kappa_\pi, \tag{6.17}$$

where again the sum ranges over all partitions $\pi = [j_1^{m_1}, j_2^{m_2}, \ldots, j_k^{m_k}]$ of the integer $n$, and $\kappa_\pi = \prod_{j \in \pi} \kappa_j$. A closed-form expression of equality (6.17) is given by the following inversion theorem.

**Theorem 6.1 (Inversion theorem).** *Let $\kappa_\alpha$ be the cumulant umbra of $\alpha$, then $\alpha \equiv \beta.\kappa_\alpha$, where $\beta$ is the Bell umbra.*

The result follows simply by observing that $\beta.\kappa_\alpha \equiv \beta.u^{<-1>}.\beta.\alpha \equiv \chi.\beta.\alpha \equiv u.\alpha \equiv \alpha$. Recalling (6.8), we have an explicit formula of (6.17) via the complete Bell polynomials, that is $a_n = \mathcal{Y}_n[(k_a)_1, \ldots, (k_a)_n]$, with $a_n$ the $n$-th moment of the umbra $\alpha$ and $(k_a)_n$ the $n$-th moment of the umbra $\kappa_\alpha$. The complete Bell polynomials $\{\mathcal{Y}_n\}$ are of binomial type. Since from the inversion Theorem 6.1 any umbra $\alpha$ could be seen as the partition umbra of its cumulant $\kappa_\alpha$, it is then possible to prove a more general result: *every polynomial sequence of binomial type is completely determined by the sequence of its formal cumulants.* On the other hand, we have proved that any polynomial sequence of binomial type can be viewed as a sequence of moments of a polynomial umbra $x.\alpha$, and viceversa. Hence from the inversion Theorem 6.1, we deduce that any polynomial sequence of binomial type is a sequence of moments of the polynomial umbra $x.\beta.\kappa_\alpha$.

The next corollary follows from equivalence (6.9) and smooths the way to a symbolic handling of Abel sequences, as it will be more clear in Section 7.

**Corollary 6.1.** *If $\kappa_\alpha$ is the cumulant umbra of $\alpha$, then for $n = 1, 2, \ldots$*

$$\alpha^n \simeq \kappa_\alpha(\kappa_\alpha + \alpha)^{n-1} \quad and \quad \kappa_\alpha^n \simeq \alpha(-1.\alpha + \alpha)^{n-1}. \tag{6.18}$$

The first of equivalences (6.18) was assumed by Shen and Rota in [56] as definition of the cumulant umbra. In terms of moments, the first of equivalences (6.18) gives $a_n = \sum_{j=0}^{n-1} \binom{n}{j} a_j \kappa_{n-j}$, largely used in statistic framework [61]. In the following, we give an example of how the computation of cumulants of Poisson random variables is simplified by using the umbral calculus.

*Example 6.4. Cumulants of Poisson random variables.* As already underlined, the umbra $\gamma.\beta.\alpha$ corresponds to a compound randomized Poisson random variable, i.e. a random sum $S_N = X_1 + \cdots + X_N$ with $N$ a randomized Poisson random variable with parameter the random variable $Y$. In particular $\alpha$ corresponds to $X$ and $\gamma$ corresponds to $Y$. Being $\chi.(\gamma.\beta.\alpha) \equiv \kappa_\gamma.\beta.\alpha$, the cumulant umbra of the composition of $\alpha$ and $\gamma$ is the composition of $\alpha$ and $\kappa_\gamma$. So from (6.10), the cumulants of a compound randomized Poisson random variable are given by $\sum_{i=1}^{n} \kappa_i B_{n,i}(a_1, a_2, \ldots, a_{n-i+1})$, where $a_i$ are the moments of the random variable $X$ and $\kappa_i$ are the cumulants of the random variable $Y$. Now set $\gamma \equiv x.u$ in $\gamma.\beta.\alpha$. This means to consider a random variable $Y$ such that $P(Y = x) = 1$. Then, the random sum $S_N$ becomes a compound Poisson random variable of parameter $x$ corresponding to the polynomial $\alpha$-partition umbra $x.\beta.\alpha$. Since the moments of $\chi.x$ are equal to 0, except the first one which equals $x$, the umbra $\alpha$ is the umbral counterpart of $X$ with cumulants $x a_n$. If $x = 1$ the cumulant of the $\alpha$-partition umbra is $\alpha$, so that the moments of $X$ are the cumulants of the corresponding compound Poisson random variable. Now, in $x.\beta.\alpha$ take $\alpha \equiv u$. The cumulants of the Bell polynomial umbra $x.\beta$ are equals to $x$ as well as for the Poisson random variable of parameter $x$. If we choose the unity umbra $u$ as umbra $\alpha$, then $\gamma.\beta.\alpha \equiv \gamma.\beta$, whose cumulant umbra is $\kappa_\gamma.\beta$, with $\kappa_\gamma$ the cumulant umbra of $\gamma$. Its probabilistic counterpart is a randomized Poisson random variable of parameter the random variable $Y$, corresponding to the umbra $\gamma$. The cumulants of a randomized Poisson random variable of parameter the random variable $Y$ are the moments of $\kappa_\gamma.\beta$, i.e. $\sum_{i=0}^{n} S(n,i) \kappa_i$, with $\kappa_i$ the cumulants of the random variable $Y$.

Actually, the umbral theory of Poisson processes can be applied to more general processes. As example, let $(X_t, t \geq 0)$ be a real-value Lévy process, i.e. a process starting from 0 and with stationary and independent increments. According to the Lévy-Khintchine formula [25], if we assume that $X_t$ has a convergent moment generating function in some neighbourhood of 0, then $E[e^{\theta X_t}] = e^{t k(\theta)}$, where $k(\theta)$ is the cumulant generating function of $X_1$. The inversion Theorem 6.1 gives the umbral version of this last equation, that is $t.\alpha \equiv t.\beta.\kappa_\alpha$. Since the umbral syntax shows its effectiveness within free probability theory (see Section 7), the last similarity smooths the way to an umbral treatment of free Levy processes.

## 6.5 Applications in Statistics

In this section, we show a variety of applications of the classical umbral calculus within statistical fields. One feature of the classical umbral calculus is that an umbra has the structure of a random variable without reference to a probability space. This brings us closer to statistical methods. Indeed we show how the classical umbral calculus can be fruitfully used in statistics, bringing out the authentically algebraic side of many techniques commonly used to manage number sequences related to random variables. This goes in parallel with the *algebraic statistics* [43, 24] that, in recent years, has usefully used the notion of toric varieties.

Our attention has been catched by the general problem of calculating algebraic expressions such as the variance of a sample mean or, more generally, moments of sampling distributions within statistical inference [75]. In the past, the field has been cursed by the difficulty of manual computations. Symbolic computations have removed many of such difficulties, leaving neverthless some issues unresolved. One of the most intriguing questions is to explain why symbolic procedures, which are straightforward in the multivariate case, turn out to be obscure in the simpler univariate one [3]. Actually, at the root of the question, there are some aspects of the combinatorics of symmetric functions that would benefit by a shift of the attention from sets to the more general notion of multiset. On the other hand, due to its generality, the umbral calculus reduces the combinatorics of symmetric functions, commonly used by statisticians, to few relations which cover a great variety of calculations. Moreover, by means of the notion of umbrae indexed by multiset, we remove the necessity of specifying if the random variables of a vector are identically distributed or not.

### 6.5.1 U-Statistics

We focus on the special auxiliary umbra $n.(\chi\alpha)$. By this symbol, we denote an auxiliary umbra such that $n.(\chi\alpha) \equiv \chi_1\alpha_1 + \cdots + \chi_n\alpha_n$, where $\chi_1, \chi_2, \ldots, \chi_n$ are uncorrelated umbrae similar to the umbra $\chi$, and $\alpha_1, \alpha_2, \ldots, \alpha_n$ are uncorrelated umbrae similar to the umbra $\alpha$. Of course, we have $n.(\chi\alpha^r) \equiv \chi_1\alpha_1^r + \cdots + \chi_n\alpha_n^r$, for non-negative integers $r$. These umbrae, and their products, are similar to well-known symmetric polynomials. Indeed, since the umbrae $\alpha_i$ for $i = 1, 2, \ldots, n$ can be rearranged without effecting the evaluation $E$, powers of $n.(\chi\alpha)$ are umbrally equivalent to the umbral elementary symmetric polynomials in the umbrae $\alpha_1, \ldots, \alpha_n$, that is $[n.(\chi\alpha)]^k \simeq k!e_k(\alpha_1, \alpha_2, \ldots, \alpha_n)$, where $e_k(\alpha_1, \alpha_2, \ldots, \alpha_n) = \sum_{1 \le j_1 < \cdots < j_k \le n} \alpha_{j_1} \cdots \alpha_{j_k}$, for $k = 1, 2, \ldots, n$. The proof is given in [15], and it relies on the role played by the umbra $\chi$ in picking out the indeterminates $\alpha_1, \ldots, \alpha_n$.

The auxiliary umbra $n.(\chi\alpha^r)$ enables us to rewrite umbral augmented symmetric polynomials in a very compact expression. Let $\lambda = (1^{r_1}, 2^{r_2}, \ldots) \vdash i$ be a partition of the integer $i \le n$. Augmented monomial symmetric polynomials in the indeterminates $\alpha_1, \alpha_2, \ldots, \alpha_n$ are defined as

$$\tilde{m}_\lambda(\alpha_1, \alpha_2, \ldots, \alpha_n) = \sum_{j_1 \neq \ldots \neq j_{r_1} \neq j_{r_1+1} \neq \ldots \neq j_{r_1+r_2} \neq \ldots} \alpha_{j_1} \cdots \alpha_{j_{r_1}} \alpha_{j_{r_1+1}}^2 \cdots \alpha_{j_{r_1+r_2}}^2 \cdots .$$

In statistical literature, a more common notation is $[1^{r_1}2^{r_2}\ldots]$, see [35]. If $\lambda = (1^{r_1}, 2^{r_2}, \ldots) \vdash i$, then

$$\tilde{m}_\lambda(\alpha_1, \alpha_2, \ldots, \alpha_n) \simeq [n.(\chi\alpha)]^{r_1} [n.(\chi\alpha^2)]^{r_2} \cdots, \qquad (6.19)$$

taking into account the role played by the umbra $\chi$ in selecting variables. The following theorem gives an umbral formulation of a fundamental and deep result in statistics (cf. [75]), lying at the core of unbiased estimation and moments of moments literature.

**Theorem 6.2.** *If $\lambda = (1^{r_1}, 2^{r_2}, \ldots) \vdash i$ with $i \leq n$, then*

$$\alpha_\lambda \simeq \frac{1}{(n)_{\nu_\lambda}} [n.(\chi\alpha)]^{r_1} [n.(\chi\alpha^2)]^{r_2} \cdots, \qquad (6.20)$$

*where $\alpha_\lambda = (\alpha_1)^{.r_1} (\alpha_2^2)^{.r_2} \ldots$ with $\alpha_1, \alpha_2, \ldots$ uncorrelated umbrae similar to $\alpha$.*

Equivalence (6.20) states how to estimate products of moments $\alpha_\lambda$ by means of only $n$ bits of information drawn from the population. In umbral terms, the population is represented by $\alpha$ and the $n$ bits are the uncorrelated umbrae $\alpha_1, \ldots, \alpha_n$, coming into $[n.(\chi\alpha^i)]^{r_i}$. Moreover, since the umbral polynomials $[n.(\chi\alpha^i)]^{r_i}$ is similar to elementary polynomials, Theorem 6.2 discloses a more general result: products of moments are umbrally equivalent to products of umbral elementary polynomials. The symmetric polynomial on the right hand side of equivalence (6.20) is named *U-statistic* of uncorrelated and similar umbrae $\alpha_1, \ldots, \alpha_n$. We take a moment to motivate this denomination. Usually an $U$-statistic has the form $U = \sum \Phi(X_{j_1}, X_{j_2}, \ldots, X_{j_k})/(n)_k$, where $X_1, \ldots, X_n$ are $n$ independent random variables, and the sum ranges in the set of all permutations $(j_1, j_2, \ldots, j_k)$ of $k$ integers with $1 \leq j_i \leq n$. If $X_1, \ldots, X_n$ have the same cumulative distribution function $F(x)$, $U$ is an unbiased estimator of the population parameter
$\theta(F) = \int \cdots \int \Phi(x_1, \ldots, x_k) dF(x_1) \cdots dF(x_k)$.

In this case, the function $\Phi$ may be assumed to be a symmetric function of its arguments. Often, in the applications, $\Phi$ is a polynomial in $X_i$'s so that the $U$-statistic is a symmetric polynomial. Hence, by virtue of the fundamental theorem on symmetric polynomials, such an $U$-statistic can be expressed as a polynomial in elementary symmetric polynomials.

For computational purposes, it would be convenient to express umbral polynomials, like $[n.(\chi\alpha)]^{r_1} [n.(\chi\alpha^2)]^{r_2} \cdots$, in terms of power sums of the data points. Being $n.\alpha^r \equiv \alpha_1^r + \cdots + \alpha_n^r$, where $\alpha_1, \ldots, \alpha_n$ are uncorrelated umbrae, similar to the umbra $\alpha$, the auxiliary umbra $n.\alpha^r$ is similar to the $r$-th power sum symmetric polynomial in the indeterminates $\alpha_1, \ldots, \alpha_n$. Expressing products of auxiliary umbrae such as $[n.(\chi\alpha^i)]^{r_i}$ in terms of $n.\alpha^j$, for some $j$, is the content of the next paragraph. These formulae allow us to express moments of sampling distributions in terms of population moments [17].

### 6.5.2 *Moments of Sampling Distributions*

Let us consider the symmetric functions commonly used in computing moments of sampling distributions, i.e. augmented monomial symmetric functions and power sums, with special care to conversion formulae. Such polynomials are classical bases of the algebra of symmetric polynomials. The well-known changes of bases involve the lattice of partitions [33]. Several packages are available aiming to implement changes of bases (see http://garsia.math.yorku.ca/MPWP/). For instance, the SF package [74] is an integrated MAPLE package devoted to symmetric functions. The use of such packages requires a good knowledge of symmetric function theory. Due to their generality, such packages are slow when applied to large variable sets.

By using the umbral notation, the change of bases between augmented symmetric functions and power sums is simplified and can be extended in a natural way to the multivariate case, taking advantage of multiset notion. In the following we summarize the necessary steps to construct such formulae in the most general case, which have applications in multivariate statistics. An extensive account is given in [15].

The starting point is the expression of moments of $n.(\chi\alpha)$ in terms of $n.\alpha$ and viceversa:

$$[n.(\chi\alpha)]^i \simeq \sum_{\lambda \vdash i} d_\lambda (\chi.\chi)_\lambda (n.\alpha)^{r_1} (n.\alpha^2)^{r_2} \cdots \quad (n.\alpha)^i \simeq \sum_{\lambda \vdash i} d_\lambda [n.(\chi\alpha)]^{r_1} [n.(\chi\alpha^2)]^{r_2} \cdots .$$

(6.21)

Equivalences (6.21) may be rewritten replacing $\alpha$ with any power $\alpha^k$. The next step is to express more general products like $[n.(\chi\alpha)]^{r_1} [n.(\chi\alpha^2)]^{r_2} \cdots$ (that are augmented symmetric polynomials) in terms of power sums. With this aim, equivalences (6.21) have to be rewritten by using set partitions instead of integer partitions. We say in advance that the final step consists in replacing the set with the more general structure of multiset.

Let $C$ be a subset of $R[A]$ with $n$ elements and let $\Pi_n$ be the set of all partitions $\pi = \{B_1, \ldots, B_k\}$ of $C$, with $k \le n$. Assume $\{\alpha_1, \ldots, \alpha_n\}$ a set of $n$ uncorrelated umbrae similar to an umbra $\alpha$. The symbol $\alpha^{.\pi}$ denotes the umbra $\alpha^{.\pi} \equiv \alpha_{i_1}^{|B_1|} \cdots \alpha_{i_k}^{|B_k|}$, where $\pi$ is a partition of the set $\{\alpha_1, \ldots, \alpha_n\}$ and $i_1, \ldots, i_k$ are distinct integers, chosen in $\{1, \ldots, n\}$. Note that $\alpha^{.\pi} \equiv \alpha_\lambda$, when $\lambda$ is the partition of the integer $n$ determined by $\pi$. Indeed, a set partition is said to be of type $\lambda = (1^{r_1}, 2^{r_2}, \ldots)$ if there are $r_1$ blocks of cardinality 1, $r_2$ blocks of cardinality 2 and so on. The number of set partitions of type $\lambda$ is $d_\lambda$ given in (6.16) replacing $j_i$ with $i$, and $m_i$ with $r_i$. By using set partitions, equivalences (6.21) may be rewritten as

$$(n.\alpha)^i \simeq \sum_{\pi \in \Pi_i} [n.(\chi\alpha)]^{r_1} [n.(\chi\alpha^2)]^{r_2} \cdots \quad [n.(\chi\alpha)]^i \simeq \sum_{\pi \in \Pi_i} (\chi.\chi)^{.\pi} (n.\alpha)^{r_1} (n.\alpha^2)^{r_2} \cdots ,$$

(6.22)

where $E[(\chi.\chi)^{.\pi}] = E\left[(\chi.\chi)^{|B_1|}\cdots(\chi''.\chi'')^{|B_k|}\right] = \prod_{i=1}^{k} x_{|B_i|}$. By using the notion of multiset, equivalences (6.22) suggest the way to generalize such formulae to the multivariate case.

Recall that a multiset $M$ is a pair $(\bar{M}, f)$, where $\bar{M} \subset R[A]$ is a set, called the support of the multiset, and $f$ is a function from $\bar{M}$ to the non-negative integers. For each $\mu \in \bar{M}$, $f(\mu)$ is called the multiplicity of $\mu$. If the support of $M$ is a finite set, say $\bar{M} = \{\mu_1, \mu_2, \ldots, \mu_k\}$, we write

$$M = \{\mu_1^{(f(\mu_1))}, \mu_2^{(f(\mu_2))}, \ldots, \mu_k^{(f(\mu_k))}\} \quad \text{or} \quad M = \{\underbrace{\mu_1, \ldots, \mu_1}_{f(\mu_1)}, \ldots, \underbrace{\mu_k, \ldots, \mu_k}_{f(\mu_k)}\}.$$

The length of the multiset $M$ is the sum of multiplicities of all elements of $\bar{M}$, that is $|M| = \sum_{\mu \in \bar{M}} f(\mu)$. From now on, we denote a multiset $(\bar{M}, f)$ simply by $M$. In the following, we set

$$\mu_M = \prod_{\mu \in \bar{M}} \mu^{f(\mu)} \quad \text{and} \quad (n.\mu)_M = \prod_{\mu \in \bar{M}} (n.\mu)^{g(\mu)}. \tag{6.23}$$

For instance, if $M = \{\alpha^{(i)}\}$ then $(n.\alpha)_M \simeq (n.\alpha)^i$ and $[n.(\chi\alpha)]_M \simeq [n.(\chi\alpha)]^i$. This notation can be easily extended to umbral polynomials. If $\lambda = (1^{r_1}, 2^{r_2}, \ldots)$ is an integer partition, set

$$P_\lambda = \{\underbrace{\alpha, \ldots, \alpha}_{r_1}, \underbrace{\alpha^2, \ldots, \alpha^2}_{r_2}, \ldots\}. \tag{6.24}$$

If $\lambda$ is the type of the set partition $\pi$, equivalences (6.22) may be rewritten as

$$(n.\alpha)_M \simeq \sum_{\pi \in \Pi_i} [n.(\chi\alpha)]_{P_\lambda} \quad \text{and} \quad [n.(\chi\alpha)]_M \simeq \sum_{\pi \in \Pi_i} (\chi.\chi)^{.\pi}(n.\alpha)_{P_\lambda}, \tag{6.25}$$

since $(n.\alpha)_{P_\lambda} \simeq (n.\alpha)^{r_1}(n.\alpha^2)^{r_2}\cdots$ and $[n.(\chi\alpha)]_{P_\lambda} \simeq [n.(\chi\alpha)]^{r_1}[n.(\chi\alpha^2)]^{r_2}\cdots$ by using notation (6.23). In equivalences (6.25), we have $M = \{\alpha^{(i)}\}$. The last step is the generalization of such equivalences to any multiset $M$. To this aim, we recall the notion of multiset subdivision. Such notion is quite natural and it is equivalent to splitting the multiset into disjoint blocks (submultisets) whose union gives the whole multiset.

A subdivision of a multiset $M$ is a multiset $S = (\bar{S}, g)$ of $k \leq |M|$ non-empty submultisets $M_i = (\bar{M}_i, f_i)$ of $M$ such that $\cup_{i=1}^{k} \bar{M}_i = \bar{M}$ and $\sum_{i=1}^{k} f_i(\mu) = f(\mu)$ for any $\mu \in \bar{M}$. Recall that a multiset $M_i = (\bar{M}_i, f_i)$ is a submultiset of $M = (\bar{M}, f)$ if $\bar{M}_i \subseteq \bar{M}$ and $f_i(\mu) \leq f(\mu), \forall \mu \in \bar{M}_i$. If $M = \{\alpha^{(i)}\}$, then subdivisions are of type

$$S_\lambda = \{\underbrace{\{\alpha\}, \ldots, \{\alpha\}}_{r_1}, \underbrace{\{\alpha^{(2)}\}, \ldots, \{\alpha^{(2)}\}}_{r_2}, \ldots\}$$

with $r_1 + 2r_2 + \cdots = i$, and we will say that the subdivision $S$ is of type $\lambda = (1^{r_1}, 2^{r_2}, \ldots) \vdash i$. The support of $S_\lambda$ is $\bar{S}_\lambda = \{\{\alpha\}, \{\alpha^{(2)}\}, \ldots\}$. With subdivisions of type

$$S = \{\underbrace{M_1\ldots,M_1}_{g(M_1)}, \underbrace{M_2,\ldots,M_2}_{g(M_2)},\ldots,\underbrace{M_k,\ldots,M_k}_{g(M_k)}\}, \tag{6.26}$$

we set $\mu_S = \prod_{M_i \in \bar{S}} \mu_{M_i}^{g(M_i)}$ and $(n.\mu)_S = \prod_{M_i \in \bar{S}} (n.\mu_{M_i})^{g(M_i)}$, extending notation (6.23). By using this notation and recalling (6.24), we have $(n.\alpha)_{P_\lambda} \equiv (n.\alpha)_{S_\lambda}$ and $[n.(\chi\alpha)]_{P_\lambda} \equiv [n.(\chi\alpha)]_{S_\lambda}$, with $\lambda = (1^{r_1},2^{r_2},\ldots) \vdash i$. Then equivalences (6.25) may be written as

$$(n.\alpha)_M \simeq \sum_{\pi\in\Pi_i} [n.(\chi\alpha)]_{S_\lambda} \quad \text{and} \quad [n.(\chi\alpha)]_M \simeq \sum_{\pi\in\Pi_i} (\chi.\chi)^{\pi}(n.\alpha)_{S_\lambda}. \tag{6.27}$$

When integer partitions are replaced by multiset subdivisions, the fundamental expectation result (6.20) becomes

$$[n.(\chi\mu)]_S \simeq (n)_{|S|}\mu^S, \tag{6.28}$$

with $S$ given in (6.26) and $\mu^S \equiv (\mu_{M_1})^{\cdot g(M_1)}\cdots(\mu'_{M_k})^{\cdot g(M_k)}$, where $\mu_{M_i}$ are uncorrelated umbral monomials. When the multiset $M = \{\alpha^{(i)}\}$ is replaced by an arbitrary multiset, one more remark allows us to remove integer partitions $\lambda$ from (6.27). Let us observe that a subdivision of the multiset $M$ may be constructed in the following way: suppose the elements of $M$ to be all distinct, build a set partition and then replace each element in any block by the original one. In this way, any subdivision corresponds to a set partition $\pi$ and we will write $S_\pi$. Note that it is $|S_\pi| = |\pi|$ and it could be $S_{\pi_1} = S_{\pi_2}$ for $\pi_1 \neq \pi_2$. Finally, equivalences (6.25) may be rewritten as follows

$$(n.\mu)_M \simeq \sum_{\pi\in\Pi_i} [n.(\chi\mu)]_{S_\pi} \quad \text{and} \quad [n.(\chi\mu)]_M \simeq \sum_{\pi\in\Pi_i} (\chi.\chi)^{\pi}(n.\mu)_{S_\pi}, \tag{6.29}$$

where $S_\pi$ is the subdivision of $M$ corresponding to the partition $\pi$ of a set $C$ such that $|C| = |M| = i$. The symbolic expression of such equivalences does not change if the multiset $M = \{\alpha^{(i)}\}$ is replaced by an arbitrary multiset.

The following example shows the effectiveness of the umbral notation in handling moments of sampling distributions.

*Example 6.5.* If $M = \{\mu_1,\mu_1,\mu_2\}$, then $(n.\mu)_M = (n.\mu_1)^2(n.\mu_2)$. In statistical terminology, moments of $(n.\mu)_M$ correspond to moments of the product of sums $\left(\sum_{i=1}^n X_i\right)^2\left(\sum_{i=1}^n Y_i\right)$, where $(X_1,Y_1),\ldots,(X_n,Y_n)$ are separately independent and identically distributed random variables. In order to apply the first part of (6.29), we need to compute all subdivisions of $M$. For instance, let us consider the subdivision $S_1 = \{\{\mu_1\},\{\mu_1,\mu_2\}\}$ of the multiset $M = \{\mu_1,\mu_1,\mu_2\}$. The support of $S_1$ consists of two multisets, $M_1 = \{\mu_1\}$ and $M_2 = \{\mu_1,\mu_2\}$, each of one with multiplicity 1 so that $[n.(\chi\mu)]_{S_1} = n.(\chi\mu_{M_1})n.(\chi\mu_{M_2})$. Since $n.(\chi\mu_{M_1}) = n.(\chi\mu_1)$ and $n.(\chi\mu_{M_2}) = n.(\chi\mu_1\mu_2)$, we have $[n.(\chi\mu)]_{S_1} = n.(\chi\mu_1)n.(\chi\mu_1\mu_2)$. Repeating the same arguments for all subdivisions of $M$, we get the following result $(n.\mu_1)^2(n.\mu_2) \simeq n.(\chi\mu_1^2\mu_2) + 2[n.(\chi\mu_1)][n.(\chi\mu_1\mu_2)] + [n.(\chi\mu_2)][n.(\chi\mu_1^2)] + [n.(\chi\mu_1)]^2[n.(\chi\mu_2)]$, that is

$$\left(\sum_{i=1}^{n} X_i\right)^2 \left(\sum_{i=1}^{n} Y_i\right) = \sum_{i=1}^{n} X_i^2 Y_i + 2 \sum_{1 \le i \ne j \le n} X_i X_j Y_j + \sum_{1 \le i \ne j \le n} X_i^2 Y_j + \sum_{1 \le i \ne j \ne k \le n} X_i X_j Y_k.$$

(6.30)

In order to evaluate the mean of the sums on the right hand side of (6.30) in terms of population moments, we have to use (6.28) and finally we obtain

$$E\left[\left(\sum_{i=1}^{n} X_i\right)^2 \left(\sum_{i=1}^{n} Y_i\right)\right] =$$
$$nE[X^2 Y] + 2(n)_2 E[X]E[XY] + (n)_2 E[X^2]E[Y] + (n)_3 E[X]^2 E[Y].$$

### 6.5.3 Products of Statistics

This paragraph is devoted to a different application of equivalences (6.29), allowing us to evaluate the mean of product of augmented polynomials in separately independent and identically distributed random variables. Suppose, for instance, we need the mean of

$$\left(\sum_{i \ne j}^{n} X_i^2 X_j\right)\left(\sum_{i=1}^{n} X_i^2 Y_i\right)^2 = S_{\{\{2,0\},\{1,0\}\}} S_{\{\{2,1\}\}} S_{\{\{2,1\}\}},$$

(6.31)

where $(X_1, Y_1), \ldots, (X_n, Y_n)$ are separately independent and identically distributed random variables and

$$S_{\{\{k_1,l_1\},\{k_2,l_2\},\ldots\}} = \sum_{i_1 \ne i_2 \ne \cdots} X_{i_1}^{k_1} Y_{i_1}^{l_1} X_{i_2}^{k_2} Y_{i_2}^{l_2} \cdots,$$

(6.32)

where we use the notation introduced in [82]. If we expand the product (6.31) as a linear combination of augmented symmetric polynomials (6.32), then we are able to apply the fundamental expectation result (6.28) and to evaluate the mean of (6.31). The umbral tools so far introduced are sufficient to do the job. Therefore, since $\sum_{i \ne j}^{n} X_i^2 X_j$ corresponds to $[n.(\chi \mu_1^2) n.(\chi \mu_1)]$, and $\sum_{i=1}^{n} X_i^2 Y_i$ corresponds to $[n.(\chi \mu_1^2 \mu_2)]$, the product (6.31) is umbrally represented by $n.(\chi_1 \mu_1^2) n.(\chi_1 \mu_1) n.(\chi_2 \mu_1^2 \mu_2) n.(\chi_3 \mu_1^2 \mu_2)$, where $\{\chi_i\}, i = 1, 2, 3$ are uncorrelated singleton umbrae. Indeed, a product of uncorrelated singleton umbrae does not "delete" the same indexed umbrae. Moreover, the sum (6.32) is umbrally represented by

$$n.(\chi \mu_1^{k_1} \mu_2^{l_1}) n.(\chi \mu_1^{k_2} \mu_2^{l_2}) \cdots.$$

(6.33)

Set $M = \{\chi_1 \mu_1^2, \chi_1 \mu_1, \chi_2 \mu_1^2 \mu_2, \chi_3 \mu_1^2 \mu_2\}$. The length of $M$ is 4, and since the monomials are all different, $M$ is a set. From the second part of (6.23), we have

$$(n.\mu)_M \simeq [n.(\chi_1 \mu_1^2) n.(\chi_1 \mu_1)][n.(\chi_2 \mu_1^2 \mu_2)][n.(\chi_3 \mu_1^2 \mu_2)].$$

Via the first part of equivalence (6.29), we have

$$[n.(\chi_1\mu_1^2)n.(\chi_1\mu_1)][n.(\chi_2\mu_1^2\mu_2)][n.(\chi_3\mu_1^2\mu_2)] \simeq \sum_{\pi \in \Pi_4}[n.(\chi\mu)]_\pi,$$

where $S_\pi = \pi$. When in one - or more - blocks of the subdivision $S_\pi$, there are at least two umbral monomials involving correlated singleton umbrae, the auxiliary umbra $[n.(\chi\mu)]_{S_\pi}$ has the evaluation equal to zero. If in every block of the subdivision $S_\pi$ there are only uncorrelated singleton umbrae, then $[n.(\chi\mu)]_{S_\pi}$ gives rise expressions like (6.33), umbrally representing (6.32).

In the next paragraph, we show how to use the same formulae in order to express any $U$-statistic in terms of power sums. In particular, we will refer to $k$-statistics, unbiased estimators of cumulants, which are among the most useful $U$-statistics.

### 6.5.4 k-statistics

The $i$-th $k$-statistic $k_i$ is the unique symmetric unbiased estimator of the cumulant $\kappa_i$ of a given statistical distribution, that is $E[k_i] = \kappa_i$, see [75]. The theory of $k$-statistics has a long history beginning with Fisher [28], who rediscovered the half-invariants theory of Thiele [78]. Fisher introduced $k$-statistics (single and multivariate) as new symmetric functions of a random sample, with the aim to estimate cumulants without using moment estimators. Dressel [23] developed a theory of more general functions, resumed later by Tukey [80], who named them polykays. Both Tukey [81] and Wishart [85] developed methods to express polykays in terms of Fisher's $k$-statistics. These methods are straightforward enough, but their execution leads to intricate computations and some cumbersome expressions, except in very simple cases. The whole subject is later described by Stuart and Ord [75] in great detail. In the Eighties, tensor notation was taken by Speed [62] – [67] and extended to polykays and single $k$-statistics. This extension reveals the coefficients defining polykays to be values of the Moebius function over the lattice of set partitions. As a consequence, Speed used the set theoretic approach to symmetric functions introduced by Doubilet [22]. In the same period, McCullagh [36, 37] simplified the tensor notation of Kaplan [32] by introducing the notion of generalized cumulants. Symbolic operators for expectation and the derivation of unbiased estimates for multiple sums were introduced by Andrews et al. [1] – [3].

In this paragraph, we provide an unifying point of view for the theory of $k$-statistics. All we need is to recover $k$-statistics in terms of power sums of the data points, as indicated by Fisher. In umbral terms, this means to express moments of the $\alpha$-cumulant umbra $\chi.\alpha$ in terms of products like $(n.\alpha)^{r_1}(n.\alpha^2)^{r_2}\cdots$ .

**Theorem 6.3 ($k$-statistics).** *If $i \le n$, then*

$$(\chi.\alpha)^i \simeq \sum_{\lambda \vdash i} \frac{(\chi.\chi)^{\nu_\lambda}}{(n)_{\nu_\lambda}} d_\lambda \sum_{\pi \in \Pi_{\nu_\lambda}} (\chi.\chi)^{\pi}(n.\alpha)_{S_\pi} \qquad (6.34)$$

*where* $\lambda = (1^{r_1}, 2^{r_2}, \ldots)$ *runs over all the partitions of the integer* $i$ *and* $S_\pi$ *is the subdivision of the multiset* $P_\lambda = \{\alpha^{(r_1)}, \alpha^{2^{(r_2)}}, \ldots\}$ *corresponding to the partition* $\pi \in \Pi_{\nu_\lambda}$.

Since $E[(\chi.\alpha)^i] = \kappa_i$, equivalence (6.34) gives the umbral expression of the $i$-th cumulant in terms of umbral power sum symmetric polynomials, i.e. the $i$-th $k$-statistic. The proof is given in [15] and relies on equivalences (6.14) and (6.20).

Products of $k$-statistics are known as *polykays*. Indeed the symmetric statistic $k_{r,\ldots,t}$ such that $E[k_{r,\ldots,t}] = \kappa_r \cdots \kappa_t$, (where $\kappa_r, \ldots, \kappa_t$ are cumulants) generalizes $k$-statistics and they were originally called *generalized k-statistics* by Dressel [23]. Being a product of cumulants, the umbral expression of a polykay is simply $k_{r,\ldots,t} \simeq (\chi.\alpha)^r \cdots (\chi'.\alpha')^t$ with $\chi, \ldots, \chi'$ uncorrelated umbrae as well as $\alpha, \ldots, \alpha'$ with $\alpha \equiv \ldots \equiv \alpha'$. The following proposition provides $k_{r,\ldots,t}$ in terms of umbral power sum symmetric polynomials.

**Theorem 6.4 (Polykays).** *If* $r + \cdots + t \le n$ *then*

$$k_{r,\ldots,t} = \sum_{(\lambda \vdash r, \ldots, \eta \vdash t)} \frac{(\chi.\chi)^{\nu_\lambda} \cdots (\chi.\chi)^{\nu_\eta} d_\lambda \cdots d_\eta}{(n)_{\nu_\lambda + \cdots + \nu_\eta}} \sum_{\pi \in \Pi_{\nu_\lambda + \cdots + \nu_\eta}} (\chi.\chi)^{\cdot\pi}(n.\alpha)_{S_\pi}, \qquad (6.35)$$

*where* $\lambda = (1^{r_1}, 2^{r_2}, \ldots)$ *runs over all the partitions of* $r$, $\eta = (1^{t_1}, 2^{t_2}, \ldots)$ *runs over all the partitions of* $t$ *and* $S_\pi$ *is the subdivision of*

$$P_{\lambda + \cdots + \eta} = \{\alpha^{(r_1 + \cdots + t_1)}, \alpha^{2^{(r_2 + \cdots + t_2)}}, \ldots\}$$

*corresponding to the partition* $\pi \in \Pi_{\nu_\lambda + \cdots + \nu_\eta}$.

In order to recover the umbral expression of multivariate $k$-statistics and multivariate polykays, we need the notion of multivariate cumulant. In umbral terms, a multivariate cumulant is the element of $R$ corresponding to $E[(\chi.\mu)_M] = \kappa_{t_1,\ldots,t_r}$, where $M = \{\mu_1^{(t_1)}, \mu_2^{(t_2)}, \ldots, \mu_r^{(t_r)}\}$ is a multiset and $(\chi.\mu)_M = (\chi.\mu_1)^{t_1}(\chi.\mu_2)^{t_2} \cdots (\chi.\mu_r)^{t_r}$. By using the notion of multivariate cumulants and equivalence (6.28), the following umbral expression for multivariate $k$-statistics is recovered.

**Theorem 6.5 (Multivariate k-statistics).** *If* $|M| = i < n$, *then*

$$(\chi.\mu)_M \simeq \sum_{\pi \in \Pi_i} \frac{(\chi.\chi)^{|\pi|}}{(n)_{|\pi|}} \sum_{\tau \in \Pi_{|\pi|}} (\chi.\chi)^{\cdot\tau}(n.\mu)_{S_\tau}, \qquad (6.36)$$

*where* $S_\tau$ *is the subdivision of* $M$ *corresponding to the partition* $\tau$ *of the set built with the blocks of* $\pi \in \Pi_i$.

Multivariate polykays were introduced by Robson in [50]. The symmetric statistic $k_{t_1 \ldots t_r, \ldots, l_1 \ldots l_m}$ such that $E[k_{t_1 \ldots t_r, \ldots, l_1 \ldots l_m}] = \kappa_{t_1 \ldots t_r} \cdots \kappa_{l_1 \ldots l_m}$, (where $\kappa_{t_1 \ldots t_r}$ is a multivariate cumulant) generalizes polykays. Being a product of uncorrelated multivariate cumulants, the umbral expression of a multivariate polykay is simply $k_{t_1 \ldots t_r, \ldots, l_1 \ldots l_m} \simeq (\chi.\mu)_T \cdots (\chi'.\mu')_L$, with $\chi$ and $\chi'$ uncorrelated as well as the umbral monomials $\mu \in T$ and $\mu' \in L$ with $T = \{\mu_1^{(t_1)}, \ldots, \mu_r^{(t_r)}\}, \ldots, L = \{\mu'^{(l_1)}_1, \ldots, \mu'^{(l_m)}_m\}$.

**Theorem 6.6 (Multivariate polykays).** *If* $|T| + \cdots + |L| < n$, *then*

$$k_{t_1 \ldots t_r, \ldots, l_1 \ldots l_m} \simeq \sum_{(\pi \in \Pi_{|T|}, \ldots, \tilde{\pi} \in \Pi_{|L|})} \frac{(\chi \cdot \chi)^{|\pi|} \cdots (\chi' \cdot \chi')^{|\tilde{\pi}|}}{(n)_{|\pi| + \cdots + |\tilde{\pi}|}} \sum_{\tau \in \Pi_{|\pi| + \cdots + |\tilde{\pi}|}} (\chi \cdot \chi)^{\cdot \tau} (n.p)_{S_\tau}, \qquad (6.37)$$

*where* $S_\tau$ *is the subdivision of the multiset obtained by the disjoint union of* $T, \ldots, L$, *with no uncorrelation labels, and corresponding to the partition* $\tau$ *of the set built with the blocks of* $\{\pi, \ldots, \tilde{\pi}\}$.

The umbral techniques, investigated in [15], have allowed us to implement a single algorithm for $k$-statistics, multivariate $k$-statistics, polykays and multivariate polykays [16]. Nevertheless, the elegance of a unifying outlook pays a price in computational costs that become comparable with those of MathStatica [52] for polykays and not competitive for univariate and multivariate $k$-statistics.

A frequently asked question is: why are these calculations relevant? High order statistics have a variety of applications. Recently, Rao et al. [47] have shown applications of high order cumulants in statistical inference and time series. Moreover, there are different areas, such as astronomy (see [46] and references therein), astrophysics [26] and biophysics [38], where one computes high order $k$-statistics in order to recognize a Gaussian population or characterizes asymptotic behavior of high order $k$-statistics when the population is Gaussian. Indeed, $k$-statistics are independent from the sample mean if and only if the population is Gaussian [35] and in such a case $k$-statistics of order greater than 2 should be nearly to zero. Since, higher order statistics require enormous amounts of data to estimate with any accuracy, for such applications, increasing speed and efficiency is a significant investment.

By forfeiting the elegant idea to produce only one algorithm for the whole subject, the umbral calculus produces different but fast algorithms for $k$-statistics and their generalizations. So we have set up radically innovative procedures for generating all these estimators, by realizing a substantial improvement of computational times compared with those available in the literature [18]. The main idea grounds on a noteworthy theoretical result obtained by the umbral syntax: *cumulants of a random variable can be obtained via cumulants of a suitable compound Poisson random variable*, as we summarize in the next paragraph.

### 6.5.5 Fast Symbolic Computation of k-statistics

Computing set partitions or multiset subdivisions requires a significant computational cost, also when these procedures are suitably optimized [17]. An improvement in the performance is achievable removing any of these procedures. By using the classical umbral calculus, we have recovered $k$-statistics by means of cumulants of compound Poisson random variables. This fact allows us to insert exponential polynomials (6.6) in formula (6.34), removing set partitions and speeding up its implementation.

In order to set up such an algorithm, we need the notion of *multiplicative inverse* of an umbra. Two umbrae are said to be *multiplicative inverse* to each other when $\alpha\gamma \equiv u$. In dealing with a saturated umbral calculus, the multiplicative inverse of an umbra is not unique, but any two multiplicative inverse umbrae of the umbra $\alpha$ are similar. From the definition, we have $a_n g_n = 1$ for all nonnegative integer $n$, i.e. $g_n = 1/a_n$, where $a_n$ and $g_n$ are moments of $\alpha$ and $\gamma$ respectively. In the following, the multiplicative inverse of an umbra $\alpha$ will be denoted by the symbol $1/\alpha$.

Let us consider the umbra $\chi.y.\beta$ introduced in [14], where $y$ is an indeterminate. This is a very special umbra, because $(\chi.y.\beta)^n \simeq y$ for all nonnegative integer $n$. The umbra $(\chi.y.\beta).\alpha \equiv \chi.(y.\beta.\alpha)$ is the cumulant umbra of the polynomial $\alpha$-partition umbra $y.\beta.\alpha$, that corresponds to a compound Poisson random variable of parameter $y$. Such an umbra is the keystone in constructing cumulants of an umbra $\alpha$. Indeed the moments of $n.\chi.(y.\beta.\alpha)$, i.e. the sum of $n$ uncorrelated cumulant umbrae of $y.\beta.\alpha$, are the polynomials $c_i(y) = \sum_{\lambda \vdash i} y^{\nu_\lambda}(n)_{\nu_\lambda} d_\lambda a_\lambda$, with $c_0(y) = 1$ and $c_i(y)$ of degree $i$ for every integer $i$. The next theorem states that the $\alpha$-cumulant umbra has moments umbrally equivalent to the umbral polynomials obtained from the polynomials $c_i(y), i = 1, 2, \ldots$, by replacing $y$ with the umbra $\chi.\chi/n.\chi$.

**Theorem 6.7.** *If* $c_i(y) = E[(n.\chi.y.\beta.\alpha)^i]$, *then* $(\chi.\alpha)^i \simeq c_i(\chi.\chi/n.\chi)$ *for* $i = 1, 2, \ldots$.

Via Theorem 6.7, cumulants of $\alpha$ can be recovered from the moments of $n.(\chi.y.\beta).\alpha$, by a suitable replacement of the indeterminate $y$. If we express the polynomials $c_i(y)$ in terms of power sums $n.\alpha^r \equiv \alpha_1^r + \cdots + \alpha_n^r$, we recover the umbral expression of $k$-statistics. To this aim, being $n.\chi.y.\beta.\alpha \equiv n.[(\chi.y.\beta)\alpha]$, we can use the following equivalence $[n.(\gamma\alpha)]^i \simeq \sum_{\lambda \vdash i} d_\lambda (\chi.\gamma)_\lambda (n.\alpha)^{r_1}(n.\alpha^2)^{r_2} \cdots$ with $\gamma$ replaced by $(\chi.y.\beta)$. This is the starting point to prove the following result, from which the fast algorithm for $k$-statistics follows.

**Theorem 6.8.** *Assume* $p_n(y) = \sum_{k=1}^{n}(-1)^{k-1}(k-1)! S(n,k)y^k$, *where* $S(n,k)$ *are the Stirling numbers of second type. For every* $\alpha \in A$ *we have*

$$(\chi.\alpha)^i \simeq \sum_{\lambda \vdash i} d_\lambda p_\lambda \left(\frac{\chi.\chi}{n.\chi}\right)(n.\alpha)^{r_1}(n.\alpha^2)^{r_2} \cdots \qquad (6.38)$$

*with* $\lambda = (1^{r_1}, 2^{r_2}, \ldots)$ *and* $p_\lambda(y) = [p_1(y)]^{r_1}[p_2(y)]^{r_2} \cdots$.

Comparing equivalence (6.38) with equivalence (6.34), having removed the summation over set partitions, the reduction of the complexity is plain.

We show how to generalize Theorem 6.8 to polykays. For the multivariate analogous, the reader is referred to [18]. For plainness, in the following we just deal with two subindexes $k_{r,t}$, the generalization to more than two being straightforward.

Let us consider a polynomial umbra whose first $k$ moments are all equal to $y$, and the reimanders are all zero. We denote this polynomial umbra by the symbol $\delta_{y,k}$ so that $\delta_{y,k}^i \simeq (\chi.y.\beta)^i$ for $i = 0, 1, 2, \ldots, k$, otherwise being zero. If $r, t$ are nonnegative integers and $k = \max\{r, t\}$, then $[n.(\delta_{y,k}\alpha)]^{r+t} \simeq \sum_{(\lambda \vdash r, \eta \vdash t)} y^{\nu_\lambda + \nu_\eta}(n)_{\nu_\lambda + \nu_\eta} d_{\lambda + \eta} \alpha_{\lambda + \eta}$. Now, let $p_{r,t}(y)$ be the polynomial obtained by evaluating the right hand side of the previous equivalence. The following theorem allows us to express products of uncorrelated cumulant umbrae by using the polynomials $p_{r,t}(y)$.

**Theorem 6.9.** *If $q_{r,t}$ is the umbral polynomial obtained via $p_{r,t}(y)$ by replacing $y^{\nu_\lambda + \nu_\eta}$ with*

$$\frac{(\chi \cdot \chi)^{\nu_\lambda}(\chi' \cdot \chi')^{\nu_\eta}}{(n \cdot \chi)^{\nu_\lambda + \nu_\eta}} \frac{d_\lambda d_\eta}{d_{\lambda + \eta}}, \tag{6.39}$$

*then $k_{r,t} \simeq (\chi \cdot \alpha)^r (\chi' \cdot \alpha')^t \simeq q_{r,t}$.*

These two last results are sufficient to express the polykay $k_{r,t}$ in terms of power sums. The steps are summarized below:

i) by using the umbra $n.(\delta_{y,k} \alpha)$, with $k = \max\{r,t\}$, the polynomials $p_{r,t}(y)$ are expressed in terms of power sums, that is

$$[n.(\delta_{y,k} \alpha)]^{r+t} \simeq p_{r,t}(y) \simeq \sum_{\xi \vdash (r+t)} d_\xi (\chi \cdot \delta_{y,k})_\xi (n \cdot \alpha)^{s_1} (n \cdot \alpha^2)^{s_2} \cdots ; \tag{6.40}$$

ii) the cumulants of the umbra $\delta_{y,k}$ are evaluated by recalling that the moments corresponding to powers greater than $k$ are zero;

iii) thanks to Theorem 6.9, occurrences of $y^{\nu_\lambda + \nu_\eta}$ in (6.40) are replaced by (6.39).

The steps $i) - iii)$ are the building blocks of the fast algorithm presented in [18] for generating polykays.

## 6.5.6 Sheppard's Corrections

In the computation of moments of a random sample, it is often worthwhile, or even necessary, to divide the given data into classes or groups. The moments computed by means of the resulting grouped frequency distribution are looked up as a first approximation to moments of the parent distribution, but they suffer from the error committed in grouping. The correction for grouping is a sum of two terms, the first depending on the length of the grouping interval, the second being a periodic function of the position [31]. This very old problem was first discussed by Thiele [78], who studied the second term missing the first, and then by Sheppard [60] who studied the first term, missing the second. Both Bruns [5] and Fisher [27] provide compelling reasons for neglecting the second term and so for using the so-called Sheppard's corrections, that nowadays are still employed [75].

For a continuous parent distribution over a finite range whose $n$-th moment is $a_n$, Sheppard's corrections have the following expression:

$$a_n = \sum_{j=0}^{n} \binom{n}{j} \left(2^{1-j} - 1\right) \mathcal{B}_j h^j \tilde{a}_{n-j}, \tag{6.41}$$

where $\{\mathcal{B}_j\}$ are the *Bernoulli numbers* [49] and $\{\tilde{a}_j\}$ are the moments of the distribution grouped in classes of length $h$.

In order to prove (6.41), Sheppard uses the Euler-MacLaurin summation assuming that $f(x)$ has high order contact with the $x$-axis at both ends. The problem is that not only the high order contact hypothesis is not always satisfied by the frequency distribution commonly used, but also it is quite impossible to know if it is satisfied by the theoretical parent distribution, before calculating the moments. Wilson [84] corrects Sheppard's corrections, by using Fourier's series and by removing the high order contact hypothesis effectively. Such formulae have some serious limits due to the involved series, whose general term draws in not only trigonometric integrals but also integrals of the unknown parent density function. Wilson ends the paper by observing that the mean value of the inserted trigonometric corrections is zero, so that the usual formulae for Sheppard's corrections are true on the average for any frequency function, for which the highest moments considered exist, without any notice of how the function behaves at the ends of its range or within it. Moreover, in the absence of any knowledge about the nature of the parent density function, the errors due to the employment of trigonometric correction terms are of the same order of magnitude as Sheppard's corrections, so it is useful to apply Sheppard's corrections in any case, despite the averages of the trigonometric terms vanish.

Sheppard's corrections to moments remain a standard topic in textbooks. Contributions on validity of these corrections as well as their proofs stop to appear around the forties, except for some sporadic papers that enrich the literature without making real innovations.

The expressions analogous to Sheppard's corrections, but involving discrete parent distributions, were first given in the Editorial of Volume 1, Number 1, of *Annals of Mathematical Statistics* (page 111). The method used to develop the general formula was extremely laborious. In a paper of 1936, Craig [7] considerably reduces and simplifies the derivation of these corrections by using the logarithm of the moment generating function, that is working on cumulants instead of moments. The noteworthy simplification in the expression of corrections, when referred to cumulants instead of moments, was first pointed out by Langdon and Ore [34] in 1929 for a continuous parent distribution over $(-\infty, \infty)$. This idea is resorted by Craig. From the corrected cumulants, he deduces the corrected moments, by using the usual formula connecting these two sequences. He supposes that $m$ consecutive values of the discrete variable in question are grouped in a frequency class of width $h$. Letting $m \to \infty$, he recovers Sheppard's corrections. The method is then extended to a bivariate random variable, recovering the result of Baten [4] with a suitable limit operation. Craig's development does not impose the high contact condition, being corrections stated on the average. At the moment Craig's results represent the most general way to find such corrections both for continuous and discrete parent distributions.

A very simple closed-form formula for Sheppard's corrections has been recovered by means of the classical umbral calculus as well as a more general closed-form formula for discrete parent distributions. The main tool is the *Bernoulli umbra*, as introduced in [55] and revised in [19].

Up to similarity, the Bernoulli umbra $\iota$ is the unique umbra such that $(\iota + 1)^{n+1} \simeq \iota^{n+1}$ for all nonnegative integers $n$. Then, the Bernoulli umbra $\iota$ turns to be the unique (up to similarity) umbra such that $E[\iota^n] = \mathcal{B}_n$, for $n = 0, 1, 2, \ldots$ where $\{\mathcal{B}_n\}$ are Bernoulli numbers. By using this definition, main properties of Bernoulli numbers can be easily proved. Here, we just recall that *Bernoulli polynomials* $\mathcal{B}_n(x) = \sum_{k=0}^n \binom{n}{k} \mathcal{B}_k x^{n-k}$ are moments of the Bernoulli polynomial umbra $x + \iota$. The Bernoulli polynomials are characterized to have an average value of 0 that is $\int_0^1 \mathcal{B}_n(x) \, dx = 0$. In umbral terms we have $E[\mathcal{B}_n(-1.\iota)] = \int_0^1 \mathcal{B}_n(x) \, dx$, for all nonnegative integers $n$, where $-1.\iota$ denotes the inverse of the Bernoulli umbra. This equality can be generalized: if $p(x) \in \mathbb{R}[x]$ and $h \in \mathbb{R} - \{0\}$, then $E\left[p(-1.(h\iota) + x)\right] = 1/h \int_x^{x+h} p(t) \, dt$. This last result allows us to prove in few steps the following result.

**Theorem 6.10 (Sheppard's correction).** *If the sequence $\{\tilde{a}_n\}$ of grouped moments is umbrally represented by the umbra $\tilde{\alpha}$ and the sequence $\{a_n\}$ of raw moments of a continuous parent distribution is umbrally represented by the umbra $\alpha$, then*

$$\alpha \equiv \tilde{\alpha} + h\left(\iota + \frac{1}{2}\right). \tag{6.42}$$

For a discrete parent distribution, assume $m$ consecutive values are grouped in a frequency class of width $h$. The $m$ smaller intervals of width $h/m$ go to make up the class width $h$ in such a way that the $m$ values of the variable represent the mid-points of the sub-intervals. In this case, equivalence (6.42) can be corrected, by using the umbral version of the *multiplication theorem* [51] for Bernoulli polynomials, that is

$$\left(x + \frac{\iota}{m}\right)^n \simeq \frac{1}{m} \sum_{k=0}^{m-1} \left(x + \frac{k}{m} + \iota\right)^n$$

with $n$ and $m$ nonnegative integers.

**Theorem 6.11 (Sheppard's correction for discrete parent distributions).** *If the sequence $\{\tilde{a}_n\}$ of grouped moments is umbrally represented by the umbra $\tilde{\alpha}$ and the sequence $\{a_n\}$ of raw moments of a discrete parent distribution is umbrally represented by the umbra $\alpha$, then*

$$\alpha \equiv \tilde{\alpha} + h\left(\iota + \frac{1}{2}\right) + \frac{h}{m}\left(-1.\iota - \frac{1}{2}\right). \tag{6.43}$$

It is interesting to compare equivalence (6.42) with equivalence (6.43). In this last equivalence, we find just one addend more, which is an umbra whose moments are $2^{-j}/(j+1)$ if $j$ is even, otherwise being zero. Obviously, if $m \to \infty$ from (6.43) we recover (6.42). These observations turn to be useful when we seek expression of raw moments $\{a_n\}$ in terms of grouped moments $\{\tilde{a}_n\}$. The unique expression is given by Craig in [7], but its structure is quite complex. We have given the following different expression:

$$a_n = \sum_{k=0}^{\lceil \frac{n}{2} \rceil} \binom{n}{2k} \left(\frac{h}{2m}\right)^{2k} \frac{1}{2k+1} \sum_{j=0}^{n-2k} \binom{n-2k}{j} \left(2^{1-j} - 1\right) \mathcal{B}_j h^j \tilde{a}_{n-2k-j}.$$

Except for the papers of Craig [7] and Baten [4], no attention was payed to multivariate generalizations of Sheppard's corrections, probably due to the complexity of the resulting formulae. The classical umbral calculus allows us to clarify the structure of multivariate formulae and gives a simple way to implement these formulae in any symbolic package.

Suppose $\mathbf{X} = (X_1, X_2, \ldots, X_j)$ a multivariate random variable with joint density function $f_{\mathbf{X}}(\mathbf{x})$ over $\mathbb{R}^j$, but we can deal with any range of type a bounded rectangle. As usual $m_{t_1 \ldots t_j}$ denotes a joint moment of $\mathbf{X}$. The moments calculated from the grouped frequencies are denoted by $\tilde{m}_{t_1 \ldots t_j}$ computed on $R_j = [-h_1/2, h_1/2] \times \cdots \times [-h_j/2, h_j/2]$, where $\{h_k\} \in \mathbb{R} - \{0\}$ are the window width for any component.

**Theorem 6.12 (Multivariate Sheppard's correction).** *If the sequence $\{\tilde{m}_{t_1 \ldots t_j}\}$ of moments, calculated from the grouped frequencies, is umbrally represented by the umbral monomial $\tilde{\mu}_M$, with $M$ a multiset of finite support $\{\mu_1, \ldots, \mu_j\}$, and the sequence $\{m_{t_1 \ldots t_j}\}$ of raw moments is umbrally represented by the umbral monomial $\mu_M$, then*

$$\mu_M \equiv \left[ \tilde{\mu} + h \left( \iota + \frac{1}{2} \right) \right]_M$$

*where $\{\iota_k\}$ are uncorrelated Bernoulli umbrae and we obtain the identity*

$$\left[ \tilde{\mu} + h \left( \iota + \frac{1}{2} \right) \right]_M = \prod_{k=1}^{j} \left[ \tilde{\mu}_k + h_k \left( \iota_k + \frac{1}{2} \right) \right]^{t_k}.$$

An analogous generalization holds also for discrete parent multivariate distributions. For more details, the reader is referred to [12].

## 6.6 Sheffer Sequences

As well-known, many polynomial sequences like Laguerre polynomials, first and second kind Meixner polynomials, Poisson-Charlier polynomials and Stirling polynomials are Sheffer sequences. Sheffer sequences can be considered the core of the umbral calculus of Roman and Rota [51]. Apart from the preliminary paper of Taylor [77], Sheffer sequences have been completely described in terms of umbrae in [19]. This new syntax has provided noteworthy computational simplifications and conceptual clarifications in many results involving Sheffer sequences.

Let $\gamma$ be an umbra with $E[\gamma] = g_1 \neq 0$, so that its generating function $f(\gamma, t)$ admits compositional inverse. The *adjoint umbra* of $\gamma$ is the umbra $\gamma^{<-1>}$-partition umbra, that is $\gamma^* = \beta.\gamma^{<-1>}$. The name parallels the adjoint of an umbral operator [51] since $\gamma.\alpha^*$ gives the umbral composition of $\gamma$ and $\alpha^{<-1>}$. Being $f(\gamma^{<-1>}, t) = f^{-1}(\gamma, t)$, the adjoint umbra has generating function $f(\gamma^*, t) = \exp[f^{-1}(\gamma, t) - 1]$. In particular the adjoint of the compositional inverse of an umbra is similar to its partition umbra, i.e. $(\gamma^{<-1>})^* \equiv \beta.\gamma$.

A polynomial umbra $\sigma_x$ is said to be a Sheffer umbra for $(\alpha, \gamma)$ if

$$\sigma_x \equiv (-1.\alpha + x.u).\gamma^*, \tag{6.44}$$

where $\gamma^*$ is the adjoint umbra of $\gamma$. In the following, we denote a Sheffer umbra by $\sigma_x^{(\alpha,\gamma)}$ in order to make explicit the dependence on $\alpha$ and $\gamma$. The generating function of $\sigma_x^{(\alpha,\gamma)}$ is the composition of $f^{-1}(\gamma,t)$ and $f(-1.\alpha + x.u, t) = e^{xt}/f(\alpha,t)$, that is

$$f(\sigma_x^{(\alpha,\gamma)}, t) = \frac{1}{f[\alpha, f^{<-1>}(\gamma,t) - 1]} e^{x[f^{<-1>}(\gamma,t)-1]}. \tag{6.45}$$

The moments of $\sigma_x^{(\alpha,\gamma)}$ result to be a Sheffer sequence. Any Sheffer umbra is uniquely determined by its moments evaluated at 0, since via equivalence (6.44) we have $\sigma_0^{(\alpha,\gamma)} \equiv -1.\alpha.\gamma^*$.

*Example 6.6. Power polynomials.* Choosing the augmentation umbra $\epsilon$ as umbra $\alpha$ and the singleton umbra $\chi$ as umbra $\gamma$, from equivalence (6.44) we have $\sigma_x^{(\epsilon,\chi)} \equiv x.u$, being $\chi^* \equiv u$. Since the sequence of polynomials $\{x^n\}$ are moments of the Sheffer umbra $\sigma_x^{(\epsilon,\chi)}$, then $\{x^n\}$ is a Sheffer sequence.

*Example 6.7. Poisson-Charlier polynomials.* Choosing the umbra $a.\beta$ as umbra $\alpha$ with $a \neq 0$ and the umbra $\chi.a.\beta$ as umbra $\gamma$, we recover the Poisson-Charlier umbra $(x.\chi - a)/a$ whose moments are the well-known Poisson-Charlier polynomials.

A polynomial umbra $\sigma_x$ is a Sheffer umbra if and only if there exists an umbra $\eta$, provided with compositional inverse, such that $\sigma_{\eta+x.u} \equiv \chi + \sigma_x$. In terms of moments, the previous equivalence gives $\sigma_{\eta+x.u}^n \simeq \sigma_x^n + n\sigma_x^{n-1}$ for all nonnegative integers $n$. This last equivalence translates a well-known result on Sheffer sequences: a polynomial sequence $\{s_n(x)\}$ is of Sheffer if and only if there exists a delta linear operator $Q$, on polynomials in $x$, shift-invariant such that $Qs_n(x) = ns_{n-1}(x)$.

Two special classes of Sheffer polynomials are often employed in the applications: the polynomial sequence of binomial type (see Section 3) and the Appell sequence. This last is a sequence of polynomials $\{p_n(x)\}$ satisfying the identity

$$\frac{d}{dx}p_n(x) = np_{n-1}(x) \quad n = 1, 2, \ldots. \tag{6.46}$$

In [19], it has been proved that polynomial sequences of binomial type are umbrally represented by a Sheffer umbra for $(\epsilon,\gamma)$, where $\gamma$ has compositional inverse and $\epsilon$ is the augmentation umbra. The umbra $\sigma_x^{(\epsilon,\gamma)}$ is called the associated umbra of $\gamma$. Moreover, the Appell sequence is umbrally represented by a Sheffer umbra for $(\alpha,\chi)$. The umbra $\sigma_x^{(\alpha,\chi)}$ is called the Appell umbra.

From (6.44), closed form formulae have been recovered for polynomial sequences of binomial type and Appell sequences. For example, a polynomial umbra $\sigma_x$ is the associated umbra of $\gamma$ if and only if $\sigma_x \equiv x.\gamma^*$, where $\gamma^*$ is the adjoint umbra of $\gamma$. So polynomial sequences of binomial type are umbrally represented by polynomial umbrae $x.\gamma^*$ with generating function $f(x.\gamma^*, t) = e^{x[f^{<-1>}(\gamma,t)-1]}$, because in equation (6.45) we have $f(\alpha,t) = f(\epsilon,t) = 1$. A polynomial sequence $\{p_n(x)\}$

is *associated* to an umbra $\gamma$ if and only if $p_n(x) \simeq (x.\gamma^*)^n$, for all nonnegative integers $n$. In particular, a sequence $\{p_n(x)\}$ is associated to the umbra $\gamma$ if and only if $\{p_n(x)\}$ is a binomial polynomial sequence. All these results are in agreement with the theory of Bell umbrae presented in Section 3.

*Example 6.8. Power polynomials.* The umbra $x.u$ is associated to the umbra $\chi$. Indeed, the polynomial sequence $\{x^n\}$ is associated to the adjoint umbra $\chi^* \equiv \beta.\chi^{<-1>} \equiv u$.

*Example 6.9. Lower factorial polynomials.* The umbra $x.u^* \equiv x.\chi$ is associated to the umbra $u$. The associated polynomial sequence is $\{(x.\chi)^n\}$, that is $\{(x)_n\}$.

*Example 6.10. Exponential polynomials.* The umbra $x.(u^{<-1>})^* \equiv x.\beta$ is associated to the umbra $u^{<-1>}$. The associated polynomial sequence is $\{(x.\beta)^n\}$, that is the sequence $\{\Phi_n(x)\}$ of exponential polynomials (6.6).

A polynomial umbra $\sigma_x$ is said to be the Appell umbra of $\alpha$ if and only if $\sigma_x \equiv -1.\alpha + x.u$. By equivalence (6.45), the generating funxtion of $(-1.\alpha + x.u)$ is $f(-1.\alpha + x.u, t) = e^{xt}/f(\alpha, t)$, being $f(\chi, t) = 1 + t$. A polynomial sequence $\{p_n(x)\}$ is an Appell sequence if and only if $p_n(x) \simeq (-1.\alpha + x.u)^n$, for all nonnegative integers $n$.

*Example 6.11. Bernoulli polynomials.* The umbra $-1.\iota + x$, where $\iota$ is the Bernoulli umbra, is the Appell umbra of $\iota$.

**Theorem 6.13 (Sheffer identity).** *A polynomial umbra $\sigma_x$ is a Sheffer umbra if and only if there exists an umbra $\eta$, provided with compositional inverse, such that*

$$\sigma_{x+y} \equiv \sigma_x + y.\eta^*. \tag{6.47}$$

Equivalence (6.47) gives the well-known Sheffer identity

$$s_n(x+y) = \sum_{k=0}^{n} \binom{n}{k} s_k(x) p_{n-k}(y) \tag{6.48}$$

that is recovered by using the binomial expansion and observing that $s_n(x+y) = E[\sigma_{x+y}^n]$, $s_k(x) = E[\sigma_x^k]$ and $p_{n-k}(y) = E[(y.\eta^*)^{n-k}]$. The Sheffer identity (6.47) for an Appell umbra $\sigma_x$ is simply $\sigma_{x+y} \equiv \sigma_x + y$. The umbral equivalent of equation (6.46) is the following result: a polynomial umbra $\sigma_x$ is an Appell umbra for some umbra $\alpha$ if and only if $\sigma_{\chi+x.u}^n \simeq \sigma_x^n + n\sigma_x^{n-1}$. Roughly speaking, the previous result says that, when in the Appell umbra the indeterminate $x$ is replaced by $\chi + x.u$, the umbra $\chi$ acts as a derivative operator.

In many special combinatorial problems, the hardest part of the solution may be the discovery of an effective recursion. Once a recursion has been established, Sheffer polynomials are often a simple and general tool for finding answers in closed form. Main contributions in this respect are due to Niederhausen [39] – [41]. Further contributions are given by Razpet [48] and Di Bucchianico and Soto y Koelemeijer [11]. We present an example of how to use umbrae in order to solve recursions.

*Example 6.12.* Suppose we are asked to solve the difference equation

$$s_n(x+1) = s_n(x) + s_{n-1}(x) \tag{6.49}$$

under the condition $\int_0^1 s_n(x)dx = 1$ for all nonnegative integers $n$. Equation (6.49) fits the Sheffer identity (6.48) if we set $y = 1$, choose the sequence $\{p_n(x)\}$ such that $p_0(x) = 1, p_1(1) = 1$ and $p_n(1) = 0$ for all $n \geq 2$ and consider the Sheffer sequence $\{n! s_n(x)\}$. The sequence $\{p_n(x)\}$ is associated to the umbra $\chi$, so we are looking for solutions of (6.49) such that $n! s_n(x) \simeq (\sigma_x^{(\alpha,\gamma)})^n$ with $\gamma^* \equiv \chi$, i.e. $\gamma \equiv u$. The condition $\int_0^1 s_n(x)dx = 1$ can be translated in umbral terms by looking for an umbra $\zeta$ such that $E[s_n(\zeta)] = 1$ for all nonnegative integers $n$. Such an umbra has generating function $\int_0^1 e^{xt}dx = (e^t - 1)/t$, so that $\zeta \equiv -1.\iota$, with $\iota$ the Bernoulli umbra. Therefore, the umbra $\alpha$ satisfies the following identity $(-1.\alpha + -1.\iota).\chi \equiv u$. Being $\beta.\chi \equiv u$, we have $-1.\alpha + -1.\iota \equiv \beta$ and so $\alpha \equiv -1.(\iota + \beta)$. Solutions of (6.49) are moments of the Sheffer umbra $(\iota + \beta + x.u).\chi$ divided by $n!$.

One may ask where this updating of Sheffer sequence theory turns to be useful. Also ignoring the resulting elegant approach, here just outlined, a first partial answer is given in the next section, where it is recalled a fundamental result allowing us an umbral approach to free probability theory.

### 6.6.1 Abel Polynomials

The main result of this section is the connection between sequences of binomial type and Abel polynomials. More precisely, any sequence of binomial type can be represented by Abel polynomials. This result plays a leading role in the conversion between moments and cumulants, as we will see in the next paragraphs. The proof given in [57] was a hybrid, based both on the early Roman-Rota version of the umbral calculus and the last version, introduced by Rota-Taylor. In [19], we have given a very simple proof by using the derivative of an umbra.

The derivative umbra $\alpha_D$ of an umbra $\alpha$ is the umbra whose moments are $(\alpha_D)^n \simeq \partial_\alpha \alpha^n \simeq n\alpha^{n-1}$ for $n = 1, 2, \ldots$. Its generating function is $f(\alpha_D, t) = 1 + tf(\alpha, t)$, since $E[\alpha_D] = 1$.

*Example 6.13.* The singleton umbra $\chi$ is the derivative umbra of the augumentation umbra $\epsilon$, that is $\epsilon_D \equiv \chi$. The unity umbra is the derivative umbra of the Bernoulli inverse umbra, that is $u \equiv (-1.\iota)_D$. The compositional inverse of the unity umbra is the derivative umbra of the $\iota$-factorial umbra, that is $u^{<-1>} \equiv (\iota.\chi)_D$.

**Theorem 6.14 (Abel representation of binomial sequences).** *If $\gamma$ is an umbra provided with a compositional inverse, then*

$$(x.\alpha_D^*)^n \simeq x(x - n.\alpha)^{n-1}, \quad n = 1, 2, \ldots. \tag{6.50}$$

The polynomials $\{x(x-n.\alpha)^{n-1}\}$ are said umbral Abel polynomials. Since the generating function of $x.\alpha_D^*$ is $\exp[x(f^{<-1>}(\alpha_D,t)-1)]$, this last is the generating function of umbral Abel polynomials.

Theorem 6.14 includes the well-known Transfer Formula [51]. In the following we recall various results usually derived by Transfer Formula, reformulated via Theorem 6.14. We start with the Lagrange inversion formula. The Lagrange inversion formula gives the coefficients of the compositional inverse of a formal power series. In umbral terms, being $\chi.\beta \equiv u$, for $n = 1,2,...$ we have $(\alpha_D^{<-1>})^n \simeq (\chi.\beta.\alpha_D^{<-1>})^n \simeq \chi(\chi-n.\alpha)^{n-1}$, by using equivalence (6.50) with $x$ replaced by $\chi$. By using the binomial expansion and recalling $\chi^{k+1} \simeq 0$ for $k = 1,2,...,n-1$, we recover $\chi(\chi-n.\alpha)^{n-1} \simeq (n.\alpha)^{n-1}$ by which the Lagrange inversion formula follows:

$$(\alpha_D^{<-1>})^n \simeq (-n.\alpha)^{n-1}, \quad n = 1,2,.... \tag{6.51}$$

So moments of the compositional inverse of $\alpha_D$ can be recovered by using the inverse of the umbra $\alpha$. Equivalence (6.51) refers to the umbra $\alpha_D$, having first moment equal to 1. If one would consider umbrae having first moment not only different from zero but also different from one, one more step is necessary. For any umbra $\alpha$ such that $E[\alpha] = a_1 \neq 0$, there exists an umbra $\gamma$ such that $\alpha/a_1 \equiv \gamma_D$. Indeed such an umbra $\gamma$ has moments $\gamma^{n-1} \simeq \alpha^n/(n a_1^n)$ for all nonnegative $n$ and generating function $f(\gamma,t) = [f(\gamma,t/a_1)-1]/t$. In particular $a_1\gamma \equiv \mathfrak{a}$, where $\mathfrak{a}$ is the umbra introduced in [13] having moments

$$E[\mathfrak{a}^n] = \frac{a_{n+1}}{a_1(n+1)} \quad n = 0,1,.... \tag{6.52}$$

with $a_{n+1}$ the $(n+1)$-th moment of $\alpha$. Hence the generalized Lagrange inversion formula becomes $\alpha.^n(\alpha^{<-1>})^n \simeq (-n.\mathfrak{a})^{n-1}$.

An application of Theorem 6.14 is the proof of the following property for Abel polynomials, known as *Abel identity*:

$$(x+y)^n \simeq \sum_{k \geq 0} \binom{n}{k} y(y-k.\alpha)^{k-1}(x+k.\alpha)^{n-k}. \tag{6.53}$$

From Abel identity (6.53), the umbral expression of the Bell exponential polynomials turns to be

$$(x.\beta.\alpha_D)^n \simeq \sum_{k \geq 0} \binom{n}{k} (k.\alpha)^{n-k} x^k. \tag{6.54}$$

The extension of equivalence (6.54) to umbrae $\alpha$ with first moment $a_1 \neq 0$ different from one is

$$(x.\beta.\alpha)^n \simeq \sum_{k \geq 0} \binom{n}{k} \alpha.^k [k.\mathfrak{a}]^{n-k} x^k.$$

In equivalence (6.54), choosing the inverse of the Bernoulli umbra $-1.\iota$ as umbra $\alpha$, then the umbral expression of Stirling numbers of second kind is recovered

$S(n,k) \simeq \binom{n}{k}(-k.\iota)^{n-k}$ for $k = 0, 1, \ldots, n$ and nonnegative integers $n$. Similarly, the umbral version of Stirling numbers of first kind results $s(n,k) \simeq \binom{n}{k}(k.\iota.\chi)^{n-k}$.

One more interesting consequence of Theorem 6.14 is the free cumulant theory, as we will show in the next section.

To this aim, we introduce the notion of quasi-free cumulant umbra. Its connection with free cumulants will be clarified in Section 7.

### 6.6.2 Quasi-free Cumulant Umbra

For a given umbra $\alpha$, the unique umbra $\mathfrak{K}_\alpha$ (up to similarity) such that

$$(-1.\mathfrak{K}_\alpha)_D \equiv \alpha_D{}^{<-1>}, \qquad (6.55)$$

is called the *quasi-free cumulant umbra* of $\alpha$. Some comments on this definition, for more details see [21]. We will prove that for $\mathfrak{K}_\alpha$ additivity and homogeneity properties hold, but semi-invariant fails. This is why we have called $\mathfrak{K}_\alpha$ quasi-free cumulant umbra. In addition, $f(\alpha_D{}^{<-1>}, t)$ is of type $1 + t h(t)$, where $n h_{n-1} = g_n$ for $n = 2, 3, \ldots$ and $g_n$ moments of $\alpha_D{}^{<-1>}$. So, $\alpha_D^{<-1>} \equiv \gamma_D$ for some umbra $\gamma$ which umbrally represents $\{h_n\}$. The quasi-free cumulant umbra of $\alpha$ is the compositional inverse of the umbra $\gamma$ having generating function $1/h(t)$, and in particular $[(-1.\mathfrak{K}_\alpha)_D]^n \simeq (-n.\alpha)^{n-1}$ from the Lagrange inversion formula (6.51).

Further results can be recovered by using definition (6.55). For example, if $\mathfrak{K}_\alpha$ is the quasi-free cumulant umbra of $\alpha$, then $\alpha \equiv \mathfrak{K}_\alpha.\beta.\alpha_D$, that in terms of generating functions becomes $f(\alpha, t) = f[\mathfrak{K}_\alpha, f(\alpha_D, t) - 1]$. Closed form formulae converting quasi-free cumulants in terms of moments and viceversa are $\mathfrak{K}_\alpha \equiv \alpha.\alpha_D^*$ and $\alpha \equiv \mathfrak{K}_\alpha.(-1.\mathfrak{K}_\alpha)_D^*$. The previous conversion formulae can be expressed via suitable parametrizations by using umbral Abel polynomials.

**Theorem 6.15 (Abel parametrization).** *If $\mathfrak{K}_\alpha$ is the quasi-free cumulant umbra of $\alpha$, then $\alpha^n \simeq \mathfrak{K}_\alpha(\mathfrak{K}_\alpha + n.\mathfrak{K}_\alpha)^{n-1}$ and $\mathfrak{K}_\alpha^n \simeq \alpha(\alpha - n.\alpha)^{n-1}$.*

*Remark 6.1.* Volume polynomials. There is a connection between umbral polynomials like $\{\alpha(\alpha + n.\alpha)^{n-1}\}$ and the $n$-volume polynomial $V_n(x)$, introduced by Pitman and Stanley [45]. Let us recall that the $n$-volume polynomial $V_n(x)$ is the following homogeneous polynomial of degree $n$:

$$V_n(x) = \frac{1}{n!} \sum_{p \in park(n)} x_p, \qquad (6.56)$$

where $x_p = x_{p_1} \cdots x_{p_n}$ and $park(n)$ is the set of all parking functions of length $n$. A *parking function* of length $n$ is a sequence $p = (p_1, \ldots, p_n)$ of $n$ positive integers, whose nondecreasing arrangement $p^\uparrow = (p_{i_1}, \ldots, p_{i_n})$ is such that $p_{i_j} \le j$. In [42] an explicit umbral expression for the volume polynomial (6.56) in the uncorrelated umbrae $\alpha_1, \ldots, \alpha_n$ similar to $\alpha$ has been proved to be $n! V_n(\alpha_1, \ldots, \alpha_n) \simeq \alpha(\alpha + n.\alpha)^{n-1}$, for all $\alpha \in A$. In particular, if $\aleph_1, \ldots, \aleph_n$ are $n$ uncorrelated umbrae similar to $\aleph_\alpha$ then $n! V_n(\aleph_1, \ldots, \aleph_n) \simeq \alpha^n$.

In order to to give an umbral expression for the free probability convolution, we need one more notation. In the Lagrange inversion formula, we have introduced the umbra $\mathfrak{a}$, whose moments are (6.52). When $E[\alpha] = a_1 = 1$, we call this umbra the primitive of the umbra $\alpha$, and instead to use the notation $\mathfrak{a}$, we use $\alpha_P$ that seems more natural. When $E[\alpha] = a_1 = 1$, we have $(\alpha_D)_P \equiv (\alpha_P)_D \equiv \alpha$. Within the theory of quasi-free cumulant umbra, we deal with umbrae like $\alpha_D$, such that $E[\alpha_D] = 1$, so the primitive of an umbra turns to be useful, for example, in order to express the free convolution by a closed-form formula.

**Theorem 6.16 (Additivity property).** *If $\aleph_\alpha, \aleph_\gamma$ and $\aleph_\xi$ are the quasi-free cumulant umbrae of $\alpha, \gamma$ and $\xi$ respectively, then $\aleph_\xi \equiv \aleph_\alpha \dotplus \aleph_\gamma$ if and only if $-1.(\xi_D{}^{<-1>}{}_P) \equiv -1.(\alpha_D{}^{<-1>}{}_P) \dotplus -1.(\gamma_D{}^{<-1>}{}_P)$.*

By virtue of Theorem 6.16, the *free convolution* of $\alpha$ and $\gamma$ has to be defined

$$(\alpha \boxplus \gamma)_D{}^{<-1>}{}_P \equiv -1.[-1.(\alpha_D{}^{<-1>}{}_P) \dotplus -1.(\gamma_D{}^{<-1>}{}_P)],$$

so that $\aleph_{\alpha \boxplus \gamma} \equiv \aleph_\alpha \dotplus \aleph_\gamma$. If $\aleph_\alpha$ is the quasi-free cumulant umbra of $\alpha$, then the homogeneity property of quasi-free cumulant umbra is $\aleph_{c\alpha} \equiv c\aleph_\alpha$, for all $c \in R$, so that $\aleph_{\alpha \boxplus cu} \equiv \aleph_\alpha + c\aleph_u$ due to Theorem 6.16. Since $\aleph_u$ and $\chi$ are not similar, semi-invariant property fails.

### 6.6.3 Boolean Cumulants

The notion of Boolean cumulants arises from considering the Boolean convolution of probability measures [70]. Within stochastic differential equations, this family of cumulants is also known as "partial cumulants". More precisely, let $M(t)$ be the ordinary generating function of a random variable $X$, that is $M(t) = 1 + \sum_{i \geq 1} a_i t^i$ where $a_i = E[X^i]$. We have $M(t) = 1/[1 - H(t)]$, where $H(t) = \sum_{i \geq 1} h_i t^i$, and $h_i$ are called Boolean cumulants of $X$.

In order to associate a sequence of Boolean cumulants to an umbra, it is necessary to characterize umbrae whose generating function is of ordinary type. It is possible cutting off the factorial terms in the exponential generating function, simply by multiplying an umbra with the Boolean unity umbra.

The *Boolean unity* umbra $\bar{u}$ has moments $E[\bar{u}^n] = n!$ and generating function $f(\bar{u}, t) = (1 - t)^{-1}$. The Boolean unity is connected to the singleton umbra through the following similarities $\bar{u} \equiv -1. -\chi$ and $-1.\bar{u} \equiv -\chi$. The introduction of the umbra $\bar{u}$ allows us a simple expression for the ordinary generating function of an umbra $\alpha$. In fact, if $\alpha$ has moments $\{a_n\}$, we will denote by $\bar{\alpha}$ an umbra similar to $\bar{u}\alpha$, having moments $n! a_n$ and generating function $f(\bar{\alpha}, t) = 1 + a_1 t + a_2 t^2 + \cdots$.

This simple device allows us to set up an umbral theory of Boolean cumulants [21] by defining an umbra $\eta_\alpha$ whose moments are the Boolean cumulants $h_i$, that is

$E[\eta_\alpha^i] = h_i$. The umbra $\eta_\alpha$ is the $\alpha$-Boolean cumulant umbra. The connection between moments $\{a_i\}$ and Boolean cumulants $\{h_i\}$ is encoded by the following equivalences

$$\bar{\eta}_\alpha \equiv \bar{u}^{<-1>}.\beta.\bar{\alpha} \quad \text{and} \quad \bar{\alpha} \equiv \bar{u}.\beta.\bar{\eta}_\alpha. \tag{6.57}$$

Observe the analogy of the first equivalence in (6.57) with $\kappa_\alpha \equiv \chi.\alpha \equiv u^{<-1>}.\beta.\alpha$, giving the $\alpha$-cumulant umbra. The equivalence $\bar{\eta}_{c\alpha} \equiv c\bar{\eta}_\alpha$ gives the homogeneity property of the $\alpha$-Boolean cumulant umbra $\eta_\alpha$.

*Example 6.14.* The unique umbra (up to similarity) having sequence of Boolean cumulants $\{1\}_{n\geq 1}$ is an umbra $\alpha$ such $\bar{\alpha} \equiv (2\bar{u})_D$. This umbra has moments $2^{n-1}$, that is the number of interval partitions $\mathcal{I}_n$.

A parametrization is available for equivalences (6.57), similarly to what happens for the quasi-free cumulant umbra (see Theorem 6.15).

**Theorem 6.17 (Parametrization).** *If $\eta_\alpha$ is the $\alpha$-Boolean cumulant umbra, then*

$$\bar{\alpha}^n \simeq \bar{\eta}_\alpha(\bar{\eta}_\alpha + 2.\bar{\alpha})^{n-1} \quad and \quad \bar{\eta}_\alpha^n \simeq \bar{\alpha}(\bar{\alpha} - 2.\bar{\alpha})^{n-1}. \tag{6.58}$$

The following theorem helps to define the Boolean probability convolution in umbral terms.

**Theorem 6.18 (Additivity property).** *If $\eta_\alpha, \eta_\gamma$ and $\eta_\xi$ are the Boolean cumulant umbrae of $\alpha, \gamma$ and $\xi$ respectively, then $\bar{\eta}_\xi \equiv \bar{\eta}_\alpha \dotplus \bar{\eta}_\gamma$ if and only if $-1.\bar{\xi} \equiv -1.\bar{\alpha} \dotplus -1.\bar{\gamma}$.*

So, we can define the *Boolean convolution* of $\alpha$ and $\gamma$ to be the umbra $\alpha \uplus \gamma$ such that

$$\overline{\alpha \uplus \gamma} \equiv -1.(-1.\bar{\alpha} \dotplus -1.\bar{\gamma}). \tag{6.59}$$

Note the parallelism with the convolution linearized by classical cumulants, that is $\alpha + \gamma \equiv -1.(-1.\alpha + -1.\gamma)$. Theorem 6.18 assures this is the unique convolution linearized by Boolean cumulants. By using Theorem 6.18 and definition (6.59), the additivity property of the Boolean cumulant umbra with respect to the Boolean convolution can be stated as $\bar{\eta}_{\alpha \uplus \gamma} \equiv \bar{\eta}_\alpha \dotplus \bar{\eta}_\gamma$. In particular, the semi-invariance property $\bar{\eta}_{\alpha \uplus cu} \equiv \bar{\eta}_\alpha \dotplus c\chi$ results being $\bar{\eta}_{cu} \equiv c\chi$.

## 6.7 Free Cumulant Theory

Free probability theory is a noncommutative probability theory introduced by Voiculescu [83], with a view to tackle certain problems on operator algebra. More precisely, a new kind of independence is defined by replacing tensor products with free products and this can help understand the Von Neumann algebras of free groups. The combinatorics underling this subject is based on the notion of noncrossing partition. Within free probability theory, noncrossing partitions are extensively used by Speicher [69, 68]. Speicher takes his lead from the definition of classical multilinear cumulants in terms of the Möbius function. However, he changes the lattice where the Möbius inversion formula is applied. Instead of using the lattice of all partitions of a finite set, he uses the smaller lattice of noncrossing partitions. As well-known, some results of noncrossing partition theory can be recovered via Lagrange inversion formula. In the previous section we have shown how the Lagrange inversion formula is strictly connected to the Abel representation of binomial sequences. This is the key to manage free cumulants in umbral terms. Indeed we get a new and very simple parametrization of free cumulants in terms of moments, not yet achievable by the formal power series language.

Let us consider a noncommutative random variable $X$, i.e. an element of an unital noncommutative algebra $\mathcal{A}$. Suppose $\phi : \mathcal{A} \to \mathbb{C}$ is an unital linear functional. The $i$-th moment of $X$ is the complex number $m_i = \phi(X^i)$ while its generating function is the formal power series $M(t) = 1 + \sum_{i \geq 1} m_i t^i$. The noncrossing (or free) cumulants of $X$ are the coefficients $r_i$ of the ordinary power series $R(t) = 1 + \sum_{i \geq 1} r_i t^i$ such that

$$M(t) = R[tM(t)]. \tag{6.60}$$

In order to associate a sequence of free cumulants to an umbra, we resume the notion of quasi-free cumulant umbra. Suppose to replace the umbra $\alpha$ with $\bar{\alpha}$, so that $f(\bar{\alpha}, t)$ is the ordinary generating function of $\alpha$ corresponding to the moment generating function $M(t)$ in (6.60). In [21], we prove that the formal power series $f(\mathfrak{R}_{\bar{\alpha}}, t)$ corresponds to the generating function $R(t)$ given in (6.60). The umbra $\mathfrak{R}_{\bar{\alpha}}$ has moments $E[\mathfrak{R}_{\bar{\alpha}}^n] = n! r_n$, where $r_n$ are the free cumulants defined by Speicher. So the umbra $\mathfrak{R}_{\bar{\alpha}}$ has been called the *free cumulant umbra* of $\alpha$. The umbral version of equality (6.60) is $\bar{\alpha} \equiv \mathfrak{R}_{\bar{\alpha}}.\beta.\bar{\alpha}_D$. Since $\mathfrak{R}_{\bar{\alpha}}$ represents the sequence $\{n! r_n\}_{n \geq 1}$, then additivity and homogeneity properties of free cumulants can be recovered by using the analogous properties of the quasi-free cumulant umbra. Recall that for the quasi-free cumulant umbra the semi-invariance property fails, but it can be easily recovered for the free cumulant umbra. Indeed, from the additivity property we have $\mathfrak{R}_{\bar{\alpha} \boxplus c\bar{u}} \equiv \mathfrak{R}_{\bar{\alpha}} \dot{+} c \mathfrak{R}_{\bar{u}}$, and being $\mathfrak{R}_{\bar{u}} \equiv \bar{u}.\bar{u}_D^* \equiv \chi$, the semi-invariance property $\mathfrak{R}_{\bar{\alpha} \boxplus c\bar{u}} \equiv \mathfrak{R}_{\bar{\alpha}} \dot{+} c\chi$ is proved.

It is interesting to underline that, by introducing the notion of *Catalan umbra*, we are able to encode the well-known relation between Catalan numbers and noncrossing partitions by using free cumulants, as the following example shows.

*Example 6.15. Catalan umbra.* The Catalan umbra is the unique umbra $\varsigma$ such that $\mathfrak{R}_{\bar{\varsigma}} \equiv \bar{u}$, that is $\bar{\varsigma} \equiv \bar{u}.(-1.\bar{u})_D^*$. So in the free setting, the Catalan umbra plays the same role played by the Bell umbra $\beta$ in the classical framework. If $C_n$ is the $n$-th Catalan number, then $\varsigma^n \simeq C_n$, since $n! \varsigma^n \simeq \bar{\varsigma}^n \simeq n! \sum_{\mu \vdash n} (n)_{\ell(\mu)-1}/m(\mu)!$. As well-known (see for instance [33]), $(n)_{\ell(\mu)-1}/m(\mu)!$ is the number of noncrossing partitions of shape $\mu$. On the other hand $|\mathcal{N}C_n| = C_n$, hence $E[\varsigma^n] = |\mathcal{N}C_n| = C_n$.

*Example 6.16. Wigner semicircle distribution.* In free probability theory, the Wigner semicircle distribution is analogous to the Gaussian random variable in the classical probability. Indeed, free cumulants of degree higher than 2 of the Wigner semicircle distribution are all zero. Suppose to denoted by $X$ the Wigner semicircle random variable. It is well-known that $E[X^{2i}] = C_i$ and $E[X^{2i+1}] = 0$ for all nonnegative integers $i$. In [20], the umbra corresponding to the Wigner semicircle distribution has been recognized in the composition $\bar{\varsigma}.\beta.\bar{\delta}$, where $\varsigma$ is the Catalan umbra and $\delta$ is the umbral counterpart of a standard Gaussian random variable (see Example 6.1).

Moments in terms of free cumulants (and viceversa) can be easily recovered by suitably updating the Abel parametrization given in Theorem 6.15.

**Theorem 6.19 (Abel parametrization).** *If $\mathfrak{K}_{\bar{\alpha}}$ is the free cumulant umbra of $\alpha$, then*

$$\bar{\alpha}^n \simeq \mathfrak{K}_{\bar{\alpha}}(\mathfrak{K}_{\bar{\alpha}} + n.\mathfrak{K}_{\bar{\alpha}})^{n-1} \quad and \quad \mathfrak{K}_{\bar{\alpha}}^n \simeq \bar{\alpha}(\bar{\alpha} - n.\bar{\alpha})^{n-1} \tag{6.61}$$

Despite this last elegant parametrization obtained thanks to the power of umbral syntax, most of the spadework remains to be done, notably the extension to the multivariate case. While the notion of classical multivariate cumulant umbra introduced in [15] can be easily extended to Boolean one, this does not happen for free cumulants, since a noncommutative context is required.

*One Formula for all Kind of Cumulants.*

We have given parametrization formulae, expressing moments in terms of classical, Boolean and free cumulants (and viceversa). We have always pointed out that, by a suitable expansion of these umbral polynomials, we recover formulae converting moments in cumulants and viceversa.

All these umbral polynomials can be viewed as special cases of the umbral polynomial $\gamma(\gamma + \zeta.\gamma)^{i-1}$, for $i = 1, 2, \ldots$ and $\zeta, \gamma \in A$, with a structure very similar to Abel polynomials. The following proposition gives an expansion of this umbral polynomial, particularly suited to be implemented in symbolic software.

**Proposition 6.2.** *If $\zeta, \gamma \in A$ then*

$$\gamma(\gamma + \zeta.\gamma)^{i-1} \simeq \sum_{\lambda \vdash i} (\zeta)_{\nu_\lambda - 1} d_\lambda \gamma_\lambda, \tag{6.62}$$

*where the sum ranges over all the integer partitions $\lambda = (1^{r_1}, 2^{r_2}, \ldots)$ of the integer $i$, $\nu_\lambda$ is the length of the partition, $\gamma_\lambda = \gamma_1^{r_1}(\gamma_2^2)^{r_2} \cdots$ with $\gamma_1, \gamma_2, \ldots$ uncorrelated umbrae similar to $\gamma$ and $d_\lambda$ is the number of set partitions of type $\lambda$.*

In order to evaluate $\gamma(\gamma + \delta.\gamma)^{i-1}$ via (6.62), we need the factorial moments of $\zeta$ and the moments of $\gamma$. Recall that, if we just have information on moments $\zeta^i$, the factorial moments can be recovered by using the well-known change of bases $(\zeta)_i \simeq \sum_{k=1}^{i} s(i,k)\zeta^k$, where $\{s(i,k)\}$ are the Stirling numbers of first kind. In particular equivalence (6.62) allows us to give any expression of cumulants (classical, Boolean, free) in terms of moments and viceversa, as shown in the following.

CLASSICAL CUMULANTS IN TERMS OF MOMENTS. Due to the first of (6.18), set $\zeta = -1.u$ and $\gamma = \alpha$. Here $E[(-1.u)_i] = (-1)_i = (-1)^i i!$.

MOMENTS IN TERMS OF CLASSICAL CUMULANTS. Due to the latter of (6.18), set $\zeta = \beta$ and $\gamma = \kappa_\alpha$. Here $E[(\beta)_i] = 1$.

BOOLEAN CUMULANTS IN TERMS OF MOMENTS. Due to the latter of (6.58), set $\delta = -2.u$ and and $\gamma = \bar{\alpha}$. Here $E[(-2.u)_i] = (-1)^i(i+1)!$.

MOMENTS IN TERMS OF BOOLEAN CUMULANTS. Due to the first of (6.58), set $\delta = 2.\bar{u}.\beta$ and $\gamma = \bar{\eta}_\alpha$. Here $E[(2.\bar{u}.\beta)_i] = E[(2.\bar{u}.\beta.\chi)^i] = (i+1)!$.

FREE CUMULANTS IN TERMS OF MOMENTS. Due to the latter of (6.61), set $\delta = -n.u$ and $\gamma = \bar{\alpha}$. Here $E[(-n.u)_i] = (-n)_i$.

MOMENTS IN TERMS OF FREE CUMULANTS. Due to the first of (6.61), set $\delta = n.u$ and $\gamma = \mathfrak{R}_{\bar{\alpha}}$. Here $E[(n.u)_i] = (n)_i$.

In 1999, Henry Crapo in "Ten abandoned golden mines "[9] has collected the expression of Gian-Carlo desires for future developments in mathematics. We have started to dig in one of these mines and soon we have found a precious vein. We hope this reading could be provoking.

# References

1. Andrews, D. F. (2001), *Asymptotic Expansions of Moments and Cumulants*, in "Stat. Comput.", 11, pp. 7-16.
2. Andrews, D. F., Stafford, J. E. (1998), *Iterated Full Partitions*, in "Stat. Comput.", 8, pp. 189-92.
3. Andrews, D. F., Stafford, J. E. (2000), *Symbolic Computation for Statistical Inference*, Oxford, Oxford University Press.
4. Baten, W. D. (1931), *Correction for the Moments of a Frequency Distribution in Two Variables*, in "Ann. Math. Stat.", 2/3, pp. 309-19.
5. Bruns, H. (1906), *Wahrscheinlichkeitsrechnung und Kollektivmasslehre*, Leipzig, Teubner.
6. Costantine, G. M., Savits, T. (1994), *A Stochastic Representation of Partition Identities*, in "SIAM J. Discrete Math.", 7, pp. 194-202.
7. Craig, C. C. (1936), *Sheppard's Corrections for a Discrete Variable*, in "Ann. Math. Stat.", 7/2, pp. 55-61.
8. Crapo, H., Senato, D. (eds.) (2001), *Algebraic Combinatorics and Computer Science: a Tribute to Gian-Carlo Rota*, Milan, Springer Italia.
9. Crapo, H. (2001), *Ten Abandoned Golden Mines*, in H. Crapo, D. Senato (eds.), *Algebraic Combinatorics and Computer Science, A Tribute to Gian-Carlo Rota*, Milan, Springer Italia, pp. 3-22.
10. D'Antona, O. (1994), *The Would-Be Method of Targeted Rings*, in B. E. Sagan, R. P. Stanley (eds.), *Mathematical Essays in Honor of Gian-Carlo Rota*, Boston, Basel, Berlin, Birkhäuser, pp. 157-72.
11. Di Bucchianico, A., Soto y Koelemeijer, G. (2001), *Solving Linear Recurrences Using Functionals*, in H. Crapo, D. Senato (eds.), *Algebraic Combinatorics and Computer Science, A Tribute to Gian-Carlo Rota*, Milan, Springer Italia, pp. 461-72.
12. Di Nardo, E. (2009), *A New Approach to Sheppards Corrections*, preprint.
13. Di Nardo, E., Senato, D. (2001), *Umbral Nature of the Poisson Random Variables*, in H. Crapo, D. Senato (eds.), *Algebraic Combinatorics and Computer Science, A Tribute to Gian-Carlo Rota*, Milan, Springer Italia, pp. 245-66.
14. Di Nardo, E., Senato, D. (2006), *An Umbral Setting for Cumulants and Factorial Moments*, in "European J. Combin.", 27/3, pp. 394-413.
15. Di Nardo, E., Guarino, G., Senato, D. (2008), *A Unifying Framework for k-statistics, Polykays and their Multivariate Generalizations*, in "Bernoulli", 14/2, pp. 440-68.
16. Di Nardo, E., Guarino, G., Senato, D. (2008), *Maple Algorithms for Polykays and Multivariate Polykays*, in "Adv. Appl. Stat.", 8/1, pp. 19-36.
17. Di Nardo, E., Guarino, G., Senato, D. (2008), *Symbolic Computation of Moments of Sampling Distributions*, in "Comp. Stat. Data Anal.", 52/11, pp. 4909-22.

18. Di Nardo, E., Guarino, G., Senato, D. (2009), *A New Method for Fast Computing Unbiased Estimators of Cumulants*, in "Statist. Comput.", 19, pp. 155-65.

19. Di Nardo, E., Niederhausen, H., Senato, D. (2008), *The Classical Umbral Calculus II: Sheffer Sequences*, arXiv:0810. 3554v1.

20. Di Nardo, E., Oliva, I. (2009), *On the Computation of Classical, Boolean and Free Cumulants*, in "Appl. Math. Comp.", 208/2, pp. 347-54.

21. Di Nardo E., Petrullo, P., Senato, D. (2009), *Cumulants, Convolutions and Volume Polynomials*, submitted.

22. Doubilet, P. (1972), *On the Foundations of Combinatorial Theory VII: Symmetric Functions Through the Theory of Distribution and Occupancy*, in "Stud. Appl. Math.", 11, pp. 377-96.

23. Dressel, P. L. (1940), *Statistical Seminvariants and their Estimates with Particular Emphasis on their Relation to Algebraic Invariants*, in "Ann. Math. Stat.", 11, pp. 33-57.

24. Drton, M., Sturmfels, B., Sullivant, S. (2008), *Lectures on Algebraic Statistics. Oberwolfach Seminars*, Boston, Basel, Berlin, Birkhäuser.

25. Feller, W. (1966), *An Introduction to Probability Theory and its Applications*, vol. 2, New York, London, Sydney, John Wiley & Sons.

26. Ferreira, P. G., Magueijo, J., Silk, J. (1997), *Cumulants as non-Gaussian Qualifiers*, in "Phys. Rev. D", 56, pp. 4592-603.

27. Fisher, R. A. (1922), *On the Mathematical Foundations of Theoretical Statistics*, in "Phil. Trans. R. Soc. Lond. A", 222, pp. 309-68.

28. Fisher, R. A. (1929), *Moments and Product Moments of Sampling Distributions*, in "Proc. London Math. Soc. (2)", 30, pp. 199-238.

29. Fréchet, M. (1940-1943), *Les probabilités associées à un système d'événements, compatibles et dépendants*, Paris, Herman.

30. Gessel, I. M. (2003), *Applications of the Classical Umbral Calculus*, "Algebra Universalis", 49/4, pp. 397-434.

31. Hald, A. (2001), *On the History of the Correction for Grouping 1873-1922*, in "Scand. J. Statist." 28/3, pp. 417-28.

32. Kaplan, E. L. (1952), *Tensor Notation and the Sampling Cumulants of k-statistics*, in "Biometrika", 39, pp. 319-23.

33. Kreweras, G. (1972), *Sur les partitions non croisée d'un cycle*, in "Discrete Math.", 1, pp. 333-50.

34. Langdon, W. L., Ore, O. (1929), *Semi-invariants and Sheppard's Correction*, in "Ann. of Math." 31/2, pp. 230-2.

35. Lukacs, E. (1955), *Applications of Faà di Bruno's Formula in Mathematical Statistics*, in "Am. Math. Mon.", 62, pp. 340-8.

36. McCullagh, P. (1984), *Tensor Notation and Cumulants of Polynomials*, "Biometrika", 71, pp. 461-76.

37. McCullagh, P. (1987), *Tensor Methods in Statistics. Monographs on Statistics and Applied Probability*, London, Chapman and Hall.

38. Müller, J. D. (2004), *Cumulant Analysis in Fluorescence Fluctuation Spectroscopy*, in "Biophys J.", 86, pp. 3981-92.

39. Niederhausen, H. (1980), *Sheffer Polynomials and Linear Recurrences*, in "Congressus Numerantium", 29, pp. 689-98.

40. Niederhausen, H. (1985), *A Formula for Explicit Solutions of Certain Linear Recursions on Polynomial Sequences*, in "Congressus Numerantium", 49, pp. 87-98.

41. Niederhausen, H. (1999), *Recursive Initial Value Problems for Sheffer Sequences*, in "Discrete Math.", 204, pp. 319-27.

42. Petrullo, P., Senato, D. (2009), *An Instance of Umbral Methods in Representation Theory: the Parking Function Module*, in "Pure Math. Appl.", in press.

43. Pistone, G., Riccomagno, E., Wynn, H. P. (2001), *Algebraic Statistics: Computational Commutative Algebra in Statistics*, London, Chapman and Hall.

44. Pitman, J. (1997), *Some Probabilistic Aspects of Set Partitions*, in "Amer. Math. Montly", 104, pp. 201-9.

45. Pitman, J., Stanley, R. P. (2002), *A Polytope Related to Empirical Distributions, Plane Trees, Parking Functions and the Associahedron*, in "Discrete Comput. Geom.", 27, pp. 603-34.

46. Prasad, S., Menicucci, N. C. (2004), *Fisher Information with Respect to Cumulants*, in "IEEE Transactions on Information Theory", 50, pp. 638-42.

47. Rao Jammalamadaka, S., Subba Rao, T., Terdik, G. (2006), *Higher Order Cumulants of Random Vectors and Applications to Statistical Inference and Time Series*, in "Sankhya", 68, pp. 326-56.

48. Razpet, M. (1990), *An Application of the Umbral Calculus*, in "J. Math. Anal. Appl.", 149, pp. 1-16.

49. Riordan, J. (1958), *An Introduction to Combinatorial Analysis*, New York, John Wiley & Sons.

50. Robson, D. S. (1957), *Applications of Multivariate Polykays to the Theory of Unbiased Ratio-type Estimation*, in "J. Amer. Statist. Assoc.", 52, pp. 511-22.

51. Roman, S. M., Rota, G.-C. (1978), *The Umbral Calculus*, in "Adv. Math.", 27, pp. 95-188.

52. Rose, C., Smith, M. D. (2002), *Mathematical Statistics with Mathematica*, (Springer Text in Statistics), New York, Berlin, Springer.

53. Rota, G.-C. (2001), *Twelve Problems in Probability No One Likes to Bring Up*, in H. Crapo, D. Senato (eds.), *Algebraic Combinatorics and Computer Science, A Tribute to Gian-Carlo Rota*, Milan, Springer Italia.

54. Rota, G.-C., Kahaner D., Odlyzko, A. (1973), *On the Foundations of Combinatorial Theory. VIII. Finite Operator Calculus*, in "Jour. Math. Anal. Appl.", 42, pp. 684-760.

55. Rota, G.-C., Taylor, B. D. (1994), *The Classical Umbral Calculus*, in "SIAM J. Math. Anal.", 25, pp. 694-711.

56. Rota, G.-C., Shen, J. (2000), *On the Combinatorics of Cumulants*, in "Jour. Comb. Theory Series A", 91, pp. 283-304.

57. Rota, G.-C., Shen, J., Taylor, B. D. (1998), *All Polynomials of Binomial Type are Represented by Abel Polynomials*, in "Ann. Scuola Norm. Sup. Pisa Cl. Sci.", 25/1, pp. 731-8.

58. Saliani, S., Senato, D. (2006), *Compactly Supported Wavelets Through the Classical Umbral Calculus*, in "Journal of Fourier Analysis and Applications", 12/1, pp. 27-36.

59. Shen, J. (1999), *Combinatorics for Wavelets: the Umbral Refinement Equation*, in "Stud. Appl. Math.", 103/2, pp. 121-47.

60. Sheppard, W. F. (1898), *On the Calculation of Most Probable Values of Frequency-Constants for Data Arranged According to Equisdistant Division of a Scale*, in "Proc. Lond. Math. Soc.", 29, pp. 353-80.

61. Smith, P. J. (1995), *A Recursive Formulation of the Old Problem of Obtaining Moments from Cumulants and Vice-versa*, "Amer. Statist.", 49, pp. 217-8.

62. Speed, T. P. (1983), *Cumulants and Partition Lattices*, in "Austral. J. Stat.", 25, pp. 378-88.

63. Speed, T. P. (1986), *Cumulants and Partition Lattices. II: Generalised k-statistics*, in "J. Aust. Math. Soc. Ser. A", 40, pp. 34-53.

64. Speed, T. P. (1986), *Cumulants and Partition Lattices. III: Multiply-indexed Arrays*, in "J. Aust. Math. Soc. Ser. A", 40, pp. 161-82.

65. Speed, T. P. (1986), *Cumulants and Partition Lattices. IV: a. s. Convergence of Generalised k-statistics*, in "J. Aust. Math. Soc. Ser. A", 41, pp. 79-94.

66. Speed, T. P., Silcock, H. L. (1988), *Cumulants and Partition Lattices. V: Calculating Generalized k-statistics*, in "J. Aust. Math. Soc. Ser. A", 44, pp. 171-96.

67. Speed, T. P., Silcock, H. L. (1988), *Cumulants and Partition Lattices. VI: Variances and Covariances of Mean Squares*, in "J. Aust. Math. Soc. Ser. A", 44, pp. 362-88.

68. Speicher, R. (1994), *Multiplicative Functions on the Lattice of Non-Crossing Partitions and Free Convolution*, in "Math. Ann.", 298, pp. 611-28.

69. Speicher, R. (1997), *Free Probability Theory and Non-Crossing Partitions*, in "Sém. Loth. Combin.", B39c.

70. Speicher, R., Woroudi, R. (1997), *Boolean Convolution*, in D. Voiculescu (ed.), "Fields Inst. Commun.", 12, AMS, pp. 267-79.

71. Stam, A. J. (1988), *Polynomials of Binomial Type and Compound Poisson Processes*, in "Jour. Math. Anal. Appl.", 130, pp. 493-508.

72. Stanley, R. (2001), *Enumerative Combinatorics II*, Cambridge, Cambridge Univerity Press.
73. Steffensen, J. F. (1950), *Interpolation*, New York, Chelsea (reprinted from 1927).
74. Stembridge, J. R. (1995), *A Maple Package for Symmetric Functions*, in "J. Symbolic Computation", 20, pp. 755-68.
75. Stuart, A., Ord, J. K. (1987), *Kendall's Advanced Theory of Statistics*, 1, London, Charles Griffin and Co.
76. Taylor, B. D. (1998), *Difference Equations Via the Classical Umbral Calculus*, in B. E. Sagan, R. P. Stanley (eds.), *Mathematical Essays in Honor of Gian-Carlo Rota*, Boston, Basel, Berlin, Birkhäuser, pp. 397-411.
77. Taylor, B. D. (2001), *Umbral Presentations for Polynomial Sequences*, in "Comput. Math. Appl.", 41, pp. 1085-98.
78. Thiele, T. N. (1897), *Elementaer Iagttagelses Laere*, København, Gyldendalske. Reprinted in English (1931), *The Theory of Observations*, in "Ann. Math. Stat.", 2, pp. 165-308.
79. Touchard, J. (1956), *Nombres exponentiels et nombres de Bernoulli*, in "Canad. J. Math.", 8, pp. 305-20.
80. Tukey, J. W. (1950), *Some Sampling Simplified*, in "J. Amer. Statist. Assoc.", 45, pp. 501-19.
81. Tukey, J. W. (1956), *Keeping Moment-like Sampling Computations Simple*, "Ann. Math. Stat.", 27, pp. 37-54.
82. Vrbik, J. (2005), *Populations Moments of Sampling Distributions*, in "Comput. Stat.", 20, pp. 611-21.
83. Voiculescu, D. (2000), *Lecture Notes on Free Probability*, in "Lecture Notes in Math.", 1738, pp. 281-349.
84. Wilson, E. B. (1927), *On the Proof of Sheppard's Corrections*, in "Proc. Natl. Acad. Sci. USA", 13/3, pp. 151-6.
85. Wishart, J. (1952), *Moment Coefficients of the k-statistics in Samples from a Finite Population*, in "Biometrika", 39, pp. 1-13.
86. Zeilberger, D. (2004), *Symbolic Moment Calculus I: Foundations and Permutation Pattern Statistics*, in "Ann. Comb.", 8, pp. 369-78.

# Chapter 7
# Two Examples of Applied Universal Algebra
## Invited Chapter

Joseph P. S. Kung

## 7.1 Prologue

In the long run, mathematics evolves organically and is independent of individual mathematicians. Multiple, and in many cases, almost simultaneous discovery
of concepts or proofs of theorems is the norm rather than the exception. Someone
will discover the concept or prove the theorem, eventually. Gian-Carlo Rota often
praised an idea, detached from its accidental discoverers, as "an idea whose time has
come". In the 1960s, the time came for the theory of Möbius functions of partially
ordered sets, and the area has flourished since then. Rota did not include his paper
*Foundations I* [4] among his most original papers. In his opinion, he only uncovered a theory already "there". Of course, there is more than a hint of *sprezzatura* in
Rota's assessment of his most influential paper, but its essential truth focuses rather
than diminishes the achievement in *Foundations I.*

Rota considered his work on Baxter algebras among his most original. This work
spans analysis, probability, and combinatorics and is summarized in the pair of papers [5]. Longer expositions can be found in [6, 8, 9]. The first and longer part of
this essay is an introduction to Baxter algebras, assuming the minimum of special-
ized knowledge. In particular, we avoid any explicit use of symmetric functions. The
only technical tool required is Möbius inversion on partition lattices.[1] The second
part gives a brief account of a theory of random variables founded on the events they
define. Rota sketched this in the 1970s but never put it into final form.

The theory of Baxter algebras reveals the algebraic structure underlying analytic
identities. Rota's development of random variables reverses this and shows how
algebraic operations (such as addition and multiplication) on random variables can

Joseph P. S. Kung
University of North Texas, Denton, TX, US

[1] Much work has been done on Baxter algebras, particularly in the last ten years or so, by M.
Aguiar, G.E. Andrews, S. de Bragança, P. Cartier, R. Diaz, A. Doohovskoy, K. Ebrahimi-Fard,
J.M. Garcia-Bondía, L. Guo, W. Keigher, W. Moreira, K. Ono, M. Páez, and others. Some authors
have renamed Baxter algebras "Rota-Baxter algebras".

E. Damiani et al. (eds.), *From Combinatorics to Philosophy,*                                    131
DOI 10.1007/978-0-387-88753-1_7, © Springer Science+Business Media, LLC 2009

be defined from lattice operations in an abstract Boolean $\sigma$-algebra. Both are good examples of applied universal algebra.

## 7.2 The Bohnenblust–Spitzer Identity

An elementary entry into the theory of Baxter operators is to consider the combinatorics of taking maximums. Let

$$x^+ = \frac{|x| + x}{2} = \max\{0, x\}.$$

Let $\mathfrak{S}_n$ be the symmetric group of all permutations acting on the set $\{1, 2, \ldots, n\}$. If $\gamma \in \mathfrak{S}_n$, let cycle$(\gamma)$ be the partition on $\{1, 2, \ldots, n\}$ whose blocks are the underlying sets of the cycles in the cycle decomposition of $\gamma$. Let $\underline{x}$ be the sequence $x_1, x_2, \ldots, x_n$ of real numbers and $S(\underline{x})$ and $T(\underline{x})$ be the multisets (of size $n!$) defined by

$$S(\underline{x}) = \{((((x_{\gamma(1)}^+ + x_{\gamma(2)})^+ + \cdots)^+ + x_{\gamma(n-1)})^+ + x_{\gamma(n)})^+ : \gamma \in \mathfrak{S}_n\}$$

$$T(\underline{x}) = \left\{ \sum_{D:\, D \in \text{cycle}(\gamma)} \left( \sum_{d:\, d \in D} x_d \right)^+ : \gamma \in \mathfrak{S}_n \right\}.$$

The *Bohnenblust–Spitzer identity*[2] [10] says that the two multisets $S(\underline{x})$ and $T(\underline{x})$ are equal. For example, let $\underline{x} = (-4, 2, 3)$. Then $S(\underline{x})$ consists of the six sums

$$(((-4)^+ + 2)^+ + 3)^+, \ ((2^+ + (-4))^+ + 3)^+, \ ((3^+ + 2)^+ + (-4))^+,$$
$$(((-4)^+ + 3)^+ + 2)^+, \ ((3^+ + (-4))^+ + 2)^+, \ ((2^+ + 3)^+ + (-4))^+,$$

arising from the permutations, written in cycle notation,

$$(1)(2)(3), \ (1\,2)(3), \ (1\,3)(2), \ (1)(2,3), \ (1,2,3), \ (1,3,2).$$

Hence, $S(\underline{x})$ is the multiset $\{5, 3, 1, 5, 2, 1\}$. On the other hand, $T(\underline{x})$ consists of the sums

$$(-4)^+ + 2^+ + 3^+, \ (-4 + 2)^+ + 3^+, \ (-4 + 3)^+ + 2^+,$$
$$(-4)^+ + (2 + 3)^+, \ (-4 + 2 + 3)^+, \ (-4 + 3 + 2)^+$$

and $T(\underline{x}) = \{5, 3, 2, 5, 1, 1\}$. The two multisets are indeed equal.

---

[2]  There are actually two multiset identities in [10]. The other identity is that the multisets $T(\underline{x})$ and

$$\left\{ \max_{k:\, 1 \leq k \leq n} \left( \sum_{i=1}^{k} x_{\gamma(i)} \right)^+ : \gamma \in \mathfrak{S}_n \right\}$$

are equal. For the example in the text, the third multiset is $\{1, 2, 5, 1, 3, 5\}$. See [10] for an elegant combinatorial proof.

## 7.3 Baxter Algebras

Let $\mathcal{B}$ be a commutative algebra over a field $\mathbb{F}$ and $\vartheta$ be an element in $\mathbb{F}$. We do not assume that $\mathcal{B}$ has an identity. An $\mathbb{F}$-linear operator $P : \mathcal{B} \to \mathcal{B}$ is a *Baxter operator* *(with parameter $\vartheta$)* if it satisfies the *Baxter identity:* for all $x$ and $y$ in $\mathcal{B}$,

$$(Px)(Py) + \vartheta P(xy) = P(x(Py)) + P((Px)y).$$

A *Baxter algebra* $(\mathcal{B}, P)$ is a pair, where $\mathcal{B}$ is an $\mathbb{F}$-algebra and $P$ is a Baxter operator on $\mathcal{B}$. A subset $\mathcal{D}$ of a Baxter algebra $(\mathcal{B}, P)$ is a *Baxter subalgebra* if $\mathcal{D}$ is an $\mathbb{F}$-subalgebra closed under $P$ (that is, if $x \in \mathcal{D}$, then $Px \in \mathcal{D}$). A function $\varphi$ from a Baxter algebra $(\mathcal{A}, Q)$ to a Baxter algebra $(\mathcal{B}, P)$ is a *Baxter algebra homomorphism* if $\varphi$ is an algebra homomorphism and commutes with the Baxter operators, that is, $\varphi Q = P \varphi$.

The following result shows that Baxter operators are abstractions of summation operators.

**Lemma.** Let $E : \mathcal{A} \to \mathcal{A}$ be an endomorphism so that the infinite sum

$$P = \sum_{i=1}^{\infty} E^i = E + E^2 + E^3 + \cdots$$

is defined. Then $P$ is a Baxter operator with parameter $-1$.

*Proof.* Observe that

$$
\begin{aligned}
P(x(Py)) &= P\left( \sum_{i=1}^{\infty} x E^i(y) \right) \\
&= \sum_{j=1}^{\infty} E^j(x) \left( \sum_{i=1}^{\infty} E^{j+i}(y) \right) \\
&= \sum_{i,j:\, j<i} E^j(x) E^i(y).
\end{aligned}
$$

Similarly,

$$(Px)(Py) = \sum_{i,j=1}^{\infty} E^i(x)E^j(y),$$

$$P(xy) = \sum_{i=1}^{\infty} E^i(x)E^i(y),$$

$$P((Px)y) = \sum_{i,j:\, j>i} E^j(x)E^i(y),$$

and Baxter's identity follows.

The proof suggests that the Baxter identity decomposes the product of two sums $(Px)(Py)$, pictured as a rectangle, into two subsums $P(x(Py))$ and $P((Px)y)$, pictured as an upper triangle and a lower triangle, with an overlap $\vartheta P(xy)$ along the diagonal. Thus, we expect natural examples of Baxter operators to have parameter $\vartheta$ equal to 1, 0, or $-1$. The Baxter operator needed to prove the Bohnenblust–Spitzer identity has parameter 1. The *indefinite integration operator I* defined on continuous functions $\mathbb{R} \to \mathbb{R}$ by

$$I(f(x)) = \int_0^x f(t)dt$$

has parameter 0. Although one can define Baxter operators with any given parameter formally, no concrete Baxter operator with parameter not equal to 1, 0, or $-1$ is known. We remark that the product rule for a derivation should perhaps be written

$$D(xy) = x(Dy) + (Dx)y - \vartheta(Dx)(Dy),$$

with $\vartheta = 0$. This idea is developed in the elegant paper [2].

## 7.4 The Standard Baxter Algebra

The most important example of a Baxter algebra is the standard algebra. Let $\mathbb{F}$ be a field, $\mathcal{A}$ an $\mathbb{F}$-algebra, and $\mathcal{A}_\infty$ the $\mathbb{F}$-algebra of all sequences $(a_i)_{1 \le i < \infty}$ with terms $a_i$ in $\mathcal{A}$, under termwise addition, multiplication, and scalar multiplication. Then the operator

$$P : (a_1, a_2, a_3, \dots) \mapsto (0, a_1, a_1 + a_2, a_1 + a_2 + a_3, \dots)$$

is a Baxter operator with parameter $-1$. This follows from the lemma with the endomorphism $E$ equal to the *shift*

$$(a_1, a_2, a_3, \dots) \mapsto (0, a_1, a_2, a_3, \dots).$$

Now let $x_{ij}$, $1 \leq i \leq n, 1 \leq j < \infty$ be infinitely many indeterminates and $\mathbb{F}(x_{ij})$ be the field of rational functions in the indeterminates $x_{ij}$. Let $\underline{x}_i$ be the sequence $(x_{i1}, x_{i2}, x_{i3}, \ldots)$. The *standard Baxter algebra* $(\mathcal{S}_n, P)$ on $n$ generators is the intersection of all Baxter subalgebras in $\mathbb{F}(x_{ij})_\infty$ containing the sequences $\underline{x}_1, \underline{x}_2, \ldots, \underline{x}_n$. The operator $P$ is a Baxter operator with parameter $-1$.

It may be useful to see some examples of elements in $\mathcal{S}_n$. The polynomial $\underline{x}_1^2 + \underline{x}_3$ is in $\mathcal{S}_n$ and it is the sequence whose $j$th term is $x_{1j}^2 + x_{3j}$. In general, if $q(X_1, X_2, \ldots, X_n)$ is a polynomial in $n$ variables, then $q(\underline{x}_1, \underline{x}_2, \ldots, \underline{x}_n)$ is the sequence with $j$th term equal to $q(x_{1j}, x_{2j}, \ldots, x_{nj})$. Writing $q(x_{1j}, x_{2j}, \ldots, x_{nj})$ in abbreviated form as $q(x_{ij})$, the operator $P$ sends $q(\underline{x}_1, \underline{x}_2, \ldots, \underline{x}_n)$ to the sequence

$$(0, q(x_{i1}), q(x_{i1}) + q(x_{i2}), q(x_{i1}) + q(x_{i2}) + q(x_{i3}),$$
$$q(x_{i1}) + q(x_{i2}) + q(x_{i3}) + q(x_{i4}), \ldots).$$

The operator $P$ replaces a sequence with the sequence of its partial sums, shifted once to the right. Note that the $j$th term in $Pq(\underline{x}_1, \underline{x}_2, \ldots, \underline{x}_n)$ is a polynomial in the variables $x_{ik}$, where $k$ is *strictly less* that $j$.

From this, we see that the standard algebra is constructed in two step. The first step, done once, is to form a polynomial algebra (with each polynomial cloned countably many times in a sequence). The second step is to add elements obtained by applying $P$ to each element and then form all polynomials in the elements already constructed. The second step is repeated infinitely many times. From the construction, we expect the standard algebra to behave like a polynomial algebra, that is, like a "free" algebra.

**Rota's theorem.** *The standard Baxter algebra $(\mathcal{S}_n, P)$ is free in the following sense: if $(\mathcal{B}, Q)$ is a Baxter algebra with parameter $-1$ generated by $n$ elements $y_1, y_2, \ldots, y_n$, then there exists a unique Baxter algebra homomorphism $\varphi : \mathcal{S}_n \to \mathcal{B}$ such that $\varphi(\underline{x}_i) = y_i$. In particular, $\mathcal{B}$ is a homomorphic image of $\mathcal{S}_n$ and an identity holding in the standard Baxter algebra $(\mathcal{S}_n, P)$ holds in all Baxter algebras with parameter $-1$ which can be generated by $n$ elements.*

The proof of Rota's theorem is technically complicated. The original account can be found in [5, 8] and an exposition is in Chapter 5 of [3]. We end this section by sketching Rota's proof. From universal algebra, we know how to "construct" free Baxter algebras abstractly.[3] To do so, we consider algebras *with a linear operator*, which are simply algebras with a specific linear operator attached. It is straightforward to construct the free $\mathbb{F}$-algebra $(\mathcal{F}_n, T)$ on $n$ generators $\xi_1, \xi_2, \ldots, \xi_n$ with a linear operator $T$. This algebra consists of all expressions that can be built recursively starting from $\xi_1, \xi_2, \ldots, \xi_n$ and the elements of $\mathbb{F}$ with the rules: if $u$ and $v$ are already built, then we can built $u + v$, $uv$, and $Tu$. For example, the expression $T(2\xi_1 T(\xi_3) + \xi_1^2) + \xi_3(T(\xi_2)))$ can be built. This forms an algebra with an operator

---

[3] A general theorem is Birkhoff's theorem ([1], p. 167): free algebras exist for any class of algebras closed under subalgebras and direct products. We shall not need this deep theorem.

with operations defined formally, so that for example, the sum of $3\xi_2$ and $T(\xi_3)$ is the formal expression $3\xi_2 + T(\xi_3)$.

A $T$-ideal in an algebra with operator $(\mathcal{A}, T)$ is an algebra ideal in $\mathcal{A}$ closed under the operator $T$. If $I$ is a $T$-ideal, then $T$ is well-defined on the quotient $\mathcal{A}/I$ as an ordinary algebra and thus, $(\mathcal{A}/I, T)$ is an algebra with operator. Extensions of the familiar homomorphism theorem and related results from ring theory hold for algebras with operators. Let $(\mathcal{A}, T)$ and $(\mathcal{B}, R)$ be algebras with operators. An algebra homomorphism $\varphi : \mathcal{A} \rightarrow \mathcal{B}$ is an algebra-with-operator homomorphism if $\varphi T = R\varphi$. It is routine to check that (a) the kernel of an algebra-with-operator homomorphism is a $T$-ideal and (b) if $\varphi$ is onto, then $(\mathcal{A}, T)/\ker \varphi \cong (\mathcal{B}, R)$. In addition, if $I$ and $J$ are $T$-ideals in $(\mathcal{A}, T)$ such that $I \subseteq J$, then there is an algebra-with-operator homomorphism $\mathcal{A}/I \rightarrow \mathcal{A}/J$.

Using this theory, we can now construct the free Baxter algebra on $n$ generators with parameter $-1$. Let $H$ be the $T$-ideal in $(\mathcal{F}_n, T)$ generated by all elements of the form

$$(Tx)(Ty) - T(xy) - T(x(Ty)) - T((Tx)y),$$

where $x, y \in \mathcal{F}_n$. The ideal $H$ is clearly the minimum $T$-ideal such that $(\mathcal{F}_n/H, T)$ is a Baxter algebra with Baxter operator $T$. Hence, it is free in the sense given in Rota's theorem.

To prove Rota's theorem, it suffices to show that the kernel of the algebra homomorphism $\varphi : \mathcal{F}_n \rightarrow \mathcal{S}_n$ defined by $\varphi(\xi_i) = \underline{x}_i$ has kernel $H$. A *monomial* in $\mathcal{F}_n$ is an expression formed not using "$+$". For example, $\xi_3 T(\xi_1 T(\xi_2))$ and $T(\xi_3 T(\xi_2^2))T(\xi_1^3)$ are monomials. Using the Baxter identity in the form

$$(Tx)(Ty) = T(xy) + T(x(Ty)) + T((Tx)y),$$

we can write every monomial in $\mathcal{F}_n$ in the form $s + r$, where $r$ is in the $T$-ideal $H$ and $s$ is a linear combination of monomial of the form

$$a(Tb), Tc, d,$$

where $a$ and $d$ are monomials formed not using the operator $T$. A linear combination of monomials in $\mathcal{F}_n$ formed not using $T$ is mapped by $\varphi$ to a polynomial $q(\underline{x}_i)$; hence, a linear combination of monomials of the form $d$ is mapped to 0, the zero polynomial, if and only if that linear combination is the zero polynomial. It remains to handle the case of a linear combination $s$ of $a(Tb)$ and $Tc$, where $a$ is a non-constant monomial. To do this, write $s = u + v$, where $u$ is the part of $s$ involving $a(Tb)$ and $v$ is the part of $s$ involving $Tb$ and suppose that $\varphi(u + v) = 0$. There can be no cancellation between terms in $\varphi(u)$ and $\varphi(v)$. The reason is that the terms $\varphi(Tc)$ in $\varphi(v)$ all has the form $P\varphi(c)$. Since $P$ shift a sequence to the right so that the $j$th term of $P\varphi(c)$ involves at most the variables $x_{ik}$, $k < j$. Thus, the terms $u$ and $v$ can be handled separately. By induction on the number of times $T$ occurs in a monomial, we can show first that $\varphi(v) = 0$ implies that $v \in H$, and finally, that $\varphi(u) = 0$ implies that $u \in H$.

The originality of Rota's theorem lies in the definition of the standard algebra, particularly, in using a infinite sequence which allows the Baxter operator to act as a combined summation and shift operator. Rota's construction shows how symbolic calculations determine analysis, or as he put it, "algebra dictates analysis".

## 7.5 A Precursor Identity

We now describe the Baxter algebra behind the Bohnenblust–Spitzer identity. Let $\mathcal{M}$ be the $\mathbb{R}$-algebra of functions $f : \mathbb{R} \to \mathbb{R}$ with finite support (that is, $f(x) = 0$ except for a finite set of real numbers), with *convolution*, defined by

$$fg(x) = \sum_{y:\, y \in \mathbb{R}} f(y)g(x-y) = \sum_{y,z:\, y+z=x} f(y)g(z),$$

as the product. Let $P : \mathcal{M} \to \mathcal{M}$ be the linear operator

$$Pf(x) = \sum_{y:\, y^+ = x} f(y).$$

Put another way, $Pf(x) = f(x)$ if $x$ is positive and $Pf(0) = \sum_{y:\, y \le 0} f(y)$.

The operator $P : \mathcal{M} \to \mathcal{M}$ is a Baxter operator with parameter 1. To prove this, observe that

$$P(fPg + gPf)(x) = \sum_{y,z:\, (y^+ + z)^+ = x \text{ or } (y+z^+)^+ = x} f(y)g(z)$$

$$(P(fg) + (Pg)(Pf))(x) = \sum_{y,z:\, (y+z)^+ = x \text{ or } y^+ + z^+ = x} f(y)g(z).$$

Thus, Baxter's identity follows if for every pair $y,z$ of real numbers, the two multisets (of size 2)

$$\{(y^+ + z)^+,\ (y + z^+)^+\}$$

and

$$\{(y+z)^+,\ y^+ + z^+\}$$

are equal. This can be checked by an easy case analysis.[4]

Let $(x_1, x_2, \ldots, x_n)$ be a sequence of real numbers and $h_i : \mathbb{R} \to \mathbb{R}$ be the functions defined by $h_i(y) = 1$ if $y = x_i$ and 0 otherwise. Then the value of the function

---

[4] It is not hard to show that if $m : \mathbb{R} \to \mathbb{R}$ is a continuous function satisfying the multiset equation

$$\{m(m(y) + z),\ m(y + m(z))\} = \{m(y + z),\ m(y) + m(z)\},$$

then $m(x) = x$, $m(x) = 0$, $m(x) = x^+$, or $m(x) = x^-$, where $x^- = \min(0, x)$.

$$\sum_{\gamma:\,\gamma\in\mathfrak{S}_n} P(h_{\gamma(n)}(Ph_{\gamma(n-1)}\cdots P(h_{\gamma(2)}Ph_{\gamma(1)})))$$

at the real number $x$ is the multiplicity of $x$ in the multiset $S(\underline{x})$. For $D$ a subset of $\{1,2,\dots,n\}$, let

$$h_D = \prod_{d:\,d\in D} h_d,$$

where the product is convolution. Then the value of

$$\sum_{\gamma:\,\gamma\in\mathfrak{S}_n}\,\prod_{D:\,D\in\mathrm{cycle}(\gamma)} Ph_D,$$

at $x$ is the multiplicity of $x$ in $T(\underline{x})$. In particular, the Bohnenblust–Spitzer identity is equivalent to the following Baxter operator identity in $\mathcal{M}$:

$$\sum_{\gamma:\,\gamma\in\mathfrak{S}_n} P(h_{\gamma(n)}(Ph_{\gamma(n-1)}\cdots P(h_{\gamma(2)}Ph_{\gamma(1)}))) = \sum_{\gamma:\,\gamma\in\mathfrak{S}_n}\,\prod_{D:\,D\in\mathrm{cycle}(\gamma)} Ph_D.$$

One way to show that this identity holds in $\mathcal{M}$ is to show that it holds in the standard Baxter algebra. To do this, we need to find a precursor of the Bohnenblust–Spitzer identity in the standard algebra.

Let $S$ be the finite set $\{1,2,\dots,n\}$, $X$ be the countably infinite set $\{1,2,3,\dots\}$, and $\mathrm{Fun}(S,X)$ the set of all functions $f: S \to X$. The *coimage* of a function $f$ in $\mathrm{Fun}(S,X)$ is the partition of $S$ whose blocks are those inverse images $f^{-1}(x)$, $x \in X$, which are non-empty. For example, if $S = \{1,2,3,4,5\}$ and $f(1) = f(3) = 11$, $f(2) = f(5) = 3$, $f(4) = 17$, then the coimage of $f$ is the partition with blocks $\{1,3\}, \{2,5\}, \{4\}$.

Let $\pi$ and $\sigma$ be partitions of the set $S$. The partition $\pi$ is *finer* than the partition $\sigma$ if every block of $\pi$ is contained in a block of $\sigma$, or equivalently, $\pi$ is obtained from $\sigma$ by further subdividing the blocks of $\sigma$. The partitions of a set $S$ are partially ordered by reverse refinement, that is,

$$\pi \le \sigma \text{ if } \pi \text{ is finer than } \sigma.$$

Under this partial order, the partitions of $S$ form a lattice $\Pi(S)$ with minimum $\hat{0}$, the partition with one-element blocks $\{a\}, a \in S$.

Let $x_{ij}$, $1 \le i \le n, 1 \le j < \infty$ be countably many variables. For each function $f: S \to X$, we associate the monomial $\mathrm{Mon}(f)$, defined by

$$\mathrm{Mon}(f) = \prod_{i:\,i\in S} x_{i,f(i)}.$$

If $\pi$ is a partition of $S$, let

$$A(\pi) = \sum_{f:\,\mathrm{coimage}(f)=\pi} \mathrm{Mon}(f),$$

$$B(\pi) = \sum_{f:\,\mathrm{coimage}(f)\geq\pi} \mathrm{Mon}(f).$$

The sums $A(\pi)$ and $B(\pi)$ are formal power series in the variables $x_{ij}$. Let $A_k(\pi)$ be the polynomial obtained by setting $x_{ij} = 0$ for all $j \geq k$ and let $\underline{A}(\pi)$ be the sequence $(A_k(\pi))_{k=1}^{\infty}$. Define $B_k(\pi)$ and $\underline{B}(\pi)$ in a similar way.

Our next task is to rewrite $\underline{A}(\pi)$ and $\underline{B}(\pi)$ as expressions in the standard Baxter algebra. We begin with $\underline{A}(\pi)$. For $1 \leq i \leq n$, let $\underline{x}_i$ be the sequence $(x_{i1}, x_{i2}, x_{i3}, \ldots)$. Then

$$P\underline{x}_1 = (0, x_{11}, x_{11} + x_{12}, x_{11} + x_{12} + x_{13}, x_{11} + x_{12} + x_{13} + x_{14}, \ldots),$$

$$\underline{x}_2 P\underline{x}_1 = (0, x_{11}x_{22}, x_{11}x_{23} + x_{12}x_{23}, x_{11}x_{24} + x_{12}x_{24} + x_{13}x_{24}, \ldots),$$

$$P(\underline{x}_2 P\underline{x}_1) = (0, 0, x_{11}x_{22}, x_{11}x_{22} + x_{11}x_{23} + x_{12}x_{23},$$
$$x_{11}x_{22} + x_{11}x_{23} + x_{12}x_{23} + x_{11}x_{24} + x_{12}x_{24} + x_{13}x_{24}, \ldots),$$

and so on. Thus, the $(j+1)$st term in

$$P(\underline{x}_n P(\underline{x}_{n-1} \cdots P(\underline{x}_2 P\underline{x}_1)))$$

is

$$\sum_{i_1,i_2,\ldots,i_n} x_{1i_1} x_{2i_2} \cdots x_{ni_n},$$

where the sum ranges over all strictly increasing length-$n$ sequences $i_1, i_2, \ldots, i_n$ with terms in $\{1, 2, \ldots, j\}$. Hence,

$$\sum_{\gamma:\,\gamma\in\mathfrak{S}_n} \left( \sum_{i_1,i_2,\ldots,i_n} x_{\gamma(1),i_1} x_{\gamma(2),i_2} \cdots x_{\gamma(n),i_n} \right) = \sum_f x_{1,f(1)} x_{2,f(2)} \cdots x_{n,f(n)},$$

where the sum on the right-hand side ranges over all one-to-one functions $f : S \to X$. Since a function $f : S \to X$ has coimage $\hat{0}$ if and only if it is one to one, we conclude that

$$A(\hat{0}) = \sum_{\gamma:\,\gamma\in\mathfrak{S}_n} P(\underline{x}_{\gamma(n)} P(\underline{x}_{\gamma(n-1)} \cdots P(\underline{x}_{\gamma(2)} P\underline{x}_{\gamma(1)})))$$

For $D$ a subset of $\{1, 2, \ldots, n\}$, let

$$\underline{x}(D) = \prod_{d:\,d\in D} \underline{x}_d \quad \text{and} \quad x(D)_j = \prod_{d:\,d\in D} x_{dj},$$

so that $x(D)_j$ is the $j$th term in the sequence $\underline{x}(D)$. Then the $(j+1)$st term of $P\underline{x}(D)$ is

$$x(D)_1 + x(D)_2 + \cdots + x(D)_j.$$

Hence, if $\pi$ is a partition of $S$ into blocks $D_1, D_2, \ldots, D_c$, a monomial in the product $(P\underline{x}(D_1))(P\underline{x}(D_2))\cdots(P\underline{x}(D_c))$ has the form

$$x(D_1)_{i_1} x(D_2)_{i_2} \cdots x(D_c)_{i_c},$$

where $i_1, i_2, \ldots, i_c$ are positive integers which are not necessarily distinct. In other words, a monomial in $(P\underline{x}(D_1))(P\underline{x}(D_2))\cdots(P\underline{x}(D_c))$ is the monomial of a function $f : S \to X$ such that coimage$(f) \geq \pi$; conversely, every monomial of a function $f$ with coimage$(f) \geq \pi$ occurs exactly once in the product. We conclude that

$$\underline{B}(\pi) = (P\underline{x}(D_1))(P\underline{x}(D_2))\cdots(P\underline{x}(D_c))$$

By definition,

$$\underline{B}(\pi) = \sum_{\sigma:\sigma \geq \pi} \underline{A}(\sigma),$$

and hence, by Möbius inversion,

$$\underline{A}(\pi) = \sum_{\sigma:\sigma \geq \pi} \mu(\pi,\sigma)\underline{B}(\sigma).$$

In particular, we have the inversion identity

$$\underline{A}(\hat{0}) = \sum_{\pi:\pi \in \Pi(D)} \mu(\hat{0},\pi)\underline{B}(\pi). \tag{$*$}$$

We can make this identity more explicit using a formula for the Möbius function of partition lattices found by Rota [4]: if $\pi$ is the partition with blocks $D_1, D_2, \ldots, D_c$, then

$$\mu(\hat{0},\pi) = (-1)^{n-c}(|D_1|-1)!(|D_2|-1)!\cdots(|D_c|-1)!.$$

In particular, $|\mu(\hat{0},\pi)|$ equals the number of permutations $\gamma$ such that cycle$(\gamma) = \pi$. Hence, we can rewrite the inversion identity $(*)$ as

$$\sum_{\gamma:\gamma \in \mathfrak{S}_n} (-1)^{c(\gamma)} \underline{B}(\pi),$$

where $c(\gamma)$ is the number of cycles in the cycle decomposition of $\gamma$. In terms of the Baxter operator $P$ in the standard algebra $(S_n, P)$,

$$\sum_{\gamma:\gamma \in \mathfrak{S}_n} P(\underline{x}_{\gamma(n)}P(\underline{x}_{\gamma(n-1)}\cdots P(\underline{x}_{\gamma(2)}P\underline{x}_{\gamma(1)})))$$
$$= \sum_{\gamma:\gamma \in \mathfrak{S}_n} (-1)^{c(\gamma)} \prod_{D:D \in \text{cycle}(\gamma)} P\underline{x}(D).$$

This is almost the identity we required, except for signs. However, recall that the Baxter operator $P$ in $\mathcal{M}$ has parameter 1. Thus, $(\mathcal{M}, -P)$ is a Baxter algebra with parameter $-1$.

Hence, by Rota's theorem, we have

$$\sum_{\gamma:\gamma\in\mathfrak{S}_n} (-1)^n P(\underline{x}_{\gamma(n)} P(\underline{x}_{\gamma(n-1)} \cdots P(\underline{x}_{\gamma(2)} P\underline{x}_{\gamma(1)})))$$

$$= \sum_{\gamma:\gamma\in\mathfrak{S}_n} (-1)^{n-c(\gamma)} \prod_{D:D\in\mathrm{cycle}(\gamma)} -P\underline{x}(D),$$

exactly the identity we required on cancelling out the signs.

## 7.6 Abstract Random Variables

The second part of this essay is about a construction of Rota which he described in a course called "Foundations of Mathematics: Logic and Probability", given by Rota at the Massachusetts Institute of Technology in 1977. This construction formalizes the practice that when working with random variables, it is the events they define, not the functions themselves, which matter.

Let $\mathcal{B}$ be a Boolean $\sigma$-algebra, with partial order $\subseteq$, meet $\wedge$, join $\vee$, and complementation $\cdot^c$. Define an *abstract random variable* $X$ to be a function $\mathbb{R} \to \mathcal{B}, t \mapsto [X \leq t]$, satisfying three properties:

RV1. If $s \leq t$, then $[X \leq s] \subseteq [X \leq t]$.

RV2. If $(s_n)_{n=1}^{\infty}$ is a sequence such that $\lim_{n\to\infty} = \infty$, then $\bigvee_{n=1}^{\infty}[X \leq s_n] = \hat{1}$; similarly, if $\lim_{n\to\infty} s_n = -\infty$, then $\bigwedge_{n=1}^{\infty}[X \leq s_n] = \hat{0}$.

RV3. If $(s_n)_{n=1}^{\infty}$ is a sequence such that $s_n > t$ and $\lim_{n\to\infty} s_n = t$, then $[X \leq t] = \bigwedge_{n=1}^{\infty}[X \leq s_n]$.

The element $[X \leq t]$ is thought of as the event defined by the condition $X \leq t$. The set of all (abstract) random variables can be partially ordered by

$$X \leq Y \text{ whenever } [X \leq t] \supseteq [Y \leq t] \text{ for all real numbers } t.$$

We can define an algebra structure on the set of random variables. Let $X$ and $Y$ be random variables. We define the sum $X + Y$ by

$$[X + Y \leq t] = \bigvee_{s: s\in\mathbb{Q}} ([X \leq s] \wedge [y \leq t - s]),$$

where $s$ ranges over all rational numbers $\mathbb{Q}$ (or any countable dense subset of $\mathbb{R}$). It is not hard to check that $X + Y$ is a random variable. The zero random variable $\mathbf{0}$ is the random variable defined by

$$[\mathbf{0} \leq t] = \begin{cases} \hat{1} & \text{if } t \geq 0, \\ \hat{0} & \text{if } t < 0. \end{cases}$$

Note that the theory thus far does not use complementation and uses only the structure of $\mathcal{B}$ as a distributive lattice.

Defining multiplication is a little more complicated. We begin with scalar multiplication by a positive scalar $\alpha$. If $X$ is a random variable, then $\alpha X$ is the random variable defined by

$$[\alpha X \le t] = [X \le t/\alpha].$$

Based on the fact that $-X \le t$ if and only if $X \ge -t$, that is, the negation of $X \le -t$ holds, we define $-X$ by

$$[-X \le t] = [X \le -t]^c.$$

We can now define multiplication by negative scalars by: $\beta X = -((-\beta)X)$ if $\beta$ is a negative real number. If $X$ is a random variable, define

$$X^+ = \begin{cases} [X \le t] & \text{if } t \ge 0, \\ \hat{0} & \text{if } t < 0 \end{cases}$$

and $X^- = (-X)^+$. Both $X^+$ and $X^-$ are *non-negative*, in the sense that $X^+ \ge \mathbf{0}$ and $X^- \ge \mathbf{0}$. This gives a (unique) decomposition $X = X^+ - X^-$ of a random variable $X$ into the difference of two non-negative random variable.

We can now define the product of two random variables. If $X$ and $Y$ are non-negative random variables, then the product $XY$ is defined by

$$[XY \le t] = \bigvee_{s:\, s \in \mathbb{Q} \text{ and } s > 0} [X \le s] \wedge [y \le t/s].$$

In general, decompose $X$ and $Y$ and define

$$XY = (X^+ - X^-)(Y^+ - Y^-) = X^+ Y^+ - X^+ Y^- - X^- Y^+ + X^- Y^-.$$

We can go further and define functional composition, expectations, and other tools required to do probability.

Replacing the Boolean $\sigma$-algebra $\mathcal{B}$ by, say, the lattice of closed subspaces of a Hilbert space, we obtain "quantum" random variables. Whether this gives a way to do quantum probability remains untested.

This reformulation of random variables does not, by itself, resolve basic problems with random variables. For example, does it offer a way to define probability densities abstractly (see Problem 2 in [7])? However, it does show that one need not use measurable functions as the basic building block of probability. In particular, it shows that the Kolmogorov axiomatization of probability using measurable functions is not necessarily the only reasonable foundation. However, most probabilists are thoroughly indoctrinated in the Kolmogorov axiomatization and regard alternative or "deviant" foundations as amusing distractions at best. This is a reason, perhaps not the main reason, for Rota's paper *Twelve Problems* [7] to have little impact so far. To Rota, the difficult part of mathematical research is not in solving problems, but finding the right problems to solve. Many of the right problems lie in the gap between how one does the subject and the axioms one professes. These are the problems "which no one likes to bring up".

We end, as we began, with the paper *Foundations I*. Only after almost completing this essay did I see the wisdom in the use of the plural "Foundations" in the title. Instead of prescribing the foundation, *Foundations I* invites its readers to think seriously about their own foundations for combinatorics. This has lead to an openness in algebraic combinatorics, far from the "personality cults" which held over some areas of combinatorics and mathematics. Many philosophers of mathematics have also arrived (independently) at this standpoint: rather than continuing the fundamentalist project of putting mathematics into the straight-jacket of one foundation, usually the philosophically and mathematically bankrupt framework given by logic and set theory, they have embraced with enthusiasm the many foundations which underlie the theory and more significantly, the practice of mathematics.

# References

1. Birkhoff, G. (1967), *Lattice Theory*, 3rd ed., Providence, American Mathematical Society.
2. Guo, L., Keigher, W. (2008), *On Differential Rota-Baxter Algebras*, in "J. Pure Appl. Algebra", 212, pp. 522-40.
3. Kung, J. P. S., Rota, G.-C., Yan, C. H. (2009), *Combinatorics. The Rota Way*, Cambridge, Cambridge University Press.
4. Rota, G.-C. (1964), *On the Foundations of Combinatorial Theory. I. Theory of Möbius Functions*, "Z. Wahrscheinlichkeitstheorie u. Verw. Gebeite", 2, pp. 340-68.
5. Rota, G.-C. (1969), *Baxter Algebras and Combinatorial Identities. I, II*, in "Bull. Amer. Math. Soc.", 75, pp. 325-34.
6. Rota, G.-C. (1994), *Baxter Operators. An Introduction*, in J. P. S. Kung (ed.), *Gian-Carlo Rota on Combinatorics*, Boston, Basel, Berlin, Birkhäuser, pp. 301-20.
7. Rota, G.-C. (2001), *Twelve Problems in Probability No One Likes to Bring Up*, in H. Crapo, D. Senato (eds.), *Algebraic Combinatorics and Computer Science, A Tribute to Gian-Carlo Rota*, Milan, Springer Italia, pp. 57-96.
8. Rota, G.-C., Smith, D. A. (1972), *Fluctuation Theory and Baxter Algebras*, in *Symposia Mathematica, Vol. IX (Convegno di Calcolo delle Probabilità, INDAM, Rome, 1971)*, London, Academic Press, pp. 179-201.
9. Rota, G.-C., Smith, D. A. (1977), *Enumeration Under Group Action*, in "Ann. Scuola Normale Superiore Pisa", 4, pp. 637-46.
10. Spitzer, F. (1956), *A Combinatorial Lemma with Applications to Probability Theory*, in "Trans. Amer. Math. Soc.", 82, pp. 323-39.

# Chapter 8

# On the Euler Characteristic of Finite Distributive Lattices

## Contributed Chapter

Emanuele Munarini

**Abstract** Some relationships between the structure of a finite distributive lattice and the algebraic or combinatorial properties of its Euler characteristic $\chi$ are investigated. Furthermore, the combinatorial meaning of $\chi$ is obtained in the particular cases of subhypergraph lattices, of dual Gödel lattices and of tree-map lattices.

## 8.1 Introduction

The Euler characteristic appears in several branches of classical mathematics, such as algebraic topology, differential topology, differential geometry and integral geometry. But the efforts made by several mathematicians in these last fifty years show that the Euler characteristic can be seen as a fundamental dimensionless quantity which can be defined in much more general contexts.

One of the first important results in this direction was obtained by G.-C. Rota in his program of bringing to light the fundamental relationships between geometry and combinatorics. Starting from the works of V. Klee [18] and H. Hadwiger [12, 13, 14, 15] on the Euler characteristic of polytopes, and from the works of G. T. Sallee [29] and E. H. Shepard [32] on valuations on convex polytopes, Rota defined the Euler characteristic of a finite distributive lattice $D$ as the unique valuation assuming value $0$ on the minimum and value $1$ on every join-irreducible element [26, 27, 28] (see also [20]).

Many other important results have been obtained following other approaches. For instance, in the attempt of categorizing $\mathbb{Z}$ in order to have a conceptual definition of negative sets, S. Schanuel defined the Euler characteristic as a generalization of cardinality (which allows negative integer values) by means of a universal property in a suitable category of polyhedral sets [31] (see also [30] and [24, 25]). Furthermore, starting from the idea that the Euler characteristic is a fundamental quantity

Emanuele Munarini
Politecnico di Milano, Italy

E. Damiani et al. (eds.), *From Combinatorics to Philosophy,*
DOI 10.1007/978-0-387-88753-1_8, © Springer Science+Business Media, LLC 2009

that deserves to be better understood, T. Leinster generalized Rota's and Schanuel's approaches defining the Euler characteristic for certain finite categories (in which the Rota-Möbius inversion holds) [22] (see also [5]).

In this paper, we explore a little more the relationships between the structure of a finite distributive lattice and the algebraic or combinatorial properties of its Euler characteristic $\chi$. Specifically, we characterize all finite distributive lattices in which $\chi$ is almost constant, constant along each level, or order-preserving. Moreover, we characterize all finite distributive lattices in which $\chi$ can assume only the values 0, 1 and 2, generalizing the concept of planar lattice. Finally, we obtain some combinatorial expressions of $\chi$ for the lattices of all subhypergraph of a finite hypergraph and for the dual of Gödel lattices, and we compute $\chi$ explicitly for the lattices of tree maps, i.e. order-preserving maps from a tree to a distributive lattice.

## 8.2 Posets and Distributive Lattices

In this section some basic definitions and properties are recalled. For further background on partially ordered sets and lattices see, for instance, [2, 6, 9, 10, 33].

In a poset $P$, an element $q$ *covers* an element $p$ when $p \leq u \leq q$ implies $p = u$ or $q = y$. An *atom* is an element covering the minimum element (when it exists). A finite poset $P$ is *ranked* when it admits a rank function, i.e. a map $r : P \to \mathbb{N}$ such that $r(p) = 0$ on every minimal element $p$ and $r(q) = r(p) + 1$ whenever $q$ covers $p$. The *height* of $P$ is its maximum rank. The *minimum* and the *maximum* element (when they exist) are denoted by $\widehat{0}$ and $\widehat{1}$, respectively.

Let $D$ be a finite distributive lattice. An element $x \in D$ is *join-irreducible* when $x \neq \widehat{0}$ and $x = u \vee v$ implies $x = u$ or $x = v$. The poset of all join-irreducible elements is the *spectrum* of $D$ and is denoted by $\mathbf{Spec}(D)$. A representation $x = x_1 \vee \cdots \vee x_k$ is *irredundant* when the elements $x_1, \ldots, x_k$ form an antichain.

An *ideal* of a poset $P$ is a subset $I$ such that $x \in I$ and $u \leq x$ implies $u \in I$. A *principal ideal* of $P$ is an ideal with exactly one maximal element, i.e. of the form $\downarrow p = \{u \in P : u \leq p\}$ for some $p \in P$. Similarly, a *filter* of $P$ is a subset $F \subseteq P$ such that $x \in F$ and $u \geq x$ implies $u \in F$. A *principal filter* of $P$ is a filter with exactly one minimal element, i.e. of the form $\uparrow p = \{u \in P : u \geq p\}$ for some $p \in P$. The set $\mathcal{J}(P)$ of all ideals of $P$, ordered by inclusion, is a distributive lattice. In particular, $\mathcal{J}(P)$ is ranked and $r(I) = |I|$ for every ideal $I$. Conversely, by *Birkhoff's representation theorem* [6, p. 139] [9, p. 61] (see also [3]), every finite distributive lattice $D$ is isomorphic to the lattice $\mathcal{J}(P)$, where $P = \mathbf{Spec}(D)$. The ideal $I$ corresponding to the element $x \in D$ is formed by all join-irreducibles $u \in D$ such that $u \leq x$.

For any ideal $I$ of $P$, let $\nabla I$ be the set of all ideals covered by $I$ and let $M(I)$ be the set of all maximal elements of $I$. Since the ideals covered by $I$ are obtained from $I$ by removing exactly one maximal element, there exists a simple bijection between $\nabla I$ and $M(I)$.

**Theorem 8.1.** *Let* $D$ *be a finite distributive lattice. Then every element* $x \in D$ *admits a unique irredundant representation* $x = x_1 \vee \cdots \vee x_k$ *where* $x_1, \ldots, x_k$ *are join-irreducible elements. Moreover, the size of this antichain is equal to the number of elements covered by* $x$, *i.e.* $k = |\nabla x|$.

*Proof.* Let $D = \mathcal{J}(P)$ and let $I$ be the ideal of $P$ corresponding to $x$. Then the irredundant representation of $x$ as a join of join-irreducible elements corresponds to the decomposition of $I$ as the union of principal ideals generated by its maximal elements.

The *Cartesian product* (or *direct product*) [9, 33] of two posets $P_1$ and $P_2$ is the poset $P_1 \times P_2$ defined as the set of all pairs $(p_1, p_2)$ with $p_1 \in P_1$ and $p_2 \in P_2$ where $(p_1, q_1) \leq (p_2, q_2)$ whenever $p_1 \leq p_2$ and $q_1 \leq q_2$. The *ordinal* (or *linear*) *sum* [6, 8, 33] of $P_1$ and $P_2$ is the poset $P_1 \oplus P_2$ on the sum (disjoint union) $P_1 + P_2$ where $p \leq q$ whenever $p, q \in P_1$ and $p \leq q$, or $p, q \in P_2$ and $p \leq q$, or $p \in P_1$ and $q \in P_2$. The Hasse diagram of $P_1 \oplus P_2$ is obtained by drawing $P_2$ above $P_1$ and by connecting every maximal element of $P_1$ with every minimal element of $P_2$. If $P_1$ has a maximum element $\widehat{1}_{P_1}$ and $P_2$ has a minimum element $\widehat{0}_{P_2}$, then the *vertical sum* [8] of $P_1$ and $P_2$ is the poset $P_1 \odot P_2$ defined by identifying the elements $\widehat{1}_{P_1}$ and $\widehat{0}_{P_2}$ in the ordinal sum $P_1 \oplus P_2$. Finally, the *dual* $P^*$ of a poset $P$ is the set $P$ ordered by setting $p \leq^* q$ whenever $p \geq q$ in $P$.

For every finite poset $P_1$ and $P_2$, and for every finite distributive lattice $D_1$ and $D_2$, we have the following dual identities:

$$\begin{aligned} \mathcal{J}(P_1 + P_2) &= \mathcal{J}(P_1) \times \mathcal{J}(P_2) & \mathbf{Spec}(D_1 \times D_2) &= \mathbf{Spec}(D_1) + \mathbf{Spec}(D_2) \\ \mathcal{J}(P_1 \oplus P_2) &= \mathcal{J}(P_1) \odot \mathcal{J}(P_2) & \mathbf{Spec}(D_1 \odot D_2) &= \mathbf{Spec}(D_1) \oplus \mathbf{Spec}(D_2) \\ \mathcal{J}(P_1 \oplus \mathbf{1} \oplus P_2) &= \mathcal{J}(P_1) \oplus \mathcal{J}(P_2) & \mathbf{Spec}(D_1 \oplus D_2) &= \mathbf{Spec}(D_1) \oplus \mathbf{1} \oplus \mathbf{Spec}(D_2). \end{aligned}$$
$$(8.1)$$

Moreover, for every finite poset $P$, we have $\mathcal{J}(P^*) = \mathcal{J}(P)^*$ and $\mathbf{Spec}(D^*) \simeq \mathbf{Spec}(D)^*$.

A map $f : P_1 \to P_2$ between two posets is *order-preserving* when $p_1 \leq q_1$ implies $f(p_1) \leq f(q_1)$, for every $p_1, q_1 \in P_1$. A map $f : D_1 \to D_2$ between two lattices is a *lattice morphism* when $f(x \wedge y) = f(x) \wedge f(y)$ and $f(x \vee y) = f(x) \vee f(y)$ for every $x, y \in D_1$. Every lattice morphism is order-preserving.

For every $n \in \mathbb{N}$, $[n]$ denotes the set $\{1, 2, \ldots, n\}$, $\mathbf{n}$ denotes the *antichain* of size $n$, $C_n$ denotes the *chain* of length $n$, and $\mathcal{B}_n$ denotes the *Boolean lattice* of order $n$. Moreover, $\mathcal{P}(X)$ denotes the set of all subsets of $X$.

## 8.3 Euler Characteristic

A (real) *valuation* on a distributive lattice $D$ is a function $v : D \to \mathbb{R}$ satisfying the identity $v(x \vee y) + v(x \wedge y) = v(x) + v(y)$ for every $x, y \in D$. When $D$ is finite, any valuation is uniquely determined by the values it takes on the set of join-irreducibles [27]. Since these values can be arbitrarily assigned, the *Euler characteristic* of $D$

can be defined as the unique valuation $\chi$ such that $\chi(\widehat{0}) = 0$ and $\chi(x) = 1$ for each join-irreducible $x$ [26, 27, 28] (see also [2, p. 182] and [3, 8, 19]). The Euler characteristic, as every valuation, satisfies the following form of the *Principle of Inclusion-Exclusion*

$$\chi(x_1 \vee x_2 \vee \cdots \vee x_n) = \sum_{\substack{S \subseteq [k] \\ S \neq \emptyset}} (-1)^{|S|-1} \chi\left(\bigwedge_{i \in S} x_i\right). \tag{8.2}$$

Moreover, it can be expressed in terms of the *Möbius function* [3, 26]. The *reduced Möbius function* of a poset $P$ with minimum $\widehat{0}$ is the unique function such that

$$\mu(\widehat{0}) = 1 \quad \text{and} \quad \sum_{u \leq x} \mu(x) = 0 \quad \text{for every } x \in P, \ x \neq \widehat{0}. \tag{8.3}$$

In particular, $\mu(x) = -1$ for every atom of $P$.

**Theorem 8.2 (Rota,[27]).** *Let $D$ be a finite distributive lattice with spectrum $P$. The Euler characteristic of an element $x \in D$ (or, of an ideal $I$ of $P$) is*

$$\chi(x) = -\sum_{p < x} \mu(p) \qquad \left( or, \quad \chi(I) = -\sum_{p \in I} \mu(p) \right)$$

*where the sum runs over all join-irreducibles $p$ below $x$ and $\mu$ is the reduced Möbius function of the* augmented spectrum $\widehat{P} = \{\widehat{0}\} \oplus P$ *obtained by adding to $P$ a new minimum element.*

Another important feature of the Euler characteristic is invariance. Since the value of $\chi(x)$ depends only on the structure of the principal ideal $\downarrow x$, we have $\chi(x) = \chi(y)$ whenever $\downarrow x \simeq \downarrow y$. Moreover, $\chi$ is invariant under the action of the group of automorphisms of $D$, that is $\chi(\varphi(x)) = \chi(x)$ for every $x \in D$ and for every $\varphi \in \mathbf{Aut}(D)$. But we also have the following useful

**Theorem 8.3.** *Let $D_1$ and $D_2$ be two finite distributive lattices and let $f : D_1 \to D_2$ be an injective lattice morphism such that $f(\widehat{0})$ has Euler characteristic $0$ and $f(x)$ has Euler characteristic $1$ for every join-irreducible $x \in D_1$. Then $f$ also preserves the Euler characteristic, i.e. $\chi(f(x)) = \chi(x)$ for every $x \in D_1$.*

*Proof.* By hypothesis, $\chi(f(\widehat{0})) = 0 = \chi(\widehat{0})$ and $\chi(f(x)) = 1 = \chi(x)$ for every atom of $D_1$. Proceeding by induction on the the rank of $D_1$, let $x \in D_1$ with $r(x) \geq 2$. If $x$ is join-irreducible, then $\chi(f(x)) = 1 = \chi(x)$ by hypothesis. If $x$ is not join-irreducible, then it covers at least two elements $x_1$ and $x_2$ such that $x = x_1 \vee x_2$. Hence, since $f$ is an injective lattice morphism, $\chi(f(x)) = \chi(f(x_1) \vee f(x_2)) = \chi(f(x_1)) + \chi(f(x_2)) - \chi(f(x_1 \wedge x_2))$. Since $x_1$, $x_2$ and $x_1 \wedge x_2$ have rank smaller then $r(x)$, by induction hypothesis, there follows that $\chi(f(x)) = \chi(x_1) + \chi(x_2) - \chi(x_1 \wedge x_2) = \chi(x_1 \vee x_2) = \chi(x)$.

As an immediate consequence of Theorem 8.3, there follows

**Theorem 8.4.** *Let $D_1$ and $D_2$ be two finite distributive lattices and let $f : D_1 \to D_2$ be an injective lattice morphism preserving the minimum element $\widehat{0}$ and all join-irreducible elements. Then $f$ also preserves the characteristic, i.e. $\chi(f(x)) = \chi(x)$ for every $x \in D_1$.*

**Proposition 8.1.** *Let $D = D_1 \times \cdots \times D_k$ be the Cartesian product of $k$ finite distributive lattices. Then $\chi(x_1, x_2, \ldots, x_k) = \chi(x_1) + \chi(x_2) + \cdots + \chi(x_k)$, for every $(x_1, x_2, \ldots, x_k) \in D$.*

*Proof.* This property will be proved in two different ways.

1. FIRST PROOF. An element $(x_1, x_2, \ldots, x_k) \in D$ is join-irreducible if and only if there exists an index $i$ such that $x_i$ is join-irreducible in $D_i$ and $x_j = \widehat{0}$ for every $j \neq i$. Hence the map $\Delta_i : D_i \to D$, defined by $\Delta_i(x) = (\widehat{0}, \ldots, \widehat{0}, x, \widehat{0}, \ldots, \widehat{0})$ where $x$ appears in position $i$, is an injective lattice morphism preserving the minimum and the join-irreducibles. So, by Theorem 8.4, $\Delta_i$ also preserves the Euler characteristic, i.e. $\chi(\widehat{0}, \ldots, \widehat{0}, x, \widehat{0}, \ldots, \widehat{0}) = \chi(x)$. Now, any element $\xi = (x_1, x_2, \ldots, x_k) \in D$ can be written as $\xi = \xi_1 \vee \xi_2 \vee \cdots \vee \xi_k$, where $\xi_i = \Delta_i(x_i)$ for $i = 1, 2, \ldots, k$. Since $\chi(\xi_i) = \chi(x_i)$ and $\xi_i \wedge \xi_j = \widehat{0}$ whenever $i \neq j$, from (8.2) there follows that

$$\chi(\xi) = \chi(\xi_1 \vee \xi_2 \vee \cdots \vee \xi_k) = \sum_{S \subseteq [k], \, S \neq \varnothing} (-1)^{|S|-1} \chi\left(\bigwedge_{i \in S} \xi_i\right) = \sum_{i=1}^{n} \chi(\xi_i) = \sum_{i=1}^{n} \chi(x_i).$$

2. SECOND PROOF. Let $D = \mathcal{J}(P)$ and $D_i = \mathcal{J}(P_i)$ for every $i$. Then $P = P_1 + \cdots + P_k$ and every its ideal is of the form $I = I_1 + \cdots + I_k$ where each $I_i$ is an ideal of $P_i$. Since the value of $\mu(p)$ is determined only by the ideal $\downarrow p$ of $\widehat{P}$, there follows that $\mu_{\widehat{P}}(p) = \mu_{\widehat{P_i}}(p)$ whenever $p \in P_i$. Hence, applying Theorem 8.2, there follows $\chi(I) = \chi(I_1) + \cdots + \chi(I_k)$.

The *coproduct* of two finite distributive lattices $D_1 = \mathcal{J}(P_1)$ and $D_2 = \mathcal{J}(P_2)$ is the distributive lattice $D_1 \otimes D_2 = \mathcal{J}(P_1 \times P_2)$. An element $x$ in $D_1 \otimes D_2$ is *decomposable* when it corresponds to an ideal $I_1 \times I_2$ with $I_1 \in \mathcal{J}(P_1)$ and $I_2 \in \mathcal{J}(P_2)$. In this case, we write $x = x_1 \otimes x_2$, where $x_1 \in D_1$ is the element corresponding to $I_1$ and $x_2 \in D_2$ is the element corresponding to $I_2$.

**Proposition 8.2.** *Let $P$ and $Q$ be two finite posets. Then $\mu_{\widehat{P \times Q}}(x, y) = -\mu_{\widehat{P}}(x)\mu_{\widehat{Q}}(y)$ for every $(x, y) \in P \times Q$.*

*Proof.* The function $\overline{\mu}$, defined by $\overline{\mu}(\widehat{0}, \widehat{0}) = 1$ and $\overline{\mu}(x, y) = -\mu_{\widehat{P}}(x)\mu_{\widehat{Q}}(y)$ for every $(x, y) \in P \times Q$, satisfies the defining property (8.3) of the Möbius function. The first condition is satisfied by definition, and for every $(x, y) \in P \times Q$ we have

$$\overline{\mu}(\widehat{0}, \widehat{0}) + \sum_{\substack{(u,v) \leq (x,y)}} \overline{\mu}(u, v) = 1 - \sum_{\substack{u \leq x \\ v \leq y}} \mu_{\widehat{P}}(x)\mu_{\widehat{Q}}(y) = 1 - \sum_{u \leq x} \mu_{\widehat{P}}(x) \sum_{v \leq y} \mu_{\widehat{Q}}(y) = 1 - 1 = 0.$$

For the uniqueness of the Möbius function, there follows that $\overline{\mu} = \mu$.

**Theorem 8.5.** *Let $D_1$ and $D_2$ be two finite distributive lattices. Then, for every decomposable element $x_1 \otimes x_2$ of $D_1 \otimes D_2$, we have $\chi(x_1 \otimes x_2) = \chi(x_1)\chi(x_2)$.*

*Proof.* Let $D_1 = \mathcal{J}(P)$ and $D_2 = \mathcal{J}(Q)$. By Theorem 8.2 and Proposition 8.2, there follows that

$$\chi(I \times J) = - \sum_{(p,q) \in P \times Q} \mu_{\widehat{P \times Q}}(p,q) = \sum_{p \in P, q \in Q} \mu_{\widehat{P}}(p)\mu_{\widehat{Q}}(q)$$

$$= \sum_{p \in P} \mu_{\widehat{P}}(p) \sum_{q \in Q} \mu_{\widehat{Q}}(q) = \chi(I)\chi(J)$$

for every ideal $I \in \mathcal{J}(P)$ and $J \in \mathcal{J}(Q)$.

Let $D_1$ and $D_2$ be two distributive lattices. Let $F$ be a principal filter of $D_1$ and let $I$ be a principal ideal of $D_2$. If $F$ and $I$ are isomorphic as lattices, i.e. if there exists a lattice isomorphism $\varphi : F \to I$, then it is possible to define a new distributive lattice $(D_1, F) \uparrow (I, D_2)$ assuming as ground set the disjoint union of the ground sets of $D_1$ and $D_2$, and by setting $x \le y$ whenever $x, y \in D_1$ and $x \le y$ in $D_1$, or $x, y \in D_2$ and $x \le y$ in $D_2$, or $x \in F$, $y \in I$ and $\varphi(x) \le y$ in $I$. If $D_2 = I \simeq F$, then $(D_1, F) \uparrow (I, D_2)$ will be simply denoted by $D_1 \uparrow F$ and $D_2$ will be called the elevated copy of $F$. The importance of this operation derives from the fact that every finite distributive lattice can be generated by a finite number of elevations of principal filters [8].

**Proposition 8.3.** *In the lattice $D = (D_1, F) \uparrow (I, D_2)$, the characteristic of an element $x' \in I$ is $\chi(x') = \chi(x) - \chi(z) + 1$, where $x = \varphi^{-1}(x')$ and $z$ is the minimum element of $F$. In particular, $\chi(x') = \chi(x)$ whenever $\chi(z) = 1$.*

*Proof.* The element $z' = \varphi(z)$ is the minimum of $I$ and is join-irreducible in $D$. Then $x' = x \vee z'$ and $x \wedge z' = z$. Hence $\chi(x') = \chi(x) + \chi(z') - \chi(z) = \chi(x) + 1 - \chi(z)$.

**Theorem 8.6.** *The Euler characteristic of a finite distributive lattice $D$ coincides with the rank function if and only if $D$ is a Boolean lattice.*

*Proof.* If $\chi = r$, then $r(x) = \chi(x) = 1$ for every join-irreducible element $x$. This implies that all join-irreducibles are atoms, i.e. that $D$ is a Boolean lattice. Viceversa, if $D = \mathcal{P}(X)$ then $\chi(I) = |I| = r(I)$ for every $I \subseteq X$.

Let $\mathbf{Im}(\chi) = \{\chi(x) : x \in D\} \subseteq \mathbb{Z}$ be the *image* of the Euler characteristic $\chi : D \to \mathbb{Z}$.

**Theorem 8.7.** *If $D$ is a finite distributive lattice with $k$ atoms, then $\{0, 1, \ldots, k\} \subseteq \mathbf{Im}(\chi)$.*

*Proof.* Let $a_1, a_2, \ldots, a_k$ be the atoms of $D$, and let $z = a_1 \vee a_2 \vee \cdots \vee a_k$ the socle of $D$. Then the interval $[\widehat{0}, z]$ is isomorphic to the Boolean lattice $\mathcal{B}_k$ and $\chi(x) = r(x)$ for each of its element.

## 8.4 Uniform Distributions of the Characteristic

### 8.4.1 $\chi$-uniform Lattices

A finite distributive lattice $D$ is $\chi$-*uniform* when the Euler characteristic is constant on every non-zero element, that is when $\mathbf{Im}(\chi) \subseteq \{0,1\}$. Finite $\chi$-uniform distributive lattices can be characterized in a very simple way.

**Theorem 8.8.** *A finite distributive lattice $D$ is $\chi$-uniform if and only if it has at most one atom (or, equivalently, if and only if its spectrum $P$ has at most one minimal element).*

*Proof.* If $D$ is $\chi$-uniform then, by Theorem 8.7, it has at most one atom. Viceversa, if $D$ has no atoms then $D = \{\widehat{0}\}$ is $\chi$-uniform. If $D$ has exactly one atom, then $\chi(\widehat{0}) = 0$ and $\chi(x) = 1$ whenever $x$ is the atom or an element covering the atom. Now, let $x \in D$ with $r(x) \geq 3$. If $x$ is not join-irreducible, then it covers at least two elements $x_1$ and $x_2$ such that $x = x_1 \vee x_2$ with $r(x_1), r(x_2), r(x_1 \wedge x_2) < r(x)$. Hence, by induction on the rank of $D$, there follows that $\chi(x) = 1 + 1 - 1 = 1$. Finally, $D = \mathcal{J}(P)$ has at most one atom if and only if $P$ has at most one minimal element.

    A *valuation* $v$ on a distributive lattice $D$ is *multiplicative* [3] when $v(x \wedge y) = v(x)v(y)$ for every $x, y \in D$.

**Theorem 8.9.** *A finite distributive lattice $D$ is $\chi$-uniform if and only if the Euler characteristic is multiplicative.*

*Proof.* If $D$ is $\chi$-uniform there follows at once that $\chi$ is multiplicative. Viceversa, if $\chi$ is multiplicative, then $D$ cannot have more then one atom. Indeed, if $a_1$ and $a_2$ are two atoms, then $0 = \chi(\widehat{0}) = \chi(a_1 \wedge a_2) = \chi(a_1)\chi(a_2) = 1$, which is absurd.

    Let $\chi_k(D)$ be the number of elements $x \in D$ such that $\chi(x) = k$. Let $e_k(x_1,\ldots,x_n)$ be the *elementary symmetric polynomial* of order $k$. ¿From $\chi$-uniformity and Proposition 8.1, we have

**Proposition 8.4.** *Let $D = D_1 \times \cdots \times D_n$ be the Cartesian product of $n$ $\chi$-uniform distributive lattices. Then the characteristic of an element $(x_1,\ldots,x_n) \in D$ is equal to the number of components $x_i \neq \widehat{0}$. Moreover, $\chi_k(D) = e_k(d_1 - 1,\ldots,d_n - 1)$, where $d_i = |D_i|$ for every $i$.*

*Remark 8.1.* Proposition 8.4 generalizes the property asserting that in the lattice $\mathcal{D}(n)$ of all divisors of $n$ the Euler characteristic $\chi(d)$ is equal to the number of all prime divisors of $d$ [27].

**Proposition 8.5.** *Let $D_1$ and $D_2$ be two finite distributive lattices. Then, the Euler characteristic of an element $x \in D_1 \oplus D_2$ is $\chi(x) = \chi_{D_1}(x)$ if $x \in D_1$ and $\chi(x) = 1$ if $x \in D_2$.*

*Proof.* Since in the lattice $D_1 \oplus D_2$ all elements in $D_2$ are greater than all elements in $D_1$, there follows that $\chi(x) = \chi_{D_1}(x)$ for every $x \in D_1$. Moreover, since in $D_1 \oplus D_2$ the minimum of $D_2$ is join-irreducible, the distribution of the characteristic on $D_2$ is as in the lattice $\mathbf{1} \oplus D_2$ with just one atom. Hence, by Theorem 8.8, all elements in $D_2$ have characteristic 1.

### 8.4.2 Order-preserving Euler Characteristic

The Euler characteristic of a finite distributive lattice $D$ defines a map $\chi : D \to \mathbb{Z}$. When $D$ is a Boolean lattice and $\mathbb{Z}$ is regarded as a chain under its natural order, this map is order-preserving (by Theorem 8.6). In general, however, it is not order-preserving.

**Proposition 8.6.** *Let $D$ be a finite distributive lattice whose Euler characteristic is an order-preserving map. Then $\chi(x) \geq 0$ for every $x \in D$ and $\chi(x) = 0$ if and only if $x = \widehat{0}$.*

*Proof.* Since $x \geq \widehat{0}$ for every $x \in D$, there follows that $\chi(x) \geq 0$. Moreover, since every element $x \neq \widehat{0}$ is always greater than or equal to an atom, there follows at once that $\chi(x) \geq 1$.

The finite distributive lattices for which $\chi$ is an order-preserving map can be completely characterized as follows.

**Theorem 8.10.** *A finite distributive lattice $D$ has an order-preserving Euler characteristic if and only if it is a Cartesian product of $\chi$-uniform distributive lattices.*

*Proof.* Let $D$ be a Cartesian product of $\chi$-uniform distributive lattices, and let $\xi_1, \xi_2 \in D$ with $\xi_1 \leq \xi_2$. Then the number of zero components of $\xi_1$ is greater then or equal to the number of zero components of $\xi_2$. So, by Proposition 8.4, $\chi(\xi_1) \geq \chi(\xi_2)$.

Viceversa, let $\chi$ be order-preserving. A join-irreducible element $x \in D$ cannot have more than one atom below it. Indeed, if $a_1$ and $a_2$ are two distinct atoms such that $a_1, a_2 \leq x$, then $a_1 \vee a_2 \leq x$ and hence $\chi(a_1 \vee a_2) \leq \chi(x)$. Since the two atoms are two disjoint join-irreducible elements, there follows that $2 \leq 1$, which is absurd. This implies that the spectrum of $D$ decomposes in the disjoint union of posets with only one minimal element, i.e. that $D$ decomposes in a Cartesian product of $\chi$-uniform lattices.

*Stone algebras* are distributive lattices equipped with a pseudocomplementation satisfying the Stone identity [9, p. 112]. A *Stone lattice* is the underlying lattice of a Stone algebra. Since finite Stone lattices are characterized as finite distributive lattices given by Cartesian product of finite distributive lattices with only one atom (dense lattices) [9, p. 113], we obtain

**Theorem 8.11.** *A finite distributive lattice $D$ has an order-preserving Euler characteristic if and only if $D$ is a Stone lattice.*

### 8.4.3 Rank $\chi$-uniform Lattices

A finite distributive lattice $D$ is *rank $\chi$-uniform* if the Euler characteristic is constant along each level. For instance, $\chi$-uniform lattices and Boolean lattices are rank $\chi$-uniform. See Figure 8.1 for some other examples.

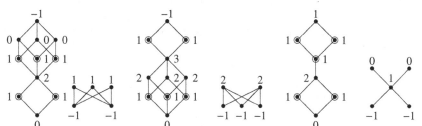

**Fig. 8.1** Some rank $\chi$-uniform lattices (with the distribution of the characteristic) and their spectrum (with the distribution of the reduced Möbius function).

**Proposition 8.7.** *Let* $P = \mathbf{m_1} \oplus \mathbf{m_2} \oplus \cdots \oplus \mathbf{m_k}$. *The Möbius function of the augmented poset* $\widehat{P}$ *is constant along the levels. More precisely, for every* $p \in \widehat{P}$ *at level* $k \geq 2$ *we have*

$$\mu(p) = \mu_k = (-1)^k (m_{k-1} - 1)(m_{k-2} - 1) \cdots (m_2 - 1)(m_1 - 1). \qquad (8.4)$$

*Proof.* The symmetry of $\widehat{P}$ implies at once that the Möbius function is constant along each level. If $\mu_k$ is the value assumed at level $k$, then $\mu_0 = 1$, $\mu_1 = -1$ and $\mu_2 = m_1 - 1$. Proceeding by induction on the rank of $\widehat{P}$, let $p$ be an element of $\widehat{P}$ of rank $k + 1$, with $k \geq 2$. From the defining property (8.3) of the Möbius function, it follows that

$$\mu_{k+1} = \mu(p) = -\sum_{u<p} \mu(u) = -\sum_{i=0}^{k} m_i \mu_i \qquad \text{(where } m_0 = 1\text{)}$$

$$= -\left( \sum_{i=2}^{k} (-1)^i m_i (m_{i-1} - 1) \cdots (m_1 - 1) - m_1 + 1 \right)$$

$$= -\sum_{i=2}^{k} (-1)^i (m_i - 1 + 1)(m_{i-1} - 1) \cdots (m_1 - 1) + m_1 - 1$$

$$= -\sum_{i=2}^{k} (-1)^i (m_i - 1)(m_{i-1} - 1) \cdots (m_1 - 1) + \sum_{i=2}^{k} (-1)^{i-1} (m_{i-1} - 1) \cdots (m_1 - 1) + m_1 - 1$$

$$= -\sum_{i=2}^{k} (-1)^i (m_i - 1)(m_{i-1} - 1) \cdots (m_1 - 1) + \sum_{i=1}^{k-1} (-1)^i (m_i - 1) \cdots (m_1 - 1) + m_1 - 1$$

$$= (-1)^{k+1} (m_k - 1) \cdots (m_1 - 1) - (m_1 - 1) + m_1 - 1$$

that is $\mu_{k+1} = (-1)^{k+1}(m_k - 1)\cdots(m_1 - 1)$.

**Proposition 8.8.** *The characteristic of the lattice* $D = B_{m_1} \odot B_{m_2} \odot \cdots \odot B_{m_k}$ *is constant along the levels. Specifically, for every element* $x$ *belonging to the i-th Boolean lattice in the vertical sum giving* $D$, *we have* $\chi(x) = 1 - (|S_x| - 1)\mu_i$, *where* $\mu_i$ *is defined by (8.4) and* $S_x$ *is the set of all maximum join-irreducible elements below* $x$.

*Proof.* By identities (8.1), the spectrum of $D$ is $P = \mathbf{m_1} \oplus \mathbf{m_2} \oplus \cdots \oplus \mathbf{m_k}$. An ideal $I$ of $P$ is of the form $I = \mathbf{m_1} \oplus \cdots \oplus \mathbf{m_{i-1}} \oplus S$, where $S$ is a subset of elements at level $i$. From Theorem 8.2, there follows that

$$\chi(I) = -\sum_{x \in I} \mu(x) = -\left(|S|\mu_i + \sum_{j=1}^{i-1} m_j\mu_j\right) = -|S|\mu_i - \sum_{j=0}^{i-1} m_j\mu_j + 1,$$

and hence, from (8.3), there follows that $\chi(I) = -|S|\mu_i + \mu_i + 1$.

**Theorem 8.12.** *A finite distributive lattice* $D$ *is rank* $\chi$-*uniform if and only if it is isomorphic to a lattice* $\mathcal{B}_{m_1} \odot \cdots \odot \mathcal{B}_{m_k} \odot H$, *where* $k \geq 0$, $m_1, \ldots, m_k \geq 2$ *and* $H$ *is a finite distributive lattice with at most one atom.*

*Proof.* Let $D$ be a rank $\chi$-uniform lattice and let $P$ be its spectrum. If $P$ is an antichain, then $D$ is a Boolean lattice. Proceeding by induction, suppose that $P = \mathbf{m_1} \oplus \cdots \oplus \mathbf{m_{i-1}} \oplus \mathbf{m_i} \oplus P_i$ with $m_1, \ldots, m_{i-1} \geq 2$ and $m_i \geq 1$. If $m_i = 1$, the theorem is proved. If $m_i \geq 2$, let $p$ be an element covering $t$ elements of rank $i$ in $P$, with $1 \leq t < m_i$. Then, since there exists at least one element $q \in P$ of rank $i$ not covered by $p$, it is possible to consider the two ideals $I_1 = \downarrow p$ and $I_2 = (\downarrow p \setminus \{p\}) \cup \{q\}$. Since they have the same size, they have the same rank in $\mathcal{J}(P)$. Since $I_1$ is principal, $\chi(I_1) = 1$, while $\chi(I_2) = -(m_1\mu_1 + \cdots + m_{i-1}\mu_{i-1} + (t+1)\mu_i) = 1 - t\mu_i$. So, $\chi(I_1) = \chi(I_2)$ if and only if $t\mu_i = 0$. However, $t \neq 0$ by assumption and $\mu_i \neq 0$ by (8.4) and the hypothesis $m_1, \ldots, m_i \geq 2$. Hence $\chi(I_1) \neq \chi(I_2)$, which is absurd. This implies that $p$ must cover all elements of $P$ with rank $i$. Proceeding in this way, at the end $P = \mathbf{m_1} \oplus \cdots \oplus \mathbf{m_k} \oplus Q$ where $Q$ is a poset with at most one minimal element.

Viceversa, if $D = \mathcal{B}_{m_1} \odot \cdots \odot \mathcal{B}_{m_k} \odot H$ with $m_1, \ldots, m_k \geq 2$ and $H$ a finite distributive lattice with at most one atom, then $D$ is rank $\chi$-uniform, by Propositions 8.8 and 8.5.

## 8.5 Pseudo-planar Distributive Lattices

A finite lattice $D$ is *planar* when its Hasse diagram is a planar graph [9, p. 51] [17]. If $D$ is distributive, it is planar if it can be obtained from the Cartesian product of two finite chains by removing a left corner and a right corner [11]. Equivalently, $D$ is planar when every element covers at most two elements [9, p. 68], or, by Theorem

8.1, when every element is the join of at most two join-irreducibles. It is easy to see that $\mathbf{Im}(\chi) \subseteq \{0,1,2\}$ for every finite planar distributive lattice. This property does not characterize planar distributive lattices (as shown in Figure 8.2). However, by Theorem 8.7, if $\mathbf{Im}(\chi) \subseteq \{0,1,2\}$ then necessarily there are at most two atoms.

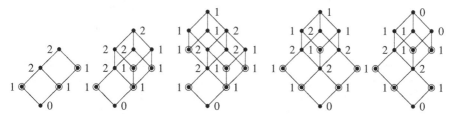

**Fig. 8.2** Some pseudo-planar distributive lattices.

A finite distributive lattice $D$ with $\mathbf{Im}(\chi) \subseteq \{0,1,2\}$ will be called *pseudo-planar*. Clearly, every $\chi$-uniform lattice and, by Proposition 8.1, every Cartesian product of two $\chi$-uniform lattices are pseudo-planar. Even though their Hasse diagram can be quite far from being planar, pseudo-planar lattices admit a characterization which generalizes one of the other characterizations of planar distributive lattices, as proved in next theorem. An element $x \in D$ will be called *pseudo-irreducible* if $\chi(x) = 1$. Clearly, every join-irreducible is pseudo-irreducible.

**Theorem 8.13.** *A finite distributive lattice $D$ is pseudo-planar if and only if every element different from the minimum is pseudo-irreducible or can be written as the join of two pseudo-irreducibles.*

*Proof.* Let $D$ be a finite distributive lattice in which every non zero element is pseudo-irreducible or can be written as the join of two pseudo-irreducibles. Clearly, $\chi(x) \in \{0,1\}$ whenever $x$ is the minimum or a join-irreducible. Proceeding by induction on the rank of $D$, let $x$ be an element with $r(x) \geq 2$ such that $x = x_1 \vee x_2$ with $x_1$ and $x_2$ pseudo-irreducibles. Then

$$\chi(x) = \chi(x_1) + \chi(x_2) - \chi(x_1 \wedge x_2) = 2 - \chi(x_1 \wedge x_2).$$

Since $r(x_1 \wedge x_2) < r(x)$, by induction hypothesis $\chi(x_1 \wedge x_2) \in \{0,1,2\}$. Hence $\chi(x) \in \{0,1,2\}$.

Viceversa, let $D$ be pseudo-planar. If $x \neq \widehat{0}$ is not pseudo-irreducible, then it covers at least two elements $x_1$ and $x_2$ with $x = x_1 \vee x_2$. If $x_1$ and $x_2$ are not both pseudo-irreducible, then there are the following two cases.

1. Only one of them is pseudo-irreducible, say $x_1$. If $\chi(x_2) = 0$, then $\chi(x_1 \wedge x_2) = 1$ and $\chi(x) = 0$ (see Figure 8.3 (a)). Now, $x_2$ covers $x_1 \wedge x_2$ and at least another element $y$. Necessarily, $\chi(y) = 0$, or $1$. If $\chi(y) = 1$ then $x = x_1 \vee y$ is the join of two pseudo-irreducibles. Otherwise, this reasoning can be reapplied to $y$, and so on. Since $D$ is finite, this process comes to an end, and this implies the existence

of a pseudo-irreducible element $z$ such that $x = x_1 \vee z$. A similar argument holds when $\chi(x_2) = 2$ (see Figure 8.3 (b)).

2. Both $x_1$ and $x_2$ are not pseudo-irreducible. Then, according to the value of $\chi(x)$, there are only the following configurations (up to vertical symmetries):

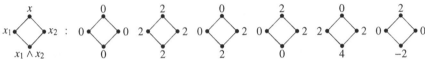

Of all these configurations, only the first two are possible. Indeed, the third and the fourth configuration lead to an infinite chain (see Figure 8.3 (c) and (d)), and this is impossible since $D$ is finite, while the last two configurations are clearly impossible for the pseudo-planarity. Reasoning as in the previous case by starting from the first two configurations, it is possible to prove (see Figure 8.3 (e) and (f)) the existence of two elements $z_1$ and $z_2$ such that $x = z_1 \vee z_2$ and $\chi(z_1) = \chi(z_2) = 1$.

In each case, $x$ can be written as the join of two pseudo-irreducibles.

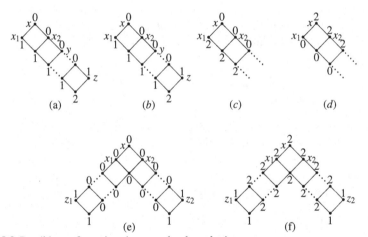

**Fig. 8.3** Possible configurations in a pseudo-planar lattice.

A subset $S$ of a distributive lattice $D$ will be called 02-*avoiding* if $\chi(x) \neq 0, 2$ for every $x \in S$. Pseudo-planar lattices can also be characterized as follows.

**Theorem 8.14.** *A finite distributive lattice $D$ is pseudo-planar if and only if it can be obtained, starting from $C_0$, with a finite number of elevations on 02-avoiding principal filters.*

*Proof.* Let $D$ be pseudo-planar and let $D = D_1 \uparrow F$ for a principal filter $F$. If $F$ contains two elements $x$ and $y$ with $\chi(x) = 0$ and $\chi(y) = 2$, then these two elements, with the corresponding elements $x'$ and $y'$ in the elevated copy $F'$ of $F$, form one of the following configurations:

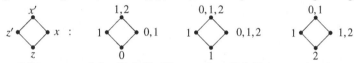

Since all these configurations are impossible, there follows that $F$ must be 02-avoiding.

Viceversa, let $D$ be pseudo-planar and let $F$ be a 02-avoiding principal filter of $D$. If $z$ is the minimum of $F$ and $z'$ is the corresponding element in the elevated copy $F'$, then for every other $x \in F$ and for the corresponding $x' \in F'$ it is $\chi(x') = \chi(x) - \chi(z) + 1$ (Proposition 8.3). Since $D$ is pseudo-planar and $F$ is 02-avoiding, according to the value of $\chi(z)$, there are only the following configurations:

In all possibile cases, $\chi(x) \in \{0, 1, 2\}$. Hence, also $D \uparrow F$ is pseudo-planar.

Similarly, one can prove the following characterization of planar distributive lattices.

**Theorem 8.15.** *A finite distributive lattice $D$ is planar if and only if it can be obtained, starting from $C_0$, with a finite number of elevations on 02-avoiding principal filters isomorphic to chains.*

Let $D$ be a finite distributive lattice. An element $x$ of $D$ is a *sphere* if $x = \widehat{0}$ or $x = x_1 \vee x_2$ for two join-irreducibles $x_1$ and $x_2$ whose meet $x_1 \wedge x_2$ is a sphere. Similarly, an element $x$ of $D$ is a *pseudo-sphere* if $x = \widehat{0}$ or $x = x_1 \vee x_2$ for two pseudo-irreducibles $x_1$ and $x_2$ whose meet $x_1 \wedge x_2$ is a pseudo-sphere. Clearly, every sphere is also a pseudo-sphere. The set of spheres (pseudo-spheres) of $D$ is univocally determined and can be obtained inductively starting from the minimum element. Moreover, proceeding by induction on the rank, it easily follows that the characteristic of a pseudo-sphere is always 0 or 2. So, join-irreducible (pseudo-irreducible) elements are never spheres (pseudo-spheres).

**Theorem 8.16.** *In a finite distributive lattice $D$ every element is a pseudo-irreducible or a pseudo-sphere if and only if $D$ is pseudo-planar.*

*Proof.* Let $D$ be pseudo-planar. The minimum is a sphere (by definition) and the atoms are join-irreducibles. Now, proceeding by induction on the rank of $D$, let $x \in D$ with $r(x) \geq 2$. If $x$ is not pseudo-irreducible, then, by Theorem 8.13, it is the join of two pseudo-irreducibles $x_1$ and $x_2$. Hence, since $\chi(x) \in \{0, 2\}$, there follows that $\chi(x_1 \wedge x_2) = 2 - \chi(x) \in \{0, 2\}$. So, by induction hypothesis, $x_1 \wedge x_2$ is a pseudo-sphere and consequently also $x$ is a pseudo-sphere.

Viceversa, if $D$ contains only pseudo-irreducibles and pseudo-spheres, then $\mathbf{Im}(\chi) \subseteq \{0, 1, 2\}$ and hence it is pseudo-planar.

In a similar way one proves

**Theorem 8.17.** *In a finite planar distributive lattice* $D$, *every element is a pseudo-irreducible or a sphere.*

In a pseudo-planar lattice not all pseudo-spheres are necessarily spheres. For instance, in the last lattice in Figure 8.2 the maximum element is not a sphere.

## 8.6 Graphs and Hypergraphs

A *hypergraph* [4] is a pair $H = \langle V, E \rangle$ where $V$ is the set of *vertices* and $E = \{e_1, \ldots, e_m\}$ is the family of *edges*, where every $e_i$ is a non-empty subset of $V$. Loops and multiple edges are allowed. A *graph* $G = \langle V, E \rangle$ is a hypergraph where every edge contains at most two vertices. A *subhypergraph* of $H$ is a pair $\langle X, Y \rangle \in \mathcal{P}(V) \times \mathcal{P}(E)$ such that all vertices of each edge $y \in Y$ are contained in $X$. Similarly, a *subgraph* of $G$ is a pair $\langle X, Y \rangle \in \mathcal{P}(V) \times \mathcal{P}(E)$ such that the endpoints of every edge $y \in Y$ are always contained in $X$. For simplicity, the subhypergraph $\langle \{x_1, \ldots, x_n\}, \{y_1, \ldots, y_m\} \rangle$ will be also written $\langle x_1 \cdots x_n, y_1 \cdots y_m \rangle$.

The set $\Sigma[H]$ of all subhypergraphs of $H$ is a distributive lattice when ordered by inclusion. Specifically, $\langle X_1, Y_1 \rangle \le \langle X_2, Y_2 \rangle$ if and only if $X_1 \subseteq X_2$ and $Y_1 \subseteq Y_2$. Moreover, $\langle X_1, Y_1 \rangle \wedge \langle X_2, Y_2 \rangle = \langle X_1 \cap X_2, Y_1 \cap Y_2 \rangle$ and $\langle X_1, Y_1 \rangle \vee \langle X_2, Y_2 \rangle = \langle X_1 \cup X_2, Y_1 \cup Y_2 \rangle$. The rank is $r\langle X, Y \rangle = |X| + |Y|$, and the height of $\Sigma[H]$ is $|V| + |E|$. In particular, also the set $\Sigma[G]$ of all subgraphs of a graph $G$ is a distributive lattice (see Figure 8.4 for an example).

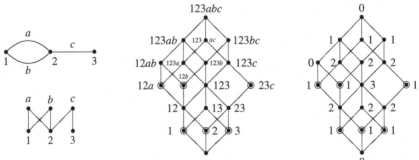

**Fig. 8.4** A graph $G$ with the incidence poset $P_G^{\varnothing}$ and the subgraph lattice $\Sigma[G]$ with the distribution of the Euler characteristic.

The *incidence poset* of a hypergraph $H = \langle V, E \rangle$ is the poset $P_H$ having the vertices and the edges of $H$ as elements so that $v < e$ whenever $v \in V$, $e \in E$ and $v \in e$. $H$ is completely determined by the poset $P_H$, which is ranked with height at most 1.

**Proposition 8.9.** *For every finite hypergraph* $H$, *the spectrum of* $\Sigma[G]$ *is the incidence poset* $P_H$ *and* $\Sigma[G] = \mathcal{J}(P_H)$.

*Proof.* From the definition of subhypergraph, we have at once that the subhypergraphs of $H$ corresponds bijectively to the ideals of $P_H$. Moreover, the join-irreducible elements of $\Sigma[H]$ are the subhypergraphs $\langle v, \varnothing \rangle$ reducing to a single vertex, and the subhypergraphs $\langle V_e, e \rangle$ reducing to a single edge (where $V_e$ denotes the set of all vertices contained in the edge $e$).

The Euler characteristic of a subhypergraph of a hypergraph $H$ does not depend on $H$, as follows from

**Theorem 8.18.** *If* $H$ *is a finite hypergraph with* $m$ *edges, then the Euler characteristic of a subhypergraph* $S \in \Sigma[H]$ *is*

$$\chi(S) = v(S) - \sum_{k=1}^{m} (k-1)\eta_k(S). \tag{8.5}$$

*where* $v(S)$ *is the number of all vertices of* $S$ *and* $\eta_k(S)$ *is the number of all edges of* $S$ *containing exactly* $k$ *vertices.*

*Proof.* If $S$ has no edges then it can be written as the disjoint union of its vertices and consequently $\chi(S) = v(S)$. If $S = \langle X, Y \rangle$ has at least one edge then it can be written as the union of all spanning subhypergraphs $\langle X, y \rangle$ with $y \in Y$. Applying (8.2), there follows that

$$\chi(S) = \chi\left(\bigcup_{y \in Y}\langle X, y \rangle\right) = \sum_{\substack{T \subseteq Y \\ T \neq \emptyset}} (-1)^{|T|-1} \chi\left(\bigcap_{y \in T}\langle X, y \rangle\right).$$

Since the intersection $\bigcap_{y \in T}\langle X, y \rangle$ reduces to $\langle X, y \rangle$ whenever $T$ is a singleton $\{y\}$ and reduces to $\langle X, \emptyset \rangle$ whenever $T$ contains at least two edges, there follows that

$$\chi(S) = \sum_{y \in Y} \chi\langle X, y \rangle + \sum_{\substack{T \subseteq Y \\ |T| \geq 2}} (-1)^{|T|-1} \chi\langle X, \emptyset \rangle.$$

A subhypergraph $\langle X, y \rangle$ can be decomposed as the disjoint union $\langle x_1, \emptyset \rangle \vee \cdots \vee \langle x_k, \emptyset \rangle \vee \langle V_y, y \rangle$, where $\{x_1, \ldots, x_k\} = X \setminus V_y$. Since all elements appearing in such a decomposition are join-irreducible in $\Sigma[H]$, there follows that $\chi\langle X, y \rangle = k + 1 = |X| - |V_y| + 1$. Hence, since $|Y| \geq 1$, we obtain the identity

$$\chi(S) = \sum_{y \in Y}(|V| - |V_y| + 1) - |X|\left(\sum_{T \subseteq Y}(-1)^{|T|} - 1 + |Y|\right) = |X| + |Y| - \sum_{y \in Y}|V_y|$$

which reduces to (8.5).

When the hypergraph reduces to a graph, Theorem 8.18 simplifies into

**Theorem 8.19.** *If $G$ is a graph, then $\chi(S) = v(S) - \eta(S) + \lambda(S)$ for every subgraph $S \in \Sigma[G]$, where $v(S)$, $\eta(S)$ and $\lambda(S)$ are the number of vertices, edges (loops included) and loops of $S$.*

*Proof.* If $S \in \Sigma[G]$, then $\eta_1(S) = \lambda(S)$, $\eta_2(S) = \eta(S) - \lambda(S)$, and $\eta_k(S) = 0$ for $k \geq 3$.

**Proposition 8.10.** *If $G$ is a simple graph, then $G$ is a tree if and only if $\chi(G) = 1$, and $G$ is unicyclic if and only if $\chi(G) = 0$.*

*Proof.* If $G$ is a simple graph then there are no loops (and no multiple edges). Hence $\chi(G) = 1$ if and only if $v(G) = \eta(G) + 1$, that is if and only if $G$ is a tree (i.e. a connected graph containing no cycles). Similarly, $\chi(G) = 0$ if and only if $v(G) = \eta(G)$, that is if and only if $G$ is a unicyclic graph (i.e. a connected graph containing exactly one cycle).

The *covering graph* of a poset $P$ is the simple graph $G_P$ where the vertices are the elements of $P$ and where two elements are adjacent when one of them is covered by the other. If $P$ has rank $1$, then the covering graph is bipartite and the partite sets are the levels $P_0$ and $P_1$. Moreover, $P$ is equivalent to the hypergraph $H_P = \langle P_0, P_1 \rangle$. Now, Theorem 8.18 can be reformulated and reproved as follows.

**Theorem 8.20.** *Let $P$ be a poset of rank $1$. Then $\chi(I) = v(G_I) - \eta(G_I)$, for every ideal $I$ of $P$, where $G_I$ is the covering graph of $I$. In particular, $\chi(I) = 1$ if and only if $G_I$ is a tree, and $\chi(I) = 0$ if and only if $G_I$ is a unicyclic graph.*

*Proof.* For every $p \in P_1$, let $d(p)$ be the number of all elements of $P_0$ covered by $p$. Then $\mu(\hat{0}) = 1$, $\mu(p) = -1$ for every $p \in P_0$ and $\mu(x) = d(x) - 1$ for every $p \in P_1$. For any ideal $I$ of $P$, let $I_0 = I \cap P_0$ and $I_1 = I \cap P_1$. Then, by Theorem 8.2, we have

$$\chi(I) = -\sum_{x \in I} \mu(p) = -\sum_{x \in I_0} \mu(p) - \sum_{p \in I_1} \mu(p) = |I_0| - \sum_{p \in I_1}(d(p) - 1) = |I| - \sum_{p \in I_1} d(p).$$

In the covering graph $G_I$ of $I$ the number of vertices is $v(G_I) = |I|$ and the number of edges is $\eta(G_I) = \sum_{p \in I_1} d(p)$. Hence $\chi(I) = v(G_I) - \eta(G_I)$.

## 8.7 Dual Gödel Lattices

Let $D$ be a finite distributive lattice and let $m : D \to \mathbb{N}$ be the map where $m(x)$ is equal to the number of all join-irreducible elements in the unique irredundant representation of the element $x \in D$. If $D = \mathcal{J}(P)$, then $m(I) = |M(I)|$ is the number of all maximal element of $I$ (Theorem 8.1). The map $m$ behaves as the Euler characteristic on $\hat{0}$ and on the join-irreducible elements. In general, however, $m$ is not a valuation and hence $m \neq \chi$. Nevertheless, the finite distributive lattices for which $m = \chi$ can be completely characterized in terms of their spectrum.

A *tree* is a poset with a unique minimal element where every principal ideal is a chain. A *forest* is a disjoint union of trees. A *cotree* is the dual of a tree and a *coforest* is the dual of a forest, i.e. the disjoint union of cotrees. In a cotree, the *children* of an element $x$ are all elements covered by $x$. The *leaves* are the minimal elements, i.e. the elements with no children.

**Theorem 8.21.** *If the Euler characteristic of a finite distributive lattice $D$ is equal to the map $m$ then the spectrum of $D$ is a coforest.*

*Proof.* Let $\chi = m$. If $x_1$ and $x_2$ are two incomparable join-irreducible elements, then

$$2 = m(x_1 \vee x_2) = \chi(x_1 \vee x_2) = \chi(x_1) + \chi(x_2) - \chi(x_1 \wedge x_2) = 2 - \chi(x_1 \wedge x_2)$$

and hence $\chi(x_1 \wedge x_2) = 0$, i.e. $m(x_1 \wedge x_2) = 0$. Since the minimum is the unique element which is the join of zero join-irreducibles, there follows that $x_1 \wedge x_2 = \widehat{0}$. So, in $D$, the meet of every two incomparable join-irreducibles is $\widehat{0}$. Now, let $P$ be the spectrum of $D$. If an element $p \in P$ is covered in $P$ by two different elements $p_1$ and $p_2$, then there were two incomparable join-irreducibles whose meet in $D$ is different from $\widehat{0}$. Since this is impossible, there follows that every principal filter of $P$ is a chain, i.e. that $P$ is a coforest.

**Theorem 8.22.** *If $\mu$ is the Möbius function of the augmentation of a finite cotree $T^*$, then $\mu(p) = d(p) - 1$, for every element $p \in T^*$, where $d(p) = |\nabla(p)|$ is the number of children of $p$.*

*Proof.* Let $\widehat{0}$ be the minimum of the augmentation of $T^*$. Consider an element $p \in T^*$ covering $p_1, \ldots, p_k$. Since $\downarrow p_i \cap \downarrow p_j = \{\widehat{0}\}$ whenever $i \neq j$, there follows that

$$\mu(p) = - \sum_{\widehat{0} \leq u < p} \mu(u) = - \left( \sum_{\widehat{0} \leq u \leq p_1} \mu(u) + \cdots + \sum_{\widehat{0} \leq u \leq p_k} \mu(u) - (k-1)\mu(\widehat{0}) \right).$$

So, applying the defining property (8.3) of the Möbius function, there follows that $\mu(p) = k - 1$.

For a graph $G$, let $\kappa(G)$ be the number of its connected components. The Euler characteristic of an ideal $I$ of a cotree has a simple geometrical interpretation in terms of its covering graph $G_I$ (similar to the one obtained in Theorem 8.20).

**Theorem 8.23.** *If $T^*$ is a finite cotree, then $\chi(I) = \kappa(G_I)$ for every $I \in \mathcal{J}(T^*)$.*

*Proof.* Theorems 8.2 and 8.22 imply

$$\chi(I) = - \sum_{x \in I} \mu(x) = - \sum_{x \in I} (d(x) - 1) = |I| - \sum_{x \in I} d(x) = \nu(G_I) - \eta(G_I).$$

Since $I$ is an ideal of a cotree, its covering graph is a forest and $\nu(G_I) - \eta(G_I) = \kappa(G_I)$.

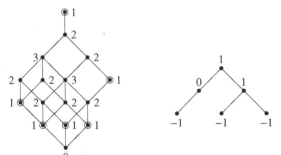

**Fig. 8.5** A lattice whose spectrum is a cotree.

Since the number of connected components of an ideal of a cotree is equal to the number of the maximum elements, from Theorems 8.21 and 8.23, there follows

**Theorem 8.24.** *The Euler characteristic of a finite distributive lattice $D$ is equal to the map $m$ if and only if the spectrum of $D$ is a coforest.*

A *Gödel algebra* is a Heyting algebra satisfying the prelinearity axiom [16] (see also [1]). A *Gödel lattice* is the underlying lattice of a Gödel algebra. Since Gödel lattices are characterized as distributive lattices whose spectrum is a forest [16], there follows at once

**Theorem 8.25.** *The Euler characteristic of a finite distributive lattice $D$ is equal to the map $m$ if and only if $D$ is the dual of a Gödel lattice.*

Let $\beta : D \to D$ be the map defined as follows. If $x$ is a join-irreducible then $\beta(x)$ is the unique element covered by $x$. More generally, if the irredundant representation of $x$ is $x_1 \vee \cdots \vee x_k$, then $\beta(x) = \beta(x_1) \vee \cdots \vee \beta(x_k)$. If $D = \mathcal{J}(P)$ then $\beta(I) = I \setminus M(I)$.

**Theorem 8.26.** *Let $D$ be a finite distributive lattice whose spectrum is a coforest. Then $\chi(x) = r(x) - r(\beta(x))$ for every $x \in D$.*

*Proof.* If $D = \mathcal{J}(T^*)$ then $r(\beta(I)) = |I \setminus M(I)| = |I| - |M(I)| = r(I) - \chi(I)$ and hence $\chi(I) = r(I) - r(\beta(I))$.

*Remark 8.2.* The map $\beta$ generalizes the *bubble sort map* defined for permutations [21, p. 108] and Theorem 8.26 generalizes a similar result obtained in [23].

## 8.8 Tree maps

Let $T$ be a finite tree and let $D$ a finite distributive lattice. The set $M_T[D]$ of all order-preserving map $f : T \to D$ can be ordered by setting $f \le g$ whenever $f(p) \le g(p)$ for every $p \in T$. In this way, $M_T[D]$ is a distributive lattice, where the minimum is the constant map $\zeta_0$ of value $\widehat{0}$ and the maximum is the constant map $\zeta_1$ of value $\widehat{1}$.

If $T$ is a chain of $k$ element, then $M_T[D]$ is equivalent to the lattice $D^{\leq k}$ of $k$-multi-chains of $D$, i.e. is the set of all $k$-tuples $(x_1, x_2, \ldots, x_k)$ such that $x_1 \leq x_2 \leq \cdots \leq x_k$, ordered componentwise. If $T = \mathbf{1} \oplus (\mathbf{k-1})$, then $M_T[D]$ is equivalent to the lattice $D^{[k]}$ whose elements are all $k$-tuples $(x, x_1, \ldots, x_{k-1})$ such that $x \leq x_1$, $\ldots$, $x \leq x_{k-1}$, ordered componentwise [10, p. 188].

**Proposition 8.11.** *If* $D$ *is a finite distributive lattice, then* $\mathbf{Spec}(M_T[D]) \simeq T^* \times \mathbf{Spec}(D)$ *and hence* $M_T[D] \simeq \mathcal{J}(T)^* \otimes D$.

*Proof.* The join-irreducibles of $M_T[D]$ are the maps $\varphi_{p,x} : T \to D$ defined by

$$\varphi_{p,x}(q) = \begin{cases} x & \text{if } q \geq p \\ \widehat{0} & \text{if } q \not\geq p \end{cases} \qquad \text{for every } q \in T, \tag{8.6}$$

where $p$ is any element of $T$ and $x$ is a join-irreducible element of $D$. Since $\varphi_{p,x} \leq \varphi_{q,y}$ if and only if $p \geq q$ and $x \leq y$, there follows that the map $\psi : \mathbf{Spec}(M_T[D]) \to T^* \times \mathbf{Spec}(D)$, defined by $\psi(f_{p,x}) = (p, x)$, is an order-preserving bijection. $\qquad \square$

For every $p \in T$ and every $x \in D$, let $\varphi_{p,x} : T \to D$ be the function defined as in (8.6). These functions will be called *simple maps*.

**Proposition 8.12.** *For every element* $p \in T$, *the map* $\Phi_p : D \to M_T[D]$, *defined by* $\Phi_p(x) = \varphi_{p,x}$, *is* $\chi$-*preserving, i.e.* $\chi(\varphi_{p,x}) = \chi(x)$ *for every* $p \in T$ *and* $x \in D$.

*Proof.* From definition (8.6), there follows that $\varphi_{p,x} \wedge \varphi_{p,y} = \varphi_{p,x \wedge y}$, $\varphi_{p,x} \vee \varphi_{p,y} = \varphi_{p,x \vee y}$ and $\varphi_{p,\widehat{0}} = \zeta_0$. Hence $\Phi_p$ is a lattice morphism preserving the minimum and the join-irreducibles (by Proposition 8.11). Since $\Phi_p$ is injective by construction, Theorem 8.4 implies that it also preserves the Euler characteristic. $\qquad \square$

**Proposition 8.13.** *Every tree map* $f \in M_T[D]$ *admits a canonical representation as a join of simple maps. Specifically,*

$$f = \bigvee_{p \in T} \varphi_{p,f(p)} = \bigvee_{p \in T} \Phi_p(f(p)).$$

*Proof.* For every element $q \in P$, from definition (8.6) there follows that

$$\bigvee_{p \in T} \varphi_{p,f(p)}(q) = \bigvee_{p \in T} \begin{cases} f(p) & q \geq p \\ \widehat{0} & q \not\geq p \end{cases} = \bigvee_{p \leq q} f(p) = f(q).$$

Let $\mathcal{C}(T)$ be the set of all chains of $T$ and let $\mathcal{C}_p(T)$ be the set of all chains of $T$ starting from $p$. Then let

$$c_p(T) = \sum_{\gamma \in \mathcal{C}_p(T)} (-1)^{|\gamma|-1}.$$

**Theorem 8.27.** *The characteristic of a tree map* $f \in M_T[D]$ *is*

$$\chi(f) = \sum_{p \in T} c_p(T)\chi(f(p)).$$

*Proof.* By Proposition 8.13, any tree map $f$ decomposes as a join of simple maps. Hence

$$\chi(f) = \chi\left(\bigvee_{p \in T} \varphi_{p,f(p)}\right) = \sum_{\substack{S \subseteq T \\ S \neq \varnothing}} (-1)^{|S|-1} \chi\left(\bigwedge_{p \in S} \varphi_{p,f(p)}\right).$$

For every $q \in T$, from (8.6) it follow that

$$\bigwedge_{p \in S} \varphi_{p,f(p)}(q) = \bigwedge_{p \in S} \begin{cases} f(p) & q \geq p \\ 0 & q \ngeq p \end{cases} = \begin{cases} \bigwedge_{p \in S} f(p) & q \geq p, \ \forall p \in S \\ 0 & \text{otherwise.} \end{cases}$$

If $S$ contains two incomparable elements, then this meet is $\zeta_0$. On the other hand, if $S$ does not contain incomparable elements, then $S$ is a chain. Since a chain has a minimum and a maximum and since $f$ is order-preserving, there follows that $\bigwedge_{p \in S} \varphi_{p,f(p)} = \varphi_{\max(S),f(\min(S))}$. So, from Proposition 8.12, there follows that

$$\chi(f) = \sum_{\substack{\gamma \in \mathcal{C}(T) \\ \gamma \neq \varnothing}} (-1)^{|\gamma|-1} \chi\left(\varphi_{\max(\gamma),f(\min(\gamma))}\right) = \sum_{\substack{\gamma \in \mathcal{C}(T) \\ \gamma \neq \varnothing}} (-1)^{|\gamma|-1} \chi(f(\min(\gamma))),$$

that is

$$\chi(f) = \sum_{p \in T} \left[ \sum_{\gamma \in \mathcal{C}_p(T)} (-1)^{|\gamma|-1} \right] \chi(f(p)) = \sum_{p \in T} c_p(T)\chi(f(p)).$$

Theorem 8.27 immediately implies

**Theorem 8.28.** *The Euler characteristic of an element of* $D^{\leq k}$ *is* $\chi(x_1, x_2, \ldots, x_k) = \chi(x_k)$, *while the Euler characteristic of an element of* $D^{[k]}$ *is*

$$\chi(x, x_1, \ldots, x_{k-1}) = \chi(x_1) + \cdots + \chi(x_{k-1}) - (k-2)\chi(x).$$

*Proof.* If $T$ is a chain with $k$ elements $p_1 < p_2 < \cdots < p_k$, then the chains in $T$ are equivalent to subsets, and hence

$$c_{p_i}(T) = \sum_{S \subseteq [i+1,k]} (-1)^{|S|} = \sum_{j=0}^{k-i} \binom{k-i}{j}(-1)^j = (1-1)^{k-i} = \delta_{ik}.$$

If $T = \{p\} \oplus \{p_1, \ldots, p_{k-1}\}$, then $c_p(T) = 1 - (k-1)$ and $c_{p_1}(T) = \cdots = c_{p_{k-1}}(T) = 1$.

# References

1. Aguzzoli, S., Gerla, B., Marra, V. (2008), *Gödel Algebras Free Over finite Distributive Lattices*, in "Ann. Pure Appl. Logic", 155, pp. 183-93.
2. Aigner, M. (1979), *Combinatorial Theory*, New York, Springer.
3. Barnabei, M., Brini, A., Rota, G.-C. (1986), *The Theory of Möbius Functions*, in "Russ. Math. Surv.", 41/3, pp. 135-88.
4. Berge, C. (1989), *Hypergraphs: The Theory of Finite Sets*, Amsterdam, North-Holland.
5. Berger, C., Leinster, T. (2008), *The Euler Characteristic of a Category as the Sum of a Divergent Series*, in "Homology, Homotopy Appl.", 10, pp. 41-51.
6. Birkhoff, G. (1966), *Lattice Theory*, Providence, Amer. Math. Soc. Colloquium Publications.
7. Crapo, H. H. (1995), *Rota's "Combinatorial Theory"*, in J. P. S. Kung (ed.), *Gian-Carlo Rota on Combinatorics*, Boston, Basel, Berlin, Birkhäuser, pp. xix-xliii.
8. Erné, M., Heitzig, J., Reinhold, J. (2002), *On the Number of Distributive Lattices*, in "Electron. J. Combin.", 9, #R24.
9. Grätzer, G. (1978), *General Lattice Theory*, Boston, Basel, Berlin, Birkhäuser.
10. Grätzer, G. (1971), *Lattice Theory. First Concepts and Distributive Lattices*, San Francisco, W. H. Freeman and Co.
11. Grätzer, G., Knapp, E. (2007), *Notes on Planar Semimodular Lattices. I. Construction*, in "Acta Sci. Math. (Szegad)", 73, pp. 445-62.
12. Hadwiger, H. (1947), *Über eine symbolisch-topologische Formel*, in "Elem. Math.", 2, pp. 35-41.
13. Hadwiger, H. (1953), *Über additive Funktionale, k-dimensionaler Eipolyeder*, in "Publ. Math. Debrecen.", 3, pp. 87-94.
14. Hadwiger, H. (1955), *Eulers Charakteristik und kombinatorische Geometrie*, in "J. Reine Angew. Math.", 194, pp. 101-10.
15. Hadwiger, H. (1960), *Zur Eulerschen Charakteristik Euklidischer Polyeder*, in "Monatsh. Math.", 64, pp. 349-54.
16. Horn, A. (1969), *Free L-algebras*, in "J. Symbolic Logic", 34, pp. 475-80.
17. Kelly, D., Rival, I. (1975), *Planar Lattices*, in "Canad. J. Math.", 27, pp. 558-66.
18. Klee, V. (1963), *The Euler Characteristic in Combinatorial Geometry*, in "Amer. Math. Monthly", 70, pp. 119-27.
19. Klain, D. A. (1997), *Kinematic Formulas for Finite Lattices*, in "Annals of Combinatorics", 1, pp. 353-66.
20. Klain, D. A., Rota, G.-C. (1997), *Introduction to Geometric Probability. Lezioni Lincee*, Cambridge, Cambridge University Press.
21. Knuth, D. E. (1973), *The Art of Computer Programming. Vol. 3: Sorting and Searching*, Reading, MA, Addison-Wesley.
22. Leinster, T. (2008), *The Euler Characteristic of a Category*, in "Doc. Math.", 13, pp. 21-49.
23. Munarini, E., Perelli Cippo, C., *Euler Characteristic and Permutation Codes*, preprint.
24. Propp, J. (2002), *Euler Measure as Generalized Cardinality*, math. CO/0203289.
25. Propp, J. (2003), *Exponentiation and Euler Measure*, "Algebra Universalis", 49, pp. 459-71.
26. Rota, G.-C. (1964), *On the Foundations of Combinatorial Theory. I. Theory of Möbius functions*, in "Z. Wahrscheinlichkeitstheorie und Verw. Gebiete", 2, pp. 340-68.
27. Rota, G.-C. (1971), *On the Combinatorics of the Euler Characteristic*, in L. Mirsky (ed.), *Studies in Pure Mathematics*, London, Academic Press, pp. 221-233.
28. Rota, G.-C. (1973), *The Valuation Ring of a Distributive Lattice*, in S. Fajtlowicz, K. Kaiser (eds.), *Proceedings of the University of Houston Lattice Theory Conference*, Houston, University of Houston, pp. 575-628.
29. Sallee, G. T. (1968), *Polytopes, Valuations, and the Euler Relation*, in "Canad. J. Math.", 20, pp. 1412-24.
30. Schanuel, S. H. (1986), *What is the Length of a Potato? An Introduction to Geometric Measure Theory*, in *Categories in Continuum Physics*, (Lecture Notes in Math., 1174), Berlin, Springer.

31. Schanuel, S. H. (1991), *Negative Sets have Euler Characteristic and Dimension*, in *Category theory (Como, 1990)*, (Lecture Notes in Math., 1488), Berlin, Springer, pp. 379-85.

32. Shephard, E. H. (1968), *Euler-type relations for convex polytopes*, in "Proc. London Math. Soc.", 18, pp. 597-606.

33. Stanley, R. P. (1997), *Enumerative Combinatorics*, Vol. 1, (Cambridge Studies in Advanced Mathematics, 49), Cambridge, Cambridge University Press.

# Chapter 9
# Rota, Probability, Algebra and Logic
## Invited Chapter

Daniele Mundici

**Abstract** Inspired by Rota's Fubini Lectures, we present the MV-algebraic extensions of various results in probability theory, first proved for boolean algebras by
De Finetti, Kolmogorov, Carathéodory, Loomis, Sikorski and others. MV-algebras
stand to Łukasiewicz infinite-valued logic as boolean algebras stand to boolean
logic. Using Elliott's classification, the correspondence between countable boolean
algebras and commutative AF C*-algebras extends to a correspondence between
countable MV-algebras and AF C*-algebras whose Murray-von Neumann order of
projections is a lattice. In this way, (faithful, invariant) MV-algebraic states are identified with (faithful, invariant) tracial states of their corresponding AF C*-algebras.
Faithful invariant states exist in all finitely presented MV-algebras. At the other extreme, working in the context of $\sigma$-complete MV-algebras we present a generalization of Carathéodory boolean algebraic probability theory.

## 9.1 MV-algebraic States and De Finetti Coherence Criterion

Suppose we are given an *arbitrary* set $E = \{X_1, \ldots, X_m\}$ together with a nonempty
subset $\mathcal{W}$ of the $m$-cube $[0, 1]^{\{X_1, \ldots, X_m\}}$. Let us agree to say that $E$ is a set of "events",
and $\mathcal{W}$ a set of "possible worlds". Already at this level of extreme generality, we
can give a precise definition of $\mathcal{W}$-coherent probability assessment of $E$:

**Definition 9.1.** A map $\beta \colon E \to [0, 1]$ is $\mathcal{W}$-*incoherent* if for some $\sigma_1, \ldots, \sigma_m \in \mathbb{R}$ we
have $\sum_{i=1}^{m} \sigma_i(\beta(X_i) - v(X_i)) < 0$ for all $v \in \mathcal{W}$. Otherwise, $\beta$ is $\mathcal{W}$-*coherent*.

The rationale behind this definition is as follows: Two players, Ada (the book-
maker) and Blaise (the bettor) wager money on the outcome of events $X_1, \ldots, X_m$ in
the set $\mathcal{W} \subseteq [0, 1]^{\{X_1, \ldots, X_m\}}$ of possible worlds. Ada proclaims her "degree of belief"

Daniele Mundici
Università di Firenze, Italy

E. Damiani et al. (eds.), *From Combinatorics to Philosophy,*
DOI 10.1007/978-0-387-88753-1_9, © Springer Science+Business Media, LLC 2009

(="betting odd") $\beta(X_i) \in [0,1]$, and Blaise chooses a "stake" $\sigma_i \in \mathbb{R}$ for his bet on $X_i$. After these preliminaries, $\sigma_i\beta(X_i)$ euro are paid, with the proviso that $-\sigma_iv(X_i)$ euro will be paid back in the possible world $v \in \mathcal{W}$ assigning the (truth-)value $v(X_i)$ to each event $X_i$. Money transfers are oriented so that "positive" means Blaise-to-Ada. In particular, for $\sigma_i < 0$, we have a reverse bet: Ada first pays Blaise $|\sigma_i|\beta(X_i)$ euro, while Blaise will pay back $|\sigma_i|v(X_i)$ in the possible world $v$. Ada's book $\beta$ would lead her to financial disaster if Blaise could choose stakes $\sigma_1,\ldots,\sigma_m$ ensuring him to win money in every $v \in \mathcal{W}$—or equivalently, to win at least one million euro in every possible world. Such a disastrous book is exactly a $\mathcal{W}$-incoherent book in the sense of Definition 9.1.

In the particular case when $\mathcal{W} \subseteq \{0,1\}^E$, Definition 9.1 yields De Finetti's celebrated no-Dutch-Book criterion for coherent probability assessments of yes-no events ([8, §7, p. 308], [9, pp. 6-7], [10, p. 87]).

For any $E$ and $\mathcal{W}$, in Theorem 9.1 below we will construct a theory $\Theta$ in infinite-valued (propositional) Łukasiewicz logic $Ł_\infty$, such that $\mathcal{W}$-coherent maps coincide with restrictions to $E$ of convex combinations of models of $\Theta$. The necessary preliminary notation and terminology are as follows:

Formulas in $Ł_\infty$ are the same strings of symbols as in boolean logic. While Łukasiewicz [24] used implication $\to$ as a main connective, together with negation $\neg$, we will use the connectives $\odot$ of conjunction and $\oplus$ of disjunction. We will write $x \to y$ as an abbreviation of $\neg x \oplus y$. *Propositional variables* are strings of symbols of the form $X_1, X_2, \ldots$, and formulas have the usual inductive definition. $\mathcal{F}_m$ will denote the set of all formulas whose variables are in the set $\{X_1, \ldots, X_m\}$.

A *(Łukasiewicz) valuation* of $\mathcal{F}_m$ is a function $v: \mathcal{F}_m \to [0,1]$ such that $v(\neg\phi) = 1 - v(\phi)$, $v(\phi \odot \psi) = \max(0, v(\phi) + v(\psi) - 1)$, and $v(\phi \oplus \psi) = \min(1, v(\phi) + v(\psi))$, for all $\phi, \psi \in \mathcal{F}_m$. We say that $v$ *satisfies a set* $\Psi \subseteq \mathcal{F}_m$ if $v(\theta) = 1$ for all $\theta \in \Psi$. The restriction map $v \mapsto v{\restriction}\{X_1, \ldots, X_m\}$ is a one-one correspondence between valuations of $\mathcal{F}_m$ and points of the $m$-cube $[0,1]^{\{X_1,\ldots,X_m\}}$. For any point $w$ in the $m$-cube, we will use the notation

$$\tilde{w}: \mathcal{F}_m \to [0,1] \quad \text{for the unique valuation of } \mathcal{F}_m \text{ extending } w. \quad (9.1)$$

A proper subset $\Theta$ of $\mathcal{F}_m$ is said to be a *(consistent) theory in the variables* $X_1,\ldots,X_m$ if $\Theta$ contains every tautology[1] $\phi \in \mathcal{F}_m$ and is closed under Modus Ponens: whenever $\psi$ and $\psi \to \chi$ are in $\Theta$ then $\chi$ is in $\Theta$. We say that a theory $\Theta \subseteq \mathcal{F}_m$ is *finitely axiomatizable* if there is a formula $\theta \in \mathcal{F}_m$ such that $\Theta$ is the smallest theory in the variables $X_1,\ldots,X_m$ containing $\theta$. In this case, following [31, p. 222-223] we write

$$\Theta = \theta^\vdash = \{\psi \in \mathcal{F}_m \mid \theta \vdash \psi\}, \quad (9.2)$$

meaning that $\Theta$ is the set of formulas in the variables $X_1,\ldots,X_m$ which are consequences[2] of $\theta$. The set of variables $X_1,\ldots,X_m$ will always be clear from the context.

---

[1] As is well known, $\phi$ is a tautology if $v(\phi) = 1$ for every valuation $v$ of $\mathcal{F}_m$.

[2] In the present case, Wójcicki's theorem [7, 4.6.7] states that $\psi$ is a syntactic consequence of $\theta$ if and only if $\psi$ is a *semantic* consequence of $\theta$ (i.e., $v(\theta) = 1 \Rightarrow v(\psi) = 1$ for all valuations $v$ of $\mathcal{F}_m$).

The assumed consistency of $\Theta$ is to the effect that $\theta$ is *consistent*, i.e., $v(\theta) = 1$ for at least one valuation $v$ of $\mathcal{F}_m$.

An *MV-algebra* is a structure $B = (B, 0, 1, \neg, \oplus, \odot)$ such that $(x \oplus y) \oplus z = x \oplus (y \oplus z)$, $x \oplus y = y \oplus x$, $x \oplus 0 = x$, $x \oplus 1 = 1$, $\neg 0 = 1$, $\neg 1 = 0$, $y \oplus \neg(y \oplus \neg x) = x \oplus \neg(x \oplus \neg y)$, $x \odot y = \neg(\neg x \oplus \neg y)$. Boolean algebras coincide with MV-algebras satisfying $x \oplus x = x$. A routine variant of [7, 4.5.5] shows that the *free m-generator MV-algebra* $Free_m$ is the MV-algebra of equivalence classes $\langle \phi \rangle$ of formulas $\phi \in \mathcal{F}_m$, two formulas $\phi, \psi \in \mathcal{F}_m$ being *equivalent* if for every valuation $v$ of $\mathcal{F}_m$, $v(\phi) = v(\psi)$. For each theory $\Theta \subseteq \mathcal{F}_m$ we define the *Lindenbaum algebra* $\mathcal{L}(\Theta)$ (*in the variables* $X_1, \ldots, X_m$) to be the quotient of $Free_m$ by the ideal $I(\Theta) \subseteq Free_m$ of all equivalence classes of formulas $\langle \phi \rangle$ such that $\neg \phi \in \Theta$. In symbols,

$$\mathcal{L}(\Theta) = \frac{Free_m}{\{\langle \phi \rangle \mid \neg \phi \in \Theta\}}. \tag{9.3}$$

Every formula $\psi$ determines a unique element of $\mathcal{L}(\Theta)$, denoted

$$\langle \psi \rangle_\Theta. \tag{9.4}$$

A *state* of an MV-algebra $B$ is a map $s \colon B \to [0, 1]$ such that $s(1) = 1$ and $s(x \oplus y) = s(x) + s(y)$ whenever $x \odot y = 0$. We say that $s$ is *faithful* if $s(x) = 0$ implies $x = 0$. We say that $s$ is *invariant* if $s(\alpha(x)) = s(x)$ for every automorphism $\alpha$ of $B$ and element $x \in B$.

**Theorem 9.1.** ([29, 32]) *For any set* $E = \{X_1, \ldots, X_m\}$ *and closed nonempty set* $\mathcal{W} \subseteq [0, 1]^E = [0, 1]^{\{1, \ldots, m\}}$, *recalling (9.1), let* $\Theta = \{\psi \in \mathcal{F}_m \mid \tilde{w}(\psi) = 1 \text{ for all } w \in \mathcal{W}\}$. *Then* $\mathcal{W}$ *coincides with the set of restrictions to* $E$ *of all valuations satisfying* $\Theta$. *Further, for any map* $\beta \colon E \to [0, 1]$ *the following conditions are equivalent:*

*(i) $\beta$ is $\mathcal{W}$-coherent.*
*(ii) $\beta$ can be extended to a convex combination of valuations satisfying $\Theta$.*
*(iii) For some state $s$ of $\mathcal{L}(\Theta)$, $\beta(X_i) = s(\langle X_i \rangle_\Theta)$, for all $i = 1, \ldots, m$.*
*(iv) For some Borel probability measure $\mu$ on $\mathcal{W}$, $\beta(X_i) = \int_{\mathcal{W}} x_i \, d\mu$, for all $i = 1, \ldots, m$.*

If in (i)$\leftrightarrow$(iii) above we assume $\mathcal{W} \subseteq \{0, 1\}^E$, we obtain De Finetti's celebrated characterization of states of boolean algebras of yes-no events, in terms of his coherence criterion 9.1, (see [8, pp. 311-312], [9, Chapter 1], [10, pp. 85-90]). De Finetti's characterization was extended by Paris [37] to several modal logics, by Kühr et al., [23] to all $[0, 1]$-valued logics whose connectives are *continuous*, including all finite-valued logics. In their paper [1], Aguzzoli, Gerla and Marra further extend De Finetti's criterion to Gödel logic [19], a logic with a *discontinuous* implication connective. By Theorem 9.1, the various kinds of "events", "possible worlds" and "coherent probability assessments" arising from all these logics can be faithfully re-interpreted in terms of valuations and theories of Łukasiewicz logic.

The main ingredient in the proof of (iii)$\leftrightarrow$(iv) is the Kroupa-Panti theorem [22, 35] stating that in every MV-algebra $A$, the integral sends (*countably additive*) regular Borel probability measures on the maximal spectrum[3] MaxSpec(A) of A, one-one onto the (*finitely additive*) states of A. The claim that De Finetti's coherence criterion only characterizes finitely additive probability measures follows from the wrong impression that the theory of finitely additive measures on boolean algebras is more general than the theory of regular Borel measures. Actually, it would be more accurate to say the reverse: finitely additive measures (= states) on boolean algebras correspond to regular Borel measures on the boolean compact spaces given by their Stone spectra [41, 18.7]. Theorem 9.1 further clarifies this point: De Finetti's coherence criterion (Definition 9.1) yields a characterization of regular probability Borel measures on *any compact space* $\mathcal{W}$.

States and observables of physical systems provide a very general source of events and possible worlds. To fix ideas, for any physical system $Q$ let $A$ be its $C^*$-algebra, $A_{sa}$ the set of self-adjoint elements of $A$, and $S^*$ the set of real-valued normalized positive linear functionals on $A_{sa}$ (see [11] and [15] for background). $Q$ is said to be classical if $A$ is commutative. For any $w \in S^*$ and $X \in A_{sa}$ the real number $w(X)$ is the expectation value of the observable $X$ when $Q$ is prepared in mode $w$, [15, p. 362]. Let $E = \{X_1, \ldots, X_m\}$ be a set of nonzero positive elements of $A_{sa}$. As explained in [15, pp. 363-369], $w$ determines the map $w': E \to [0,1]$ given by $w'(X_i) = w(X_i)/\|X_i\|$, where $\|X_i\|$ is the norm of $X_i$. Intuitively, $X_i$ says

"the observable $X_i$ is expected to have a high value",

and $w'$ evaluates how true $X_i$ is. In particular, when $X_i$ is a yes-no observable, $w'(X_i)$ is the "truth-value of the event $X_i$ in the possible world $w'$". Again, Theorem 9.1 shows that coherent probability assessments of these events and possible worlds can be faithfully re-interpreted within Łukasiewicz logic.

Given the pervasive role of states of MV-algebras, one is naturally led to develop a theory of MV-algebraic conditionals.

Keisler has justified the definition of conditional probability by showing that it provides a bridge between probability and logic.

Rényi was one of several mathematicians who were star-struck by conditional probability: He did much work aimed at reshaping the foundations of probability in terms of conditional probability, but his ideas have not caught on. They were ahead of their time, as we will see in the next brief digression. Rényi's main axiom for conditional probability can be recast as follows (I hope Rényi will not turn in his grave for my restatement of his axiom in a form that he might not recognize): $P((A|B)|C) = P(A|B \cap C)$. [40, p. 74]

Our approach to conditional probability provides a bridge between probability and Łukasiewicz infinite-valued logic. For any two (events described by) formulas $\psi$ and $\theta$ we do not consider $\psi|\theta$ as a formula in any logic, and we are not concerned with Rényi's main axiom for conditional probability. Rather, following [31, 3.1-3.2], we define a conditional as an operator acting on the whole logic $\text{Ł}_\infty$:

**Definition 9.2.** A *conditional* is a map $\mathcal{P}: \theta \mapsto \mathcal{P}_\theta$ such that, for every $m = 1, 2, \ldots$ and every consistent formula $\theta \subseteq \mathcal{F}_m$, $\mathcal{P}_\theta$ is a state of the MV-algebra $\mathcal{L}(\theta^+)$ of (9.2)-(9.3). We say that $\mathcal{P}$ is *invariant* if for any two consistent formulas $\phi \in \mathcal{F}_m$, $\psi \in \mathcal{F}_n$,

---

[3] see Definition 9.3 below.

and isomorphism $\eta$ of $\mathcal{L}(\phi^{\vdash})$ onto $\mathcal{L}(\psi^{\vdash})$, we have $\mathcal{P}_\phi = \mathcal{P}_\psi \circ \eta$, where $\circ$ denotes composition. $\mathcal{P}$ is said to be *faithful* if so is every state $\mathcal{P}_\theta$.

For every formula $\alpha \in \mathcal{F}_m$ we will write $\mathcal{P}_\theta(\alpha)$ as an abbreviation of $\mathcal{P}_\theta(\langle \alpha \rangle_{\theta^{\vdash}})$, and say that $\mathcal{P}_\theta(\alpha)$ is *the probability of $\alpha$ given $\theta$*. The main result of [31] is

**Theorem 9.2.** *Łukasiewicz propositional logic Ł$_\infty$ has a faithful invariant conditional $\mathcal{P}^*$.*

It follows that $\mathcal{P}^*$ is invariant under equivalent reformulations of the same event:

**Corollary 9.1.** *Let $\psi$ be a formula in the variables $X_1, \ldots, X_n$. Then*

$$\mathcal{P}^*_{(\psi \leftrightarrow \psi)}(\psi) = \mathcal{P}^*_{(\psi \leftrightarrow X_{n+1})}(X_{n+1}).$$

*Thus, $\mathcal{P}^*$ does not make any distinction between*

*(i) the* unconditional *probability of (the event described by) formula $\psi$, i.e., the probability of $\psi$ given the uninformative tautology $\psi \leftrightarrow \psi$, and*
*(ii) the probability of the fresh variable $X_{n+1}$ given the information that $X_{n+1}$ is equivalent to $\psi$.*

We naturally say that formula $\alpha$ is $\mathcal{P}^*$-*independent of $\theta$* if the probability of $\alpha$ given $\theta$ coincides with the unconditional probability of $\alpha$. In view of Corollary 9.1, we can equivalently write

$$\mathcal{P}^*_\theta(\alpha) = \mathcal{P}^*_{\theta \leftrightarrow \theta}(\alpha) = \mathcal{P}^*_{\alpha \leftrightarrow \alpha}(\alpha) = \mathcal{P}^*_{X \leftrightarrow \alpha}(X),$$

where $X$ is a fresh variable.

A related, purely algebraic result, is the Haar theorem for finitely presented[4] MV-algebras:

**Theorem 9.3.** [30] *Every finitely presented MV-algebra has an invariant faithful state.*

## 9.2 MV-algebraic and $C^*$-algebraic States After Elliott Classification

I will lay my cards on the table: a revision of the notion of a sample space is my ultimate concern. I hasten to add that I am not about to put forth concrete proposals for carrying out such a revision. We will, however, be guided by a belief that has been a guiding principle of the mathematics of this century. Analysis will play second fiddle to algebra. The algebraic structure sooner or later comes to dominate, whether or not it is recognized when a subject is born. Algebra dictates the analysis. [40, p. 56]

---

[4] A *finitely presented* MV-algebra is the quotient of a finitely generated free MV-algebra by a finitely generated ideal.

The key role of MV-algebras for a revised notion of "sample space" is the main topic of this paper. Again, C*-algebras provide a useful framework. Following [13, 17], by a *(unital) AF C*-algebra* we understand the norm-closure of the union of an ascending sequence $F_1 \subseteq F_2 \subseteq \ldots$ of finite-dimensional C*-algebras, all with the same unit.

On the one hand, AF C*-algebras provide a rigorous description of systems having infinitely many degrees of freedom, such as those considered in quantum statistical mechanics, [3]. On the other hand, they are deeply related to MV-algebras. To see this, we prepare some material on partially ordered (always abelian) groups.

Following [18], an *order-unit* in a partially ordered group $G$ is an element $u \geq 0$ such that for each $x \in G$ there is an integer $n \geq 0$ such that $x \leq nu$. A *state* of $(G, u)$ is an order-preserving homomorphism $\sigma : G \to \mathbb{R}$ such that $\sigma(u) = 1$. A *unital $\ell$-group* is a lattice-ordered abelian group (for short, $\ell$-group) with a distinguished order-unit.

**Theorem 9.4.** [26, 3.9] *For any $\ell$-group $G$ with order-unit $u$, let $\Gamma(G, u)$ be the unit interval $[0, u] = \{g \in G \mid 0 \leq g \leq u\}$ equipped with the operations $\neg g = u - g$, $g \oplus h = u \wedge (g + h)$, $g \odot h = 0 \vee (g + h - u)$. Further, for every $\ell$-group $G'$ with order-unit $u'$, and $\ell$-group homomorphism $\psi : G \to G'$ such that $\psi(u) = u'$, let $\Gamma(\psi)$ be the restriction of $\psi$ to the unit interval of $G$. Then $\Gamma$ is a categorical equivalence between unital $\ell$-groups and MV-algebras.*                                                                              $\square$

**Corollary 9.2.** *(i) For any state $\tau$ of $(G, u)$, its restriction $t$ to the MV-algebra $B = \Gamma(G, u) = [0, u]$ is a state of $B$. (ii) Conversely, given a state $s : B \to [0, 1]$, there is a unique extension $\sigma$ of $s$ to a state of $(G, u)$.*

*Proof.* (i) Skipping all trivialities, let $x, y \in B$. By [26, 3.8.2] we can write $x + y = (x \oplus y) + (x \odot y)$. Thus in case $x \odot y = 0$ we have $t(x \oplus y) = \tau(x \oplus y) = \tau(x + y) = \tau(x) + \tau(y) = t(x) + t(y)$. (ii) The properties of $\Gamma$, together with the above identity, ensure that $s$ is a partially defined addition in $(G, u)$ obeying the assumptions of [21, Proposition 1.5], whence the desired conclusion easily follows.

## $K_0$ of AF C*-algebras

For any AF C*-algebra $A$ let $\wp$ be the family of all finitely generated projective left $A$-modules; in other words, $P \in \wp$ iff[5] $P$ is a direct summand of a free left $A$-module of finite rank. We say that $P, Q \in \wp$ are *stably isomorphic* if there is $T \in \wp$ such that $P + T \cong Q + T$. Denoting by $[P]$ the stable isomorphism class of $P$, we define $K_0(A)^+ = \{[P] \mid P \in \wp\}$. $K_0(A)^+$ is then turned into a cancellative abelian monoid by writing $[P] + [Q] = [P + Q]$ for all $P, Q \in \wp$. By formally adjoining additive inverses we obtain an abelian group $K_0(A)$, also known as the *Grothendieck group* of $A$. Each element of $K_0(A)$ has the form $[P] - [Q]$ for some $P, Q \in \wp$. Two elements $[P] - [Q]$

---

[5] "iff" is an abbreviation of "if and only if".

and $[S] - [T]$ are equal in $K_0(A)$ iff $[P + T] = [S + Q]$. Defining $[P] \sqsubseteq [Q]$ iff $[Q] - [P] \in K_0(A)^+$, we make $K_0(A)$ into a denumerable partially ordered abelian group. The stable isomorphism class $[A]$ of the left $A$-module $A$ is an order-unit in $K_0(A)$.

The following result is the core of Elliott's classification, [14]:

**Theorem 9.5.** [2, 13, 17] *Let $A$ and $B$ be AF $C^*$-algebras.*

*(i) $A$ is isomorphic to $B$ iff $(K_0(A), [A])$ and $(K_0(B), [B])$ are isomorphic as partially ordered groups with order-unit.*

*(ii) For every order-preserving homomorphism (resp., isomorphism) $\gamma : K_0(A) \to K_0(B)$ such that $\gamma([A]) = [B]$, there is a $C^*$-algebra morphism (resp., isomorphism) $\delta : A \to B$ such that $\gamma = K_0(\delta)$.*

## *The Interval $[[0], [A]]$ as a Set of Unitary Equivalence Classes of Projections*

For any $C^*$-algebra $A$, let $\mathrm{Proj}_A$ denote the set of projections of $A$, and $1_A$ the unit element of $A$. A main consequence of Elliott's classification is that all the information about $(K_0(A), [A])$ is already contained in its unit interval $[[0], [A]]$. As will be shown by Theorem 9.6, the latter is more conveniently realized as a set of equivalence classes of projections of $A$, rather than stable isomorphism classes of finitely generated projective $A$-modules. To this purpose, following [2, p. 29], let us say that two projections $p, q \in \mathrm{Proj}_A$ are *(Murray-von Neumann) equivalent* if there exists an element $v \in A$ such that $vv^* = p$ and $v^*v = q$. By [2, p. 53], in every AF $C^*$-algebra $A$, Murray-von Neumann equivalence coincides with unitary equivalence. (Recall that $p, q \in \mathrm{Proj}_A$ are said to be *unitarily equivalent* if $p = wqw^*$ for some unitary element $w$ in $A$; the map $x \mapsto wxw^*$ is called a *unitary automorphism* of $A$). Let $[p]$ be the equivalence class of $p$, and $D(A)$ be the set of equivalence classes of projections of $A$. Then the *Murray-von Neumann order* $\leq$ on $D(A)$ is defined by stipulating that $[p] \leq [q]$ iff $p$ is equivalent to a projection $r$ such that $rq = r$. *Elliott's partial addition*, denoted $+$, is the partially defined operation on $D(A)$ given by adding two projections whenever they are orthogonal. Whenever defined, Elliott's partial addition is associative, commutative, and monotone with respect to the Murray-von Neumann order. Further, $[1_A - p]$ is the only element $[q]$ of $D(A)$ such that $[q] + [p] = [1_A]$, whence, trivially, $[1_A - p]$ is *residual* in the following sense:

For every $p \in \mathrm{Proj}_A$, $[1_A - p]$ is the $\leq$-smallest element $[q]$ of the partial monoid $D(A)$ such that $[q] + [p] = [1_A]$.

**Theorem 9.6 ([2, 13, 17]).** *There is an order-isomorphism of $D(A)$ onto the unit interval $[[0], [A]]$ of $K_0(A)$, which is also an isomorphism of partial monoids.*

We will henceforth make no distinction between Elliott's partial monoid $D(A)$ and the unit interval $[[0],[A]]$ of $(K_0(A),[A])$.

**Theorem 9.7 ([26, 33]).** *Let A be an AF C\*-algebra.*

*(i) Elliott's partial addition can be extended to an associative commutative monotone residual operation $\oplus$ defined on all pairs of $D(A)$ iff $D(A)$ is a lattice with respect to the Murray-von Neumann order, iff $K_0(A)$ is an $\ell$-group.*

*(ii) Whenever such extension $\oplus$ exists, it is unique.*

*(iii) Assume that $D(A)$ is a lattice with respect to the Murray-von Neumann order. For each $[p] \in D(A)$ let $\neg[p] = [1_A - p]$. Let further $[p] \odot [q] = \neg(\neg[p] \oplus \neg[q])$. Then the structure $(D(A),[0],[1_A],\neg,\oplus,\odot)$ is isomorphic to the MV-algebra $B = \Gamma(K_0(A),[A])$.*

Projections abound in every *commutative* AF C\*-algebra $A$: as a matter of fact, by Gelfand theorem, $A$ is isomorphic to the C\*-algebra $C(X)$ of all continuous complex-valued functions on $X$, for $X$ a separable and boolean topological space. Projections also abound in every *noncommutative* AF C\*-algebra $B$, because the linear span of projections is dense in $B$. The intuition that AF C\*-algebras are "noncommutative boolean algebras" is made precise in the following

**Theorem 9.8.** [26, 3.12] . *The map $A \mapsto \Gamma(K_0(A),[A])$ provides a one-one correspondence between isomorphism classes of AF C\*-algebras whose Murray-von Neumann order is a lattice, and isomorphism classes of countable MV-algebras. Commutative AF C\*-algebras correspond to countable boolean algebras.*

Recall that a *tracial state* on a $C^*$-algebra $A$ is a unit-preserving positive linear functional satisfying $s(aa^*) = s(a^*a)$ for all $a \in A$. We say that $s$ is *faithful* if for each positive nonzero element $a \in A$ we have $s(a) > 0$.

Following [25] we say that a map $m: \mathrm{Proj}_A \to [0,1]$ is a *measure on the projections* of $A$ if $m(1_A) = 1$ and $m(p) + m(q) = m(p + q)$, whenever $pq = 0$. We say that $m$ is *faithful* if $m(p) > 0$ for all $0 \neq p \in \mathrm{Proj}_A$; $m$ is *invariant* if $m(p) = m(\alpha(p))$ for every projection $p$ and every automorphism $\alpha$ of $A$. We let $\mathfrak{I}_A$ denote the set of invariant measures on the projections of $A$.

The following result was stated without proof in [27, 4.3]:

**Theorem 9.9.** *Let A be an AF C\*-algebra whose Murray-von Neumann order is a lattice. Let $B = \Gamma(K_0(A),[A])$ be its corresponding MV-algebra, as given by Theorem 9.8. Let $\mathfrak{S}_B$ be the set of states of B, and $\mathfrak{T}_A$ the set of tracial states of A. We then have:*

1. *Defining the map $^*: \mathfrak{I}_A \to \mathfrak{S}_B$ by $m^*([p]) = m(p)$ for all $p \in \mathrm{Proj}_A$, it follows that $^*$ maps $\mathfrak{I}_A$ one-one onto the set of invariant states of B. Faithful invariant measures on the projections of A correspond to faithful invariant states.*

2. *Defining the map $^\natural: \mathfrak{T}_A \to \mathfrak{S}_B$ by $s^\natural([p]) = s(p)$ for all $p \in \mathrm{Proj}_A$, it follows that $^\natural$ maps $\mathfrak{T}_A$ one-one onto $\mathfrak{S}_B$. Faithful tracial states of A correspond to faithful states of B.*

*Proof.* (1) By Theorem 9.6, we can write $(B, \oplus, 0) = (D(A), \oplus, 0)$. Elements of $B$ are Murray-von Neumann equivalence (=unitary equivalence) classes of projections of $A$. The $\neg$ operation in $B$ is the operation $\neg[p] = [1_A - p]$. By Theorem 9.7, the $\oplus$ operation is the only associative commutative monotone residual extension of Elliott's partial addition $+$. The Murray-von Neumann order $\preceq$ is definable by $[p] \preceq [q]$ iff $\neg[p] \oplus [q] = [1_A] = 1$ iff $[p] \odot \neg[q] = [0] = 0$. Given projections $p$ and $q$ in $A$, the condition $0 = [p] \odot [q]$ is equivalent to the existence of a projection $r \perp p$ such that $[r] = [q]$. Indeed,

$$pr = 0 \leftrightarrow p \leq (1_A - r) \leftrightarrow [p] \preceq \neg[r] = \neg[q] \leftrightarrow 0 = [p] \odot [r] \leftrightarrow 0 = [p] \odot [q].$$

Let $m \in \mathfrak{I}_A$. Let $p, q \in \text{Proj}_A$ satisfy $[p] = [q]$. Then there is a unitary automorphism sending $p$ into $q$, and hence, $m(p) = m(q)$. Thus $m$ determines a well defined map $m^* \colon B \to [0,1]$, given by $m^*([p]) = m(p)$ for all $p \in \text{Proj}_A$. It is easily seen that $m^*(1) = 1$. If $a, b \in B$ and $a \odot b = 0$, then $a = [p]$ and $b = [q]$ for suitable orthogonal projections $p, q \in A$. Now $p + q$ is still a projection, and $m(p + q) = m(p) + m(q)$. The pair $([p], [q])$ is in the domain of Elliott's partial addition and, by Theorem 9.5, the equivalence class $[p] + [q] = [p + q] \in D(A)$ coincides with the element $a \oplus b$ of $B$. This concludes the proof that $m^* \in \mathfrak{S}_B$. We will now prove that $m^*$ is invariant under every automorphism $\beta$ of $B$. By Theorems 9.4 and 9.5, we can write $\beta = \Gamma(\gamma)$ and $\gamma = K_0(\alpha)$ for some automorphism $\gamma$ of $(K_0(A), [A])$ and automorphism $\alpha$ of $A$. Further, by definition of the functors $\Gamma$ and $K_0$, for every $[p] \in B$ we have $\beta([p]) = [\alpha(p)]$, independently of the representative $p \in \text{Proj}(A)$ chosen for $[p]$. Since $m$ is invariant, so is $m^*$. In conclusion, $m^*$ is an invariant state of $B$. Faithfulness of $m^*$ follows from the faithfulness of $m$ by [17, Ex. 19J].

Conversely, if $s$ is an invariant state of $B$, then $s$ determines a $[0,1]$-valued function $s_*$ on unitary equivalence classes of projections in $A$, because unitarily equivalent projections are mapped by $K_0$ into the same element of $B$. Trivially, $s_*(1_A) = 1$. To prove that $s_*$ is additive on orthogonal projections $p, q \in A$, we first note that $[p] \odot [q]$ is the zero element of $B$. It follows that $s_*(p) + s_*(q) = s([p]) + s([q]) = s([p] \oplus [q]) = s([p] + [q]) = s([p + q]) = s_*(p + q)$, as desired. Arguing as in the first part, and again making use of Theorems 9.4 and 9.5, we see that the invariance of $s_*$ follows from the invariance of $s$. Therefore, $s_* \in \mathfrak{I}_A$. The faithfulness of $s_*$ follows from the faithfulness of $s$, by [17, Ex. 19J].

(2) By Theorem 9.7, $K_0(A)$ is an $\ell$-group. Further, the element $u = [1_A] = [A]$ is an order-unit in $K_0(A)$. By [18, p. 310] or [2, p. 57], the $K_0$-functor induces a 1-1 correspondence between $\mathfrak{I}_A$ and the set of states of $(K_0(A), [A])$. Recalling Theorem 9.5 we can write $1 = u = [A]$, and $(B, 0, 1, \neg, \oplus, \odot) = [0, u] \subseteq K_0(A)$. By Corollary 9.2 for any state $\tau$ of $(K_0(A), [A])$, its restriction $t$ to the unit interval $[0, u]$ is a state of $B$. Conversely, given a state $s \in \mathfrak{S}_B$, there is a unique extension $\sigma$ of $s$ to a state of $(K_0(A), [A])$. Thus, the functors $K_0$ and $\Gamma$ induce the desired correspondence $\natural$ between $\mathfrak{I}_A$ and $\mathfrak{S}_B$. Preservation of faithfulness is ensured by the well known properties of $K_0$ and $\Gamma$ (see, e.g., [17, p. 160] and [26, Section 3]).

See [34] for further results jointly involving MV-algebras and AF C*-algebras.

## 9.3 MV-algebraic $\sigma$-states and Carathéodory Probability Theory

A Boolean $\sigma$-algebra can be defined as a distributive lattice satisfying certain additional technical conditions. A theorem of Loomis and Sikorski insures that every Boolean $\sigma$-algebra can be represented as the quotient of a Boolean $\sigma$-algebra of events in a sample space by a $\sigma$-ideal. Pointless probability deals with an abstract Boolean $\sigma$-algebra $\Pi$ and with a real-valued function defined on $\Pi$ which imitates the definition of probability. The problem is to define an algebra of random variables.

The setup is not as artificial as it may appear. Among probabilists, mention of sample points in an argument has always been bad form. A fully probabilistic argument must be pointless. [40, p. 60]

An MV-algebra $M$ is said to be $\sigma$-*complete* if its underlying lattice [7, 1.1.5] is $\sigma$-complete, i.e., every non-empty countable subset of $M$ has a supremum in $M$. Given two $\sigma$-complete MV-algebras $N$ and $M$, a homomorphism $\eta \colon N \to M$ is said to be a $\sigma$-*homomorphism* if $\eta(\bigvee_i a_i) = \bigvee_i(\eta(a_i))$ for all sequences $a_1, a_2, \ldots \in N$.

With the trivial exception of its finite members, the class of $\sigma$-complete MV-algebras does not correspond via $K_0$ to any class of AF C*-algebras. Still, this class provides an interesting generalization of Carathéodory boolean algebraic probability theory. Every $\sigma$-algebra of sets and more generally, every boolean $\sigma$-algebra, is a $\sigma$-complete MV-algebra. As a further example of a $\sigma$-complete MV-algebra, given a set $\Omega \neq \emptyset$, a *tribe* over $\Omega$ is an MV-algebra $\mathcal{F}$ of $[0, 1]$-valued functions on $\Omega$ such that for each sequence $f_1, f_2, \ldots \in \mathcal{F}$ the pointwise supremum $f = \sup_i f_i$ belongs to $\mathcal{F}$.

**Definition 9.3.** For every MV-algebra $B$, we denote by MaxSpec($B$) the set of all maximal ideals of $B$. The family of subsets of MaxSpec($B$) of the form $F_a = \{\mathfrak{m} \in \text{MaxSpec}(B) \mid a \in \mathfrak{m}\}$, with $a$ ranging over all elements of $B$, is a basis of closed sets for the *spectral* (or, *hull-kernel*) topology of MaxSpec($B$).

The following result, independently proved in [28] and [12], generalizes the classical Loomis-Sikorski theorem:

**Theorem 9.10.** *Let $M$ be a $\sigma$-complete MV-algebra. Let $\Omega$ be the set of maximal ideals of $M$ equipped with the spectral topology. Then there exists a tribe $\mathcal{F}$ over $\Omega$ and a $\sigma$-homomorphism $\eta$ of $\mathcal{F}$ onto $M$ with the following properties: For each $a \in M$ there is precisely one continuous function $\bar{a} \in \mathcal{F}$ such that $\eta(\bar{a}) = a$. A function in $\mathcal{F}$ has the same image as $\bar{a}$ iff it differs from $\bar{a}$ on a meager subset of $\Omega$.*

Let $\Omega$ be an (always nonempty) set and $S$ a $\sigma$-algebra of subsets of $\Omega$. Let $\mathcal{F}_S$ be the set of all $S$-measurable functions from $\Omega$ to $[0,1]$. Then $\mathcal{F}_S$ is a tribe and, trivially, all $[0, 1]$-valued constant functions over $\Omega$ are members of $\mathcal{F}_S$. Conversely, given a subset $Y$ of $\Omega$ let $\chi_Y \colon \Omega \to \{0, 1\}$ denote the characteristic function of $Y$. We then have:

**Theorem 9.11.** [5, 3.2, 3.3]. *Let $\mathcal{F}$ be a tribe over $\Omega$. Let $S_{\mathcal{F}}$ be the family of subsets $Y$ of $\Omega$ such that $\chi_Y \in \mathcal{F}$. Then $S_{\mathcal{F}}$ is a $\sigma$-algebra of subsets of $\Omega$ and each function in $\mathcal{F}$ is $S_{\mathcal{F}}$-measurable. If, in addition, $\mathcal{F}$ contains all $[0, 1]$-valued constant functions then $\mathcal{F}$ coincides with the set of all $S_{\mathcal{F}}$-measurable functions $f \colon \Omega \to [0, 1]$.*

A sequence $a_1, a_2, \ldots$ of elements in a partially ordered set $(Y, \leq)$ is said to be *monotone* if $a_1 \leq a_2 \leq \ldots$. The notation $a_n \nearrow a$ stands for "$a_1, a_2, \ldots$ is a monotone sequence whose supremum is $a$".

In the light of Theorem 9.11, the following is a generalization of the notion of probability measure on a $\sigma$-algebra of sets:

**Definition 9.4.** A $\sigma$-*state* of a tribe $\mathcal{F}$ is a map $m \colon \mathcal{F} \to [0,1]$ satisfying the following conditions: (i) $m(1) = 1$; (ii) if $a, b, c \in \mathcal{F}$ and $a = b + c$, then $m(a) = m(b) + m(c)$, and (iii) for all $a, a_1, a_2, \ldots \in \mathcal{F}$, if $a_n \nearrow a$, then $m(a_n) \nearrow m(a)$.

**Theorem 9.12.** ([4, Theorem 2.6(c)], [5], [38, 8.1]). *Let $\mathcal{F}$ be a tribe over $\Omega$ with a $\sigma$-state m. Let $S$ be the $\sigma$-algebra of all subsets $Y$ of $\Omega$ such that $\chi_Y \in \mathcal{F}$. Let the map $P \colon S \to [0,1]$ be defined by $P(Y) = m(\chi_Y)$. Then $P$ is a probability measure on $S$, and $m(f) = \int_\Omega f \, \mathrm{d}P$ for every $f \in \mathcal{F}$.*

As a trivial converse, for any probability space $(\Omega, S, P)$ let $\mathcal{F}$ be the tribe of all $S$-measurable $[0,1]$-valued functions over $\Omega$. Let the $\sigma$-state $m_P \colon \mathcal{F} \to [0,1]$ be given by $m_P(f) = \int_\Omega f \, \mathrm{d}P$. Then $\mathcal{F}$ is a tribe over $\Omega$ containing all $[0,1]$-valued constant functions. Applying Theorem 9.12 to $(\Omega, \mathcal{F}, m_P)$ we recover the probability space $(\Omega, S, P)$.

## MV-algebraic States and Observables

> Carathéodory and von Neumann have shown beyond reasonable doubt by their algebraization of probability that in probabilistic reasoning the notion of a sample point is a psychological prop. To think probabilistically means to think in terms of events in a $\sigma$-algebra. To think probabilistically is to think pointlessly. [40, p. 79]

In Carathéodory's approach [6] to probability theory, one disregards sets of zero measure, and considers instead boolean $\sigma$-algebras equipped with a distinguished strictly positive probability measure. We will give only a brief sketch of the MV-algebraic generalization of Carathéodory theory. We refer to [39] for a comprehensive account and for proofs.

First of all, generalizing Definition 9.4, for any $\sigma$-complete MV-algebra $M$ we define a $\sigma$-*state*[6] to be a map $m \colon M \to [0,1]$ satisfying the following conditions for all $a, b, c, a_n, b_n \in M$: (i) $m(1) = 1$; (ii) whenever $b, c \in M$ and $b \odot c = 0$ it follows that $m(b \oplus c) = m(b) + m(c)$, and (iii) if $a_n \nearrow a$, then $m(a_n) \nearrow m(a)$. We say that $m$ is *faithful* (or *strictly positive*) if $m(x) \neq 0$ whenever $0 \neq x \in M$.

Combining Theorems 9.10 and 9.12 we immediately obtain:

**Theorem 9.13.** *Let $M$ be a $\sigma$-complete MV-algebra with a $\sigma$-state m. Let $\Omega$ be the space of maximal ideals of $M$ equipped with the spectral topology. Let the tribe $\mathcal{F}$ over $\Omega$, and the $\sigma$-homomorphism $\eta \colon \mathcal{F} \to M$ be as in Theorem 9.10. Then the*

---

[6] called "state" in [39].

*composite map $m \circ \eta$ is a $\sigma$-state on $\mathcal{F}$. Let $(\Omega, S, P)$ be obtained from $(\Omega, \mathcal{F}, m \circ \eta)$ as in Theorem 9.12. Let the map $^- \colon M \to \mathcal{F}$ send each $a \in M$ into the unique continuous function $\bar{a} \in \mathcal{F}$ such that $\eta(\bar{a}) = a$. Then for all $a \in M$, $m(a) = \int_\Omega \bar{a} \, dP$.*

Generalizing the classical notion of "probability algebra", or "boolean $\sigma$-algebra with a normalized positive measure" (see [16], and [20, p. 64]), by a *probability MV-algebra* we understand a pair $(M, m)$ where $M$ is a $\sigma$-complete MV-algebra and $m$ is a faithful $\sigma$-state on $M$.

**Proposition 9.1.** *Let $\mathcal{F}$ be a tribe over $\Omega$ with a $\sigma$-state $m$. Then the set $J$ of all $f \in \mathcal{F}$ such that $m(f) = 0$ is an ideal of the MV-algebra $\mathcal{F}$, and is closed under countable suprema. The quotient MV-algebra $M = \mathcal{F} / J$ is a $\sigma$-complete MV-algebra, and the quotient map $\theta \colon \mathcal{F} \to M$ is a $\sigma$-homomorphism. The map $\widecheck{m} \colon M \to [0, 1]$ defined by $\widecheck{m}(\theta(f)) = m(f)$ for each element $\theta(f) \in M$ is a faithful $\sigma$-state on $M$, and $(M, \widecheck{m})$ is a probability MV-algebra.*

A variant of Theorem 9.13 shows that the construction of Proposition 9.1 yields the most general possible example of a probability MV-algebra.

In the point-free version of probability, the algebraic counterpart of a random variable is not given by a real-valued function over the sample space $\Omega$, but by a function from the $\sigma$-algebra $\mathcal{B}(\mathbb{R})$ of Borel sets of $\mathbb{R}$ into the boolean $\sigma$-algebra of events. The explicit construction of sums and products of such random variables in terms of boolean operations was successfully carried out by Carathéodory.

As a generalization of his theory, for any $\sigma$-complete MV-algebra $M$ we define an *n-dimensional observable* of $M$ to be a map $x \colon \mathcal{B}(\mathbb{R}^n) \to M$ satisfying the following conditions:   (i) $x(\mathbb{R}^n) = 1$;   (ii) whenever $X, Y \in \mathcal{B}(\mathbb{R}^n)$ and $X \cap Y = \emptyset$, then $x(X) \odot x(Y) = 0$ and $x(X \cup Y) = x(X) \oplus x(Y)$;   (iii) for all $X, X_1, X_2, \ldots \in \mathcal{B}(\mathbb{R}^n)$, if $X_n \nearrow X$, then $x(X_n) \nearrow x(X)$. When $n = 1$ we simply say that $x$ is an *observable*.

In the particular case when $M$ is a boolean $\sigma$-algebra, this definition boils down to Sikorski's notion of real homomorphism [42, p. 152ff], and also coincides with Varadarajan's definition of observable [43, p. 14ff]. As an example, let $(\Omega, S, P)$ be a probability space, $\xi \colon \Omega \to \mathbb{R}$ a random variable, and $\mathcal{F}$ the tribe of all $[0, 1]$-valued $S$-measurable functions on $\Omega$. Then the map $x \colon \mathcal{B}(\mathbb{R}) \to \mathcal{F}$ defined by $x(X) = \chi_{\xi^{-1}(X)}$ is an observable.

The proof of the following proposition is immediate.

**Proposition 9.2.** *Let $M$ be a $\sigma$-complete MV-algebra with an n-dimensional observable $x \colon \mathcal{B}(\mathbb{R}^n) \to M$ and a $\sigma$-state $m \colon M \to [0, 1]$. Let the map $m_x \colon \mathcal{B}(\mathbb{R}^n) \to [0, 1]$ be defined by $m_x(X) = (m \circ x)(X) = m(x(X))$, for all $X \in \mathcal{B}(\mathbb{R}^n)$. Then $m_x$ is a probability measure on $\mathcal{B}(\mathbb{R}^n)$.*

The map $m_x$ has the same role as the probability distribution of a random variable in the classical Kolmogorov theory. Thus for instance, given a a probability MV-algebra $(M, m)$ and an observable $x \colon M \to \mathcal{B}(\mathbb{R})$ of $M$, we say that $x$ is *integrable* if the *expectation* $E(x) = \int_{\mathbb{R}} t \, dm_x(t)$ exists. We say that $x$ is *square integrable* if the *variance* $\sigma^2(x) = \int_{\mathbb{R}} (t - E(x))^2 \, dm_x(t)$ exists.

## Independence and the MV-algebraic Central Limit Theorem

In the context of $\sigma$-complete MV-algebras, the notion of independence has the following more traditional formulation than the one given at the end of Section 9.1:

Suppose we are given two random variables $\xi$ and $\eta$ in a probability space $(\Omega, S, P)$. We then have

a random vector $T\colon \Omega \to \mathbb{R}^2$ defined by $T(\omega) = (\xi(\omega), \eta(\omega))$ for each $\omega \in \Omega$;

a $\sigma$-homomorphism $h\colon \mathcal{B}(\mathbb{R}^2) \to S$ defined by $X \mapsto T^{-1}(X)$;

a probability distribution $P_T\colon \mathcal{B}(\mathbb{R}^2) \to [0,1]$ given by $P_T(X) = P(T^{-1}(X)) = P(h(X))$.

We say that $\xi$ and $\eta$ are *independent* if for all Borel sets $X, Y \subseteq \mathbb{R}$, $P(\xi^{-1}(X) \cap \eta^{-1}(Y)) = P(\xi^{-1}(X)) \cdot P(\eta^{-1}(Y))$. Recalling the notation of Proposition 9.2, since $\xi^{-1}(X) \cap \eta^{-1}(Y) = T^{-1}(X \times Y)$, we can write

$$P_T(X \times Y) = P_\xi(X) \cdot P_\eta(Y) = (P_\xi \times P_\eta)(X \times Y),$$

where $P_\xi \times P_\eta$ is the *product measure* of $P_\xi = P \circ \xi^{-1}$ and $P_\eta = P \circ \eta^{-1}$. Since $P_\xi \times P_\eta$ coincides with $P_T$ on rectangles, these two measures automatically coincide over all of $\mathcal{B}(\mathbb{R}^2)$. It follows that $P_T = P_\xi \times P_\eta$. Since $P_T = P \circ h$, we can say that $\xi$ and $\eta$ are independent iff there exists a $\sigma$-homomorphism $h\colon \mathcal{B}(\mathbb{R}^2) \to S$ such that $P \circ h = P_\xi \times P_\eta$. The generalization to $n$ random variables is straightforward.

Now let $(M, m)$ be a probability MV algebra. We say that the observables $x_1, \dots, x_n$ are *independent* if there exists an $n$-dimensional observable $h\colon \mathcal{B}(\mathbb{R}^n) \to M$ such that for all $Z_1, \dots, Z_n \in \mathcal{B}(\mathbb{R})$,

$$m(h(Z_1 \times \dots \times Z_n)) = m(x_1(Z_1)) \cdot \dots \cdot m(x_n(Z_n)) = m_{x_1}(Z_1) \cdot \dots \cdot m_{x_n}(Z_n).$$

We say that $h$ is the *joint observable* of $x_1, \dots, x_1$ in $(M, m)$. While $h$ is not uniquely determined, for any two joint observables $h_1$ and $h_2$ we have the identity $m \circ h_1 = m \circ h_2$ on all sets of the form $Z_1 \times \dots \times Z_n$, whenever $Z_1, \dots, Z_n \in \mathcal{B}(\mathbb{R})$. Since $m \circ h_1$ and $m \circ h_2$ are probability measures on $\mathcal{B}(\mathbb{R}^n)$ they must coincide over all of $\mathcal{B}(\mathbb{R}^n)$.

The *addition* $x + y$ of two independent observables $x, y$ in $(M, m)$ is defined by stipulating that, for all $X \in \mathcal{B}(\mathbb{R})$, $(x + y)(X) = h(\text{plus}^{-1}(X))$, where $\text{plus}\colon \mathbb{R}^2 \to \mathbb{R}$ denotes the usual addition operation, and $h$ is the joint observable of $x$ and $y$. Similarly, *product* is defined by $(x \cdot y)(X) = h(\text{times}^{-1}(X))$, where $\text{times}\colon \mathbb{R}^2 \to \mathbb{R}$ denotes ordinary multiplication. For any independent observables $x_1, \dots, x_n$, let $h\colon \mathcal{B}(\mathbb{R}^n) \to M$ be their joint observable. Let the map $\text{average}\colon \mathbb{R}^n \to \mathbb{R}$ be defined by $\text{average}(t_1, \dots, t_n) = \frac{1}{n} \sum_{i=1}^{n} t_i$, for all $t_1, \dots, t_n \in \mathbb{R}$. Then the *arithmetical mean* of $x_1, \dots, x_n$ is defined by

$$\frac{1}{n} \sum_{i=1}^{n} x_i = h \circ \text{average}^{-1}.$$

A sequence $x_1, x_2, \ldots$ of observables is said to be *independent* if every finite subsequence $x_1, x_2, \ldots x_n$ is independent.

**Theorem 9.14.** (Central Limit Theorem, [38, Theorem 9.2.6]) *Given a probability MV-algebra $(M, m)$, let $x_1, x_2, \ldots$ be an independent sequence of square integrable observables having the same probability distribution $m_{x_1} = m_{x_2} = \ldots$. Let a be their common expectation, and $\sigma^2 > 0$ their common variance. Then for all $t \in \mathbb{R}$,*

$$\lim_{n \to \infty} m\left( \frac{x_1 + \cdots + x_n - na}{\sigma \sqrt{n}} (\,]-\infty, t[\,) \right) = \frac{1}{\sqrt{2\pi}} \int_{-\infty}^{t} e^{-\frac{u^2}{2}} \, du.$$

# References

1. Aguzzoli, S., Gerla, B., Marra, V. (2008), *De Finetti's no-Dutch-Book Criterion for Gödel Logic*, in "Studia Logica", 90, special issue, Shier Ju et al. (eds.), *Many-valued Logic and Cognition*, pp. 25-41.
2. Blackadar, B. (1987), *K-Theory for Operator Algebras*, New York, Springer.
3. Bratteli, O., Robinson, D. W. (1979), *Operator Algebras and Quantum Statistical Mechanics I, II*, Berlin, Springer.
4. Butnariu, D. (1987), *Values and Cores of Fuzzy Games with Infinitely Many Players*, in "J. Game Theory", 16, pp. 43-68.
5. Butnariu, D., Klement, E. P. (1993), *Triangular Norm-based Measures and Games with Fuzzy Coalitions*, Dordrecht, Kluwer.
6. Carathéodory, C. (1986), *Mass und Integral und ihre Algebraisierung*, Boston, Basel, Berlin, Birkhäuser, 1956, English translation: *Algebraic Theory of Measure and Integration*, 2nd ed., New York, Chelsea.
7. Cignoli, R., D'Ottaviano, I. M. L., Mundici, D. (2000), *Algebraic Foundations of Many-Valued Reasoning*, Dordrecht, Kluwer - New York, Springer.
8. De Finetti, B. (1993), *Sul significato soggettivo della probabilità*, in "Fundamenta Mathematicae", 17 (1931), pp. 298-329. Translated into English as *On the Subjective Meaning of Probability*, in P. Monari and D. Cocchi (eds.), *Probabilità e induzione*, Bologna, Clueb, pp. 291-321.
9. De Finetti, B. (1980), *La prévision: ses lois logiques, ses sources subjectives*, in "Annales de l'Institut H. Poincaré", 7 (1937), pp. 1-68. Translated into English by H. E. Kyburg Jr., as *Foresight: Its Logical Laws, its Subjective Sources*, in H. E. Kyburg Jr. and H. E. Smokler (eds.), *Studies in Subjective Probability*, 2nd ed., New York, Krieger, pp. 53-118.
10. De Finetti, B. (1974), *Theory of Probability*, vol. 1, Chichester, John Wiley & Sons.
11. Dixmier, J. (1977), *C\*-algebras*, Amsterdam, North-Holland.
12. Dvurečenskij, A. (1999), *Loomis-Sikorski Theorem for $\sigma$-complete MV-algebras and $\ell$-groups*, in "J. Australian Math. Soc. Ser. A", 67, pp. 1-17.
13. Effros, E. G. (1981), *Dimensions and C\*-algebras*, in "CBMS Regional Conf. Series in Math.", 46, Providence, RI, American Mathematical Society.
14. Elliott, G. A. (1976), *On the Classification of Inductive Limits of Sequences of Semisimple Finite Dimensional Algebras*, in "J. Algebra", 38, pp. 29-44.
15. Emch, G. G. (1984), *Mathematical and Conceptual Foundations of 20th Century Physics*, Amsterdam, North-Holland.
16. Fremlin, D. H. (1989), *Measure Algebras*, in J. D. Monk (ed.), *Handbook of Boolean Algebras*, Vol. 3, Amsterdam, North-Holland.
17. Goodearl, K. R. (1982), *Notes on Real and Complex C\*-Algebras*, (Shiva Mathematics Series, 5), Nantwich, Shiva Publishing.

18. Goodearl, K. R. (1986), *Partially Ordered Abelian Groups with Interpolation*, in "AMS Math. Surveys and Monographs", 20.
19. Hájek, P. (1998), *Metamathematics of Fuzzy Logic*, Dordrecht, Kluwer.
20. Halmos, P. R. (1974), *Lectures on Boolean Algebras*, New York, Springer.
21. Handelman, D., Higgs, D., Lawrence, J. (1980), *Directed Abelian Groups, Countably Continuous Rings, and Rickart C\*-algebras*, in "J. London Math. Soc.", 21, pp. 193-202.
22. Kroupa, T. (2006), *Every State on Semisimple MV-algebra is Integral*, in "Fuzzy Sets and Systems", 157, pp. 2771-82.
23. Kühr, J., Mundici, D. (2007), *De Finetti Theorem and Borel States in [0,1]-valued Algebraic Logic*, in "International Journal of Approximate Reasoning", 46, pp. 605-16.
24. Lukasiewicz, J., Tarski, A. (1983), *Investigations into the Sentential Calculus*, in A Tarski, *Logic, Semantics, Metamathematics*, Oxford, Clarendon Press, 1956. Reprinted Indianapolis, Hackett, pp. 38-59.
25. Maeda, S. (1990), *Probability Measures on Projections in von Neumann Algebras*, in "Reviews in Math. Phys.", 1, pp. 235-90.
26. Mundici, D. (1986), *Interpretation of AF C\*-algebras in Łukasiewicz Sentential Calculus*, in "J. Functional Analysis", 65, pp. 15-63.
27. Mundici, D. (1995), *Averaging the Truth-value in Łukasiewicz Logic*, in "Studia Logica", 55, pp. 113-27.
28. Mundici, D. (1999), *Tensor Products and the Loomis-Sikorski Theorem for MV-algebras*, in "Advances in Applied Mathematics", 22, pp. 227-48.**22**, pp. 227-248 (1999).
29. Mundici, D. (2006), *Bookmaking Over Infinite-valued Events*, in "International J. Approximate Reasoning", 43, pp. 223-40.
30. Mundici, D. (2008), *The Haar Theorem for Lattice-ordered Abelian Groups with Order-unit*, in "Discrete and Continuous Dynamical Systems", 21, pp. 537-49.
31. Mundici, D. (2008), *Faithful and Invariant Conditional Probability in Łukasiewicz logic*, in D. Makinson, J. Malinowski and H. Wansing (eds.), *Proceedings of the Conference Trends in Logic IV, Torun, Poland, 2006*, New York, Springer, pp. 213-32.
32. Mundici, D., *Interpretation of De Finetti Coherence Criterion in Łukasiewicz logic*, in "Annals of Pure and Applied Logic", in print.
33. Mundici, D., Panti, G. (1993), *Extending Addition in Elliott's Local Semigroup*, in "J. Functional Analysis", 117, pp. 461-72.
34. Mundici, D., Tsinakis,C. (2008), *Gödel Incompleteness in AF C\*-algebras*, in "Forum Mathematicum", 20, pp. 1071-84.
35. Panti, G. (2008), *Invariant Measures in Free MV-algebras*, in "Communications in Algebra", 36, pp. 2849-61.
36. Pap, E. (ed.) (2002), *Handbook of Measure Theory, I, II*, Amsterdam, North-Holland.
37. Paris, J. (2001), *A Note on the Dutch Book Method*, in G. De Cooman, T. Fine, T. Seidenfeld (eds.), *Proceedings of the Second International Symposium on Imprecise Probabilities and their Applications, ISIPTA 2001*, Ithaca, NY, Shaker, pp. 301-6 (http://www. maths. man. ac. uk/DeptWeb/Homepages/jbp/)
38. Riečan, B., Neubrunn, T. (1997), *Integral, Measure, and Ordering*, Dordrecht, Kluwer.
39. Riečan, B., Mundici, D. (2002), *Probability on MV-algebras*, in E. Pap (ed.), *Handbook of Measure Theory*, Amsterdam, North-Holland, Vol. II, pp. 869-909.
40. Rota, G.-C. (2001), *Twelve Problems in Probability No One Likes to Bring Up*, in H. Crapo, D. Senato (eds.), *Algebraic Combinatorics and Computer Science, A Tribute to Gian-Carlo Rota*, Milan, Springer Italia, pp. 57-93.
41. Semadeni, Z. (1971), *Banach Spaces of Continuous Functions*, Vol. I, Warsaw, PWN-Polish Scientific Publishers.
42. Sikorski, R. (1960), *Boolean Algebras*, Berlin, Springer.
43. Varadarajan, V. (1968), *Geometry of Quantum Theory*, Vol. 1., Princeton, Van Nostrand.
44. Tarski, A. (1983), *Logic, Semantics, Metamathematics*, Oxford, Clarendon Press, 1956. Reprinted Indianapolis, Hackett.

# Chapter 10
# A Symbolic Treatment of Abel Polynomials
## Contributed Chapter

Pasquale Petrullo

**Abstract** We study the umbral polynomials $A_n^{(k)}(x,\alpha) = x(x - k.\alpha)^{n-1}$, by means of which a wide range of formal power series identities, including Lagrange inversion formula, can be usefully manipulated. We apply this syntax within cumulant theory, and show how moments and its formal cumulants (classical, free and Boolean) are represented by polynomials $A_n^{(k)}(\alpha,\gamma)$ for suitable choices of umbrae $\alpha$ and $\gamma$.

## 10.1 Introduction

Classical umbral calculus was formalized by Rota and Taylor [12] in 1994, with the aim of making rigorous a powerful notation, largely used by the mathematicians of the nineteenth century, involving the manipulation of sequences of numbers and polynomials. The main tools of classical umbral calculus are the notion of "umbra"and a linear functional $E$, called "evaluation", mapping the powers $\alpha^0, \alpha^1, \alpha^2, \ldots$ of an umbra $\alpha$ into a sequence $1, a_1, a_2, \ldots$ of elements of some ring $R$. In this case, such a sequence is said to be "represented"by $\alpha$, and we abbreviate $E[\alpha^n] = a_n$ by $\alpha^n \simeq a_n$. By extending coefficientwise the action of $E$ to the formal power series expansion of $e^{\alpha z}$, we obtain the "generating function"of $\alpha$, that is

$$f(\alpha, z) = E[e^{\alpha z}] = \sum_{n \geq 0} E[\alpha^n] \frac{z^n}{n!} = 1 + \sum_{n \geq 1} a_n \frac{z^n}{n!},$$

or shortly $e^{\alpha z} \simeq f(\alpha, z)$. The umbra $n.\alpha$ can be then defined to have generating function $f(\alpha, z)^n$, so that $e^{(n.\alpha)z} \simeq f(\alpha, z)^n$.

As is well known, Abel polynomials are polynomials of type $x(x - na)^{n-1}$, where $x$ is a variable, $a$ a complex number and $n$ a positive integer. By replacing $-na$ with the umbra $-n.\alpha$, Rota, Shen and Taylor [11] have defined the umbral analogous of

Pasquale Petrullo
Università degli Studi della Basilicata, Italy

E. Damiani et al. (eds.), *From Combinatorics to Philosophy,*
DOI 10.1007/978-0-387-88753-1_10, © Springer Science+Business Media, LLC 2009

Abel polynomials, $x(x - n.\alpha)^{n-1}$, and have proved that a sequence of polynomials $(p_n(x))_{n \geq 0}$, $p_n(x)$ having degree $n$, $p_0(x) = 1$ and $p_1(x) = x$, is of binomial type if and only if there exists an umbra $\alpha$ such that $x(x - n.\alpha)^{n-1} \simeq p_n(x)$ for all $n$. This paper is motivated by the following observation. Since

$$1 + \sum_{n \geq 1} x(x - na)^{n-1} \frac{z^n}{n!} = e^{x \bar{f}(z)},$$

where $\bar{f}(z) = [ze^{az}]^{<-1>}$, then we derive the identity

$$(-na)^{n-1} = n![z^n][ze^{az}]^{<-1>},$$

whose umbral analogous can be stated as follows:

$$(-n.\alpha)^{n-1} \simeq n![z^n][ze^{\alpha z}]^{<-1>}.$$

Now, by assuming $[ze^{\alpha z}]^{<-1>} \simeq [zf(\alpha, z)]^{<-1>}$ and evaluating both sides we obtain an unexpected form of the well known Lagrange inversion formula:

$$\frac{1}{n}[z^{n-1}]f(\alpha, z)^{-n} = [z^n][zf(\alpha, z)]^{<-1>}.$$

As we will show, this is only the first relation involving umbral Abel polynomials and Lagrange inversion formula. We also define "generalized umbral Abel polynomials" $x(x + k.\alpha)^{n-1}$, by means of which a wide range of formal power series identities can be usefully encoded.

In the second part of the paper, we apply this methods within cumulant theory, by extending some results of Rota and Shen [10] and Di Nardo, Petrullo and Senato [2]. Let $a = (a_n)_{n \geq 1}$ and $a' = (a'_n)_{n \geq 1}$ be sequences of complex numbers satisfying $a_n = n!a'_n$. Consider $k_a = (k_n)_{n \geq 1}$, $r_{a'} = (r_n)_{n \geq 1}$ and $s_{a'} = (r_n)_{n \geq 1}$ such that

$$M(z) = e^{K(z)-1} = \frac{1}{z}\left[\frac{z}{R(z)}\right]^{<-1>} = \frac{1}{1 - S(z)},$$

where $M(z) = 1 + \sum_{n \geq 1} a_n \frac{z^n}{n!} = 1 + \sum_{n \geq 1} a'_n z^n$, $K(z) = 1 + \sum_{n \geq 1} k_n \frac{z^n}{n!}$, $R(z) = 1 + \sum_{n \geq 1} r_n z^n$ and $S(z) = \sum_{n \geq 1} s_n z^n$. Then, $k_n$ is known as the $n$-th (formal) classical cumulant of $a$, while $r_n$ and $s_n$ are the $n$-th (formal) free and Boolean cumulant of $a'$, respectively. In probability theory, when $a$ is the sequence of moments of a probability measure, then its classical cumulants seem to characterize the probability distribution better than moments. Free cumulants are a noncommutative analogous arising in Voiculescu free probability theory [16]. Boolean cumulants are known in physics, in the context of stochastic differential equations, as "partial cumulants". Often, in literature the term "semi-invariant"is preferred to "cumulant". Each family of cumulants linearizes a certain convolution. More precisely, the classical convolution of two sequences $a$ and $b$, is a sequence $a \star b$ such that $k_{a \star b} = k_a + k_b$, the sum being componentwise. Analogously, free convolution $a \boxplus b$ and Boolean convolution $a \uplus b$ are defined by $r_{a \boxplus b} = r_a + r_b$ and $s_{a \uplus b} = s_a + s_b$. Classical convolution corresponds to the convolution of probability measures, and free convolution is its free analogous. The connection among classical, free and Boolean cumulants, not so clearly encoded in the language of formal power series, becomes more evident from a combinatorial point of view. Indeed, the formulae connecting $a$ and $k_a$ are obtained via Möbius inversion formula on the lattice of all the partitions of a finite set [5]. In 1994, Speicher [13] recovered the formulae of free cumulants by restricting the Möbius inversion to the lattice of all the noncrossing partitions. Finally, Boolean cumulants obey to the Möbius inversion on the lattice of interval partitions [15]. In particular, such a lattice turns out to be isomorphic to the Boolean lattice of all subsets of a finite set, whence the name "Boolean cumulants". Generalized umbral Abel polynomials provide a lithe syntax, thanks to which not only the formulae connecting moments and cumulants, but also the formulae connecting each family of cumulants to the others, can be easily obtained by evaluating polynomials $\gamma(\gamma + k.\alpha)^{n-1}$, for suitable umbrae $\alpha$ and $\gamma$ (Theorem 4.1, Corollary 4.3, Theorem 4.4, and Corollary 4.5). In this context, we can study the convolutions $a \star b$, $a \boxplus b$ and $a \uplus b$ by means of a general class of umbrae $\alpha_{(k)}\gamma$, called Abel-type convolutions, that in the cases $k = -1, -2, -n$ represent $((a \star b)_n)_{n \geq 1}$, $(n!(a \uplus b)_n)_{n \geq 1}$ and $(n!(a \boxplus b)_n)_{n \geq 1}$ respectively (Theorem 4.7).

The paper is structured as follows. Section 2 is devoted to Abel polynomials $A_n(x, \alpha)$ and their connections with Lagrange inversion formula. In Section 3, we introduce generalized umbral Abel polynomials $A_n^{(k)}(x, \alpha)$, and prove some useful umbral identities. In Section 4, we apply this setting within cumulant theory.

## 10.2 Abel polynomials, Lagrange inversion formula and sequences of binomial type

For basic definition and results concerning classical umbral calculus, we refer to Di Nardo and Senato [4]. In this section and in the following, we assume

$$E : \mathbb{C}[x][A \cup B] \to \mathbb{C}[x],$$

where $E$ is the linear functional evaluation, and $A$ and $B$ denote the base alphabet of umbrae and the auxiliary alphabet, respectively. Moreover, given $\alpha \in A$ we denote by $f(\alpha, z)$ the generating function of $\alpha$, that is the formal power series

$$f(\alpha, z) = E[e^{\alpha z}] = 1 + \sum_{n \geq 1} E[\alpha^n] \frac{z^n}{n!}.$$

As is well known, *Lagrange inversion formula*, in one of its equivalent forms, states that

$$[z^n]f\left([zg(z)]^{<-1>}\right) = \frac{1}{n}[z^{n-1}]f'(z)\left(\frac{1}{g(z)}\right)^n, \tag{10.1}$$

for all $f(z), g(z) \in \mathbb{C}[[z]]$ with $g(0) = 1$.

If $a \in \mathbb{C}$, then the polynomial $A_n(x,a) = x(x - na)^{n-1}$ of $\mathbb{C}[x]$ is known as an *Abel polynomial*. A generating function of the sequence $(A_n(x,a))_{n\geq 1}$ is given by

$$1 + \sum_{n\geq 1} A_n(x,a) \frac{z^n}{n!} = e^{x[ze^{az}]^{<-1>}}, \tag{10.2}$$

from which

$$(x.\beta.a_D{}^{<-1>})^n \simeq A_n(x,a),$$

where $a_D{}^{<-1>}$ denotes $(a.u)_D{}^{<-1>}$, $\alpha_D$ and $\alpha^{<-1>}$ being the derivative and the compositional inverse of $\alpha$ respectively. We replace $x$ and $a$ in (10.2) with two umbrae $\gamma$ and $\alpha$, and consider the umbral polynomials $A_n(\gamma, \alpha)$ such that

$$u + \sum_{n\geq 1} A_n(\gamma, \alpha) \frac{z^n}{n!} \simeq e^{\gamma[ze^{\alpha z}]^{<-1>}},$$

where $u$ is the unity umbra. It can be shown that

$$A_n(\gamma, \alpha) \simeq \gamma(\gamma - n.\alpha)^{n-1},$$

where $\gamma - n.\alpha$ denotes $\gamma + (-n.\alpha)$. Moreover, if we assume $[ze^{\alpha z}]^{<-1>} \simeq [zf(\alpha, z)]^{<-1>}$, then we obtain

$$(\gamma.\beta.\alpha_D{}^{<-1>})^n \simeq A_n(\gamma, \alpha). \tag{10.3}$$

Identity (10.3) is the umbral form of (10.1). Indeed, by assuming $f(\alpha, z) = f(z)$ and $f(\gamma, z) = g(z)$, then

$$(\gamma.\beta.\alpha_D{}^{<-1>})^n \simeq n![z^n]f\left([zg(z)]^{<-1>}\right),$$

and

$$A_n(\gamma, \alpha) \simeq \sum_{i=0}^{n-1} \binom{n-1}{i} \gamma^{i+1} (-n.\alpha)^{n-1-i} \simeq (n-1)![z^{n-1}]f'(z)\left(\frac{1}{g(z)}\right)^n.$$

Furthermore, setting $\gamma = \chi$ in (10.3), $\chi$ being the singleton umbra, we recover

$$(\alpha_D{}^{<-1>})^n \simeq (-n.\alpha)^{n-1}, \tag{10.4}$$

and finally

$$[z^n][zf(\alpha, z)]^{<-1>} = \frac{1}{n}[z^{n-1}]\left(\frac{1}{f(\alpha, z)}\right)^n,$$

which is (10.1) with $f(z) = z$ and $g(z) = f(\alpha, z)$. If we rewrite (10.4) as

$$n![z^n][ze^{\alpha z}]^{<-1>} \simeq (-n.a)^{n-1},$$

we gain the umbral analogous of

$$n![z^n][ze^{az}]^{<-1>} = (-na)^{n-1}.$$

We may replace only $a$ with $\alpha$ in (10.2), having $A_n(x, \alpha) \simeq x(x - n.\alpha)^{n-1}$, and

$$(x.\beta.\alpha_D^{<-1>})^n \simeq A_n(x, \alpha). \tag{10.5}$$

The polynomials $A_n(x, \alpha)$ were introduced by Rota, Shen and Taylor [11] in connection with sequences of polynomials of binomial type. Recall that, a sequence $(p_n(x))_{n \geq 0}$ of polynomials of $\mathbb{C}[x]$, with $p_n(x)$ having degree $n$, $p_0(x) = 1$ and $p_1(x) = x$, is said to be of *binomial type* if and only if

$$p_n(x + y) = \sum_{h=0}^{n} \binom{n}{h} p_h(x) p_{n-h}(y),$$

or equivalently (see [9]) if and only if there exists a formal power series $f(z)$, with $f(0) = 1$ and

$$1 + \sum_{n \geq 1} p_n(x) \frac{z^n}{n!} = e^{x[zf(z)]}.$$

In particular, since $zf(z)$ admits compositional inverse, we take $\alpha$ such that $[zf(\alpha, z)]^{<-1>} = zf(z)$, obtaining

$$p_n(x) \simeq x(x - n.\alpha)^{n-1}.$$

Such a formula says that "all polynomials of binomial type are represented by Abel polynomials" $x(x - n.\alpha)^{n-1}$, for some invertible umbra $\alpha$. This is exactly the result of Rota, Shen and Taylor.

Finally, we would like to remark that this umbral syntax point out how Lagrange inversion formula and sequences of binomial type essentially obey the same basic law involving Abel polynomials, that is identity (10.3). By starting from a formal definition of "Sheffer umbra", Di Nardo, Niederhausen and Senato [1] have given a proof of (10.5) (which they refer to as "Abel representation of binomial sequences") in a more general context related to Sheffer sequences. Such a proof can be adapted to prove (10.3) by simply setting $x = \gamma$. This way one obtains a simple proof of the Lagrange inversion formula.

## 10.3 Generalized Abel polynomials

Given $\alpha \in A$, the *noncrossing Fourier transform* or *Lagrange involution* of $\alpha$ is an auxiliary umbra $\mathfrak{L}_\alpha$ such that

$$\mathfrak{L}_\alpha \equiv \alpha_D^{<-1>}{}_P, \tag{10.6}$$

$\alpha_P$ being an auxiliary umbra such that $(n+1)\alpha_P{}^n \simeq \alpha^{n+1}$ for all $n \geq 1$. We remark that

$$\mathfrak{L}_{\mathfrak{L}_\alpha} \equiv \alpha, \tag{10.7}$$

from which the term "involution".

Moreover, since

$$f(\mathfrak{L}_\alpha, z) = \frac{1}{z}[zf(\alpha, z)]^{<-1>},$$

then the generating function of $\mathfrak{L}_\alpha$ reduces to the Fourier transform defined by Nica and Speicher [7]. Furthermore, we denote by $\alpha^{\downarrow}$ an auxiliary umbra such that $\alpha^{\downarrow n} \simeq \alpha^{n-1}$ for $n \geq 1$. Its generating function is

$$f(\alpha^{\downarrow}, z) = 1 + \int f(\alpha, z)\,dz.$$

We consider the following generalization of the Abel polynomial $A_n(x, \alpha)$. Let $k$ be an integer and define $A_n^{(k)}(x, \alpha)$ to be an umbral polynomial such that

$$A_n^{(k)}(x, \alpha) \simeq x(x + k.\alpha)^{n-1}.$$

From now on we refer to $A_n^{(k)}(x, \alpha)$ as a *generalized umbral Abel polynomial*. Of course, $A_n(x, \alpha) \simeq A_n^{(-n)}(x, \alpha)$. We reserve the notation $A_n^{(k)}(\alpha)$ to the special case $A_n^{(k)}(\alpha, \alpha)$, that is

$$A_n^{(k)}(\alpha) \simeq \alpha(\alpha + k.\alpha)^{n-1}.$$

A combinatorial treatment of $A_n^{(n)}(\alpha) \simeq \alpha(\alpha + n.\alpha)^{n-1}$ involving parking functions and noncrossing partitions has been given in [8]. Note that, since $0.\alpha \equiv \varepsilon$, then we have $A_n^{(0)}(\alpha) \simeq \alpha^n$. We are interest in umbral equivalences of type $\gamma^n \simeq A_n^{(k)}(\alpha)$. The next theorem shows how to invert them.

**Theorem 1 (First Abel Inversion Theorem)** *Let $\alpha$ and $\gamma$ be two umbrae. Then*

$$\gamma^n \simeq A_n^{(k)}(\alpha), \text{ for } n = 1, 2, \ldots$$

*if and only if*

$$\alpha^n \simeq A_n^{(-k)}(\gamma, \alpha) \text{ for } n = 1, 2, \ldots.$$

*Proof.* If $\gamma^n \simeq A_n^{(k)}(\alpha)$, since $k.\alpha - k.\alpha \equiv \varepsilon$, then we have

$$\alpha^n \simeq \alpha(\alpha + k.\alpha - k.\alpha)^{n-1} \simeq \sum_{i=0}^{n-1} A_{i+1}^{(k)}(\alpha)(-k.\alpha)^{n-1-i} \simeq A_n^{(-k)}(\gamma, \alpha).$$

Viceversa, if $\alpha^n \simeq A_n^{(-k)}(\gamma, \alpha)$ then

$$\gamma^n \simeq \gamma(\gamma - k.\alpha + k.\alpha)^{n-1} \simeq \sum_{i=0}^{n-1} A_{i+1}^{(-k)}(\gamma, \alpha)(k.\alpha)^{n-1-i} \simeq A_n^{(k)}(\alpha).$$

The proof of the following lemma is here omitted. It can be carried out by following the same steps of Theorem 3.1 in [8].

**Lemma 2** *If $\alpha$ and $\gamma$ are umbrae, then we have*

$$\alpha(\alpha + \gamma.\alpha)^{n-1} \simeq \sum_{\mu \vdash n} d_\mu(\gamma)_{\ell(\mu)-1}\alpha_\mu, \tag{10.8}$$

*where*

1. *the sum ranges over all the (integer) partitions $\mu = (\mu_1, \ldots, \mu_l)$ of $n$ with $l = \ell(\mu)$ parts,*
2. *$d_\mu = n!/[m(\mu)!\mu!]$, with $\mu! = \mu_1! \cdots \mu_l!$ and $m(\mu)! = m_1(\mu)! \cdots m_n(\mu)!$, $m_i(\mu)$ being the multiplicity of $i$ as a part of $\mu$,*
3. *$(\gamma)_k = \gamma(\gamma - 1) \cdots (\gamma - k + 1)$ and $\alpha_\mu \equiv \alpha_1^{\mu_1} \cdots \alpha_l^{\mu_l}$, the $\alpha_i$'s being uncorrelated umbrae similar to $\alpha$.*

Identity (10.8) provides the following useful equivalences.

**Proposition 1** *If $\alpha$ is an umbra, then we have*

$$A_n^{(k)}(\alpha) \simeq \sum_{\mu \vdash n} d_\mu(k)_{\ell(\mu)-1}\alpha_\mu. \tag{10.9}$$

*Moreover, $k \neq -1$ implies*

$$A_n^{(k)}(\alpha) \simeq \frac{[(k+1).\alpha]^n}{(k+1)}, \tag{10.10}$$

*while $k = -1$ gives*

$$A_n^{(-1)}(\alpha) \simeq (\chi.\alpha)^n. \tag{10.11}$$

*Proof.* By setting $\gamma = k$ in (10.8) we have (10.9). Recall that,

$$(\gamma.\alpha)^n \simeq \sum_{\mu \vdash n} d_\mu(\gamma)_{\ell(\mu)}\alpha_\mu. \tag{10.12}$$

If $k \neq -1$, then by comparing (10.9) and (10.12) with $\gamma = k + 1$, we deduce (10.10). Finally, since $(\chi)_n \simeq (-1)^{n-1}(n-1)! = (-1)_{n-1}$, then from (10.12) we gain (10.11).

**Proposition 2** *If $\alpha$ is an umbra, then we have*

$$\gamma^n \simeq A_n^{(k)}(\alpha), \text{ for } n = 1, 2, \ldots \iff \gamma \equiv (k.\chi)^{\downarrow}.\beta.\alpha. \tag{10.13}$$

*Proof.* Since $(\gamma.\beta)_n \simeq \gamma^n$ for all $\gamma \in A$, and $(k.\chi)^n \simeq (k)_n$, we have $[(k.\chi)^{\downarrow}]^n \simeq (k)_{n-1}$ and we prove (10.13) by means of (10.9) and (10.12).

**Theorem 3** *If $\alpha$ is an umbra, then*

$$A_n^{(k)}(-1.\alpha) \simeq -A_n^{(-(k+2))}(\alpha). \tag{10.14}$$

*Proof.* If $k \neq -1$, then equivalence (10.14) follows by applying (10.10). Let now $k = -1$.

Since $-(-\chi)_n \simeq (\chi)_n$, by means of (10.11) we gain

$$A_n^{(-1)}(-1.\alpha) \simeq (\chi. - 1.\alpha)^n \simeq (-\chi.\alpha)^n \simeq -\sum_{\mu \vdash n} d_\mu(\chi)\ell_{(\mu)}\alpha_\mu \simeq -A_n^{(-1)}(\alpha).$$

In closing this section we study polynomials $A_n^{(n+k)}(\gamma, \alpha)$, namely polynomials whose upper parameter depends on the degree. From (10.4), (10.6) and (10.10) we obtain

$$A_n^{(0)}(\mathfrak{L}_\alpha) \simeq A_n^{(n)}(-1.\alpha).$$

The following theorem generalizes this result.

**Theorem 4** *Let $\alpha$ be an umbra and let $\mathfrak{L}_\alpha$ be defined as in (10.6). Then*

$$A_n^{(k)}(\mathfrak{L}_\alpha) \simeq A_n^{(n+k)}(-1.\alpha) \simeq -A_n^{(-(n+k+2))}(\alpha). \tag{10.15}$$

*Proof.* Second equivalence in (10.15) is a consequence of (10.14). Let $k \neq -1$. From (10.10), first equivalence can be stated in the following generating function form:

$$[z^m]\{[zf(\alpha,z)]^{<-1>}\}^{k+1} = \frac{k+1}{m}[z^{m-k-1}]\left(\frac{1}{f(\alpha,z)}\right)^m, \tag{10.16}$$

with $m = n + k + 1$. Let $k = -1$. Then, again from (10.10), and by means of (10.11), first equivalence in (10.15) says that

$$[z^n]\log\left(\frac{1}{z}[zf(\alpha,z)]^{<-1>}\right) = \frac{1}{n}[z^n]\left(\frac{1}{f(\alpha,z)}\right)^n. \tag{10.17}$$

Both (10.16) and (10.17) can be derived from (10.1) (see [14]) and the theorem is proved.

It is well known (see for instance [14]) that identity (10.17) has to be considered the correct analogous of (10.16) when $k + 1 = 0$. We remark that first equivalence in (10.15) underlines even more this circumstance. Now, we are able to give an inversion rule for polynomials $A_n^{(n+k)}(\gamma, \alpha)$.

**Theorem 5 (Second Abel Inversion Theorem)** *Let $\alpha$ and $\gamma$ be umbrae. Then*

$$\gamma^n \simeq A_n^{(n+k)}(\alpha), \text{ for } n = 1, 2, \ldots$$

*if and only if*

$$\alpha^n \simeq A_n^{(-(n+k))}(\gamma, \mathfrak{L}_{-1.\alpha}), \text{ for } n = 1, 2, \ldots.$$

*Proof.* Let $\gamma^n \simeq A_n^{(n+k)}(\alpha)$. From (10.15) we have

$$\gamma^n \simeq A_n^{(k)}(\mathfrak{L}_{-1.\alpha}) \text{ and } \alpha^n \simeq A_n^{(-n)}(\mathfrak{L}_{-1.\alpha}). \tag{10.18}$$

First Abel Inversion Theorem gives $(\mathfrak{L}_{-1.\alpha})^n \simeq A_n^{(-k)}(\gamma, \mathfrak{L}_{-1.\alpha})$, from which we have $A_n^{(-n)}(\mathfrak{L}_{-1.\alpha}) \simeq A_n^{(-(n+k))}(\gamma, \mathfrak{L}_{-1.\alpha})$, so that first part of the theorem is proved. Viceversa, let $\delta$ be an umbra such that $\delta^n \simeq A_n^{(-k)}(\gamma, \mathfrak{L}_{-1.\alpha})$.

Then, we have

$$\alpha^n \simeq A_n^{(-n)}(\delta, \mathfrak{L}_{-1.\alpha}). \tag{10.19}$$

By virtue of (10.3) and (10.6), identity (10.19) says that $\alpha \equiv \delta.\beta.(-1.\alpha)_D$. Therefore, $\alpha^n \simeq A_n^{(-n)}(\mathfrak{L}_{-1.\alpha})$ implies $\alpha \equiv \mathfrak{L}_{-1.\alpha}.\beta.(-1.\alpha)_D$, and by comparing the two similarities we have $\delta \equiv \mathfrak{L}_{-1.\alpha}$, that is $(\mathfrak{L}_{-1.\alpha})^n \simeq A_n^{(-k)}(\gamma, \mathfrak{L}_{-1.\alpha})$. First Abel Inversion Theorem returns first equivalence in (10.18), and (10.15) completes the proof.

## 10.4 Cumulants and convolutions

In this section we apply the results of Section 3 within cumulant theory. If $a = (a_n)_{n \geq 1}$ is a sequence of complex numbers, then we will denote by $a' = (a'_n)_{n \geq 1}$ the sequence such that $a_n = n! a'_n$. Let $M(z)$ be the following formal power series,

$$M(z) = 1 + \sum_{n \geq 1} a_n \frac{z^n}{n!} = 1 + \sum_{n \geq 1} a'_n z^n.$$

Consider the sequences $k_a = (k_n)_{n \geq 1}$, $r_{a'} = (r_n)_{n \geq 1}$ and $s_{a'} = (s_n)_{n \geq 1}$ defined by

$$M(z) = e^{K(z)-1} = \frac{1}{z} \left[ \frac{z}{R(z)} \right]^{<-1>} = \frac{1}{1 - S(z)},$$

where

$$K(z) = 1 + \sum_{n \geq 1} k_n \frac{z^n}{n!}, \ R(z) = 1 + \sum_{n \geq 1} r_n z^n, \text{ and } S(z) = \sum_{n \geq 1} s_n z^n.$$

The complex number $k_n$ is known as the $n$-th (formal) *classical cumulant* of $a$. The numbers $s_n$ and $r_n$ are named *Boolean cumulant* and *free* (or *noncrossing*) *cumulant* respectively of $a'$. Sometimes, we will write $k_n(a), r_n(a')$ and $s_n(a')$ to explicit the sequence each cumulant refers to. When $a$ is the sequence of moments of some probability measure $P$, its sequence of cumulants $k_a$ has a meaning which is not purely formal. In fact, $k_1$ is the mean of $P$, and $k_2$ is the variance. Free cumulants $r_{a'}$ are a noncommutative analogous of classical cumulants. They occurs in Voiculescu free probability theory, see [16]. Boolean cumulants are known in physics in the context of partial differential equations as "partial cumulants". The connection among classical, free and Boolean cumulants, not so clearly encoded in the language of formal power series, becomes more evident from a combinatorial point of view. Indeed, by means of multiplicative functions [5], it can be shown that the formulae connecting the sequences $a$ and $k_a$ are obtained via Möbius inversion formula on the lattice of all the partitions of a finite set. In 1994, Speicher [13] restricted the Möbius inversion to the lattice of all the noncrossing partitions, recovering the formulae of free cumulants. Finally, Boolean cumulants obey to the Möbius inversion on the lattice of the interval partitions [15].

Let $\alpha$ be an umbra, and assume $\alpha^n \simeq a_n = n! a'_n$. Consider the umbrae $\kappa_\alpha, \eta_\alpha$, and $\mathfrak{R}_\alpha$

such that

$$\alpha \equiv \beta.\kappa_\alpha, \tag{10.20}$$

$$\alpha \equiv \bar{u}.\beta.\eta_\alpha, \tag{10.21}$$

$$\alpha \equiv \mathfrak{L}_{-1}.\mathfrak{K}_\alpha, \tag{10.22}$$

where $\bar{u} \equiv -1. -\chi$ is the Boolean unity. Now, we have $\kappa_\alpha{}^n \simeq k_n(a)$, $\eta_\alpha{}^n \simeq n!s_n(a')$, and $\mathfrak{K}_\alpha{}^n \simeq n!r_n(a')$. The following theorem highlights the role of the polynomials $A_n^{(k)}(x, \alpha)$ in this context.

**Theorem 6 (Abel parametrization of cumulants)** *If $\alpha$ is an umbra and $\kappa_\alpha, \eta_\alpha$ and $\mathfrak{K}_\alpha$ are the umbrae introduced in (10.20), (10.21) and (10.22), then we have*

$$\alpha^n \simeq A_n^{(1)}(\kappa_\alpha, \alpha) \quad and \quad \kappa_\alpha{}^n \simeq A_n^{(-1)}(\alpha),$$
$$\alpha^n \simeq A_n^{(2)}(\eta_\alpha, \alpha) \quad and \quad \eta_\alpha{}^n \simeq A_n^{(-2)}(\alpha),$$
$$\alpha^n \simeq A_n^{(n)}(\mathfrak{K}_\alpha) \quad and \quad \mathfrak{K}_\alpha{}^n \simeq A_n^{(-n)}(\alpha).$$

*Proof.* From (10.11) and (10.20) we have $\kappa_\alpha{}^n \simeq (\chi.\alpha)^n \simeq A_n^{(-1)}(\alpha)$. First Abel Inversion Theorem gives $\alpha^n \simeq A_n^{(1)}(\kappa_\alpha, \alpha)$. Identity (10.21) provides $\eta_\alpha \equiv \bar{u}^{<-1>}.\beta\alpha$. Since $\bar{u}^{<-1>} \equiv (-2.\chi)^{\downarrow}$, second pair of equivalences follows from (10.13) and First Abel Inversion Theorem. By combining (10.15) and (10.22), we have $\alpha^n \simeq (\mathfrak{L}_{-1}.\mathfrak{K}_\alpha)^n \simeq A_n^{(n)}(\mathfrak{K}_\alpha)$. By virtue of (10.7) and (10.22) we obtain $\mathfrak{K}_\alpha \equiv -1.\mathfrak{L}_\alpha$. Finally, by means of (10.15) we have $\mathfrak{K}_\alpha{}^n \simeq A_n^{(0)}(-1.\mathfrak{L}_\alpha) \simeq -A_n^{(-2)}(\mathfrak{L}_\alpha) \simeq A_n^{(-n)}(\alpha)$.

The equivalence $\alpha^n \simeq \kappa_\alpha(\kappa_\alpha + \alpha)^{n-1}$ was first proved by Rota and Shen [10]. Abel parametrization extends such a result to Boolean and free cumulants. These relations can be found also in [2], with a slightly different proof. By using Abel parametrization, the formulae connecting a sequence with its cumulants can be easily deduced.

**Corollary 7** *Let $a = (a_n)_{n\geq 1}$ be a sequence of complex numbers and let $k_n, s_n$ and $r_n$ be its $n$-th classical, Boolean and free cumulants respectively. Then, we have*

$$a_n = \sum d_\mu k_\mu \quad and \quad k_n = \sum d_\mu (-1)^{\ell(\mu)-1}(\ell(\mu) - 1)!a_\mu,$$
$$a_n = \sum \frac{\ell(\mu)!}{m(\mu)!}s_\mu \quad and \quad s_n = \sum \frac{\ell(\mu)!}{m(\mu)!}(-1)^{\ell(\mu)-1}a_\mu,$$
$$a_n = \sum \frac{(n)_{\ell(\mu)-1}}{m(\mu)!}r_\mu \quad and \quad r_n = \sum \frac{(-n)_{\ell(\mu)-1}}{m(\mu)!}a_\mu,$$

*where the sums range over all the partitions $\mu = (\mu_1, \ldots, \mu_l)$ of $n$, and $f_\mu = f_{\mu_1} \cdots f_{\mu_l}$.*

*Proof.* By assuming Abel parametrization of cumulants, from (10.20) and (10.21) we have $\alpha^n \simeq \kappa_\alpha(\kappa_\alpha + \beta.\kappa_\alpha)^{n-1} \simeq \eta_\alpha(\eta_\alpha + 2.\bar{u}.\beta.\eta_\alpha)^{n-1}$. If $\alpha^n \simeq a_n$ and $\kappa_\alpha{}^n \simeq k_n$, since $(\beta)_n \simeq 1$ and $(-1)_{n-1} = (-1)^{n-1}(n-1)!$, then we prove the first pair of identities by means (10.8). If $\alpha^n \simeq n!a_n$ and $\eta_\alpha{}^n \simeq n!s_n$, since $(2.\bar{u}.\beta)_{n-1} \simeq n!$ and $(-2)_{n-1} = (-1)^{n-1}n!$, the second pair of identities is proved again by means of (10.8). Finally, if $\alpha^n \simeq n!a_n$ and $\mathfrak{K}_\alpha{}^n \simeq n!r_n$, then the third pair of formulae follows directly from (10.9).

The following theorem completes the Abel parametrization.

**Theorem 8 (Mixed Abel parametrization of cumulants)** *If $\alpha$ is an umbra and $\kappa_\alpha, \eta_\alpha$ and $\Re_\alpha$ are the umbrae introduced in (10.20), (10.21) and (10.22), then we have*

$$
\begin{aligned}
\kappa_\alpha{}^n &\simeq A_n^{(1)}(\eta_\alpha, \alpha) \simeq A_n^{(n-1)}(\Re_\alpha),\\
\eta_\alpha{}^n &\simeq A_n^{(-1)}(\kappa_\alpha, \alpha) \simeq A_n^{(n-2)}(\Re_\alpha),\\
\Re_\alpha{}^n &\simeq A_n^{(1-n)}(\kappa_\alpha, \alpha) \simeq A_n^{(2-n)}(\eta_\alpha, \alpha).
\end{aligned}
$$

*Proof.* Starting from the Abel parametrization, let $\alpha'$ similar and uncorrelated to $\alpha$. Thus we have $-1.\alpha \equiv -2.\alpha + \alpha'$ and $-2.\alpha \equiv -1.\alpha - 1.\alpha'$. In this way

$$
\kappa_\alpha{}^n \simeq \alpha(\alpha - 2.\alpha + \alpha')^{n-1} \simeq A_n^{(1)}(\eta_\alpha, \alpha)
$$

and

$$
\eta_\alpha{}^n \simeq \alpha(\alpha - 1.\alpha - 1.\alpha')^{n-1} \simeq A_n^{(-1)}(\kappa_\alpha, \alpha).
$$

The other equivalences can be deduced by applying Second Abel Inversion Theorem and (10.22). $\square$

Perhaps the reader has noticed the duality between $\alpha$ and $\Re_\alpha$ arising from Abel parametrization. We have $\kappa_\alpha{}^n \simeq A_n^{(-1)}(\alpha) \simeq A_n^{(n-1)}(\Re_\alpha)$ and $\eta_\alpha{}^n \simeq A_n^{(-2)}(\alpha) \simeq A_n^{(n-2)}(\Re_\alpha)$. This is due to (10.15), and by this we deduce the general rule

$$
A_n^{(-k)}(\alpha) \simeq A_n^{(n-k)}(\Re_\alpha).
$$

Now, the formulae connecting $k_a$, $r_{a'}$ and $s_{a'}$ easily follow. We remark that $k_a$, $r_{a'}$ and $s_{a'}$ refer to different sequences of moments. Combinatorial formulae expressing $k_a$ in terms of $r_a$, and $s_a$ in terms of $k_a$ can be found in [6].

**Corollary 9** *Let $a = (a_n)_{n\geq 1}$ and $a' = (a'_n)_{n\geq 1}$ be sequences of complex numbers such that $a_n = n! a'_n$. Let $k_n = k_n(a)$, $s_n = s_n(a')$ and $r_n = r_n(a')$, then the following identities hold:*

$$
\begin{aligned}
k_n &= \sum \tfrac{n!}{m(\mu)!}(\ell(\mu) - 1)! s_\mu = \sum \tfrac{n!}{m(\mu)!}(n-1)_{\ell(\mu)-1} r_\mu,\\
s_n &= \sum \tfrac{(-1)^{\ell(\mu)-1}}{\mu! m(\mu)!} k_\mu = \sum \tfrac{(n-2)_{\ell(\mu)-1}}{m(\mu)!} r_\mu,\\
r_n &= \sum \tfrac{(1-n)_{\ell(\mu)-1}}{\mu! m(\mu)!} k_\mu = \sum \tfrac{(n-2)_{\ell(\mu)-1}}{m(\mu)!}(-1)^{\ell(\mu)-1} s_\mu,
\end{aligned}
$$

*where the sums are extended over all the partitions $\mu$ of $n$.*

*Proof.* Analogously to the proof of Corollary 7. $\square$

Each kind of cumulant linearizes a certain convolution. To be precise, given $a = (a_n)_{n\geq 1}$, $b = (b_n)_{n\geq 1}$, the *classical convolution* of $a$ and $b$ is defined to be a sequence $a \star b$ such that

$$
k_n(a \star b) = k_n(a) + k_n(b).
$$

Analogously, the *Boolean convolution* $a \uplus b$ and the *free convolution* $a \boxplus b$ of $a$ and $b$, are sequences enjoying the properties

$$
s_n(a \uplus b) = s_n(a) + s_n(b) \text{ and } r_n(a \boxplus b) = r_n(a) + r_n(b).
$$

Classical convolution gives the moments of the convolution of two probability measures with sequence of moments $a$ and $b$. Free convolution is its free analogous. Let us consider two umbrae $\alpha$ and $\gamma$ such that $\alpha^n \simeq a_n = n!a_n'$ and $\gamma^n \simeq b_n = n!b_n'$. Define $\alpha \star \gamma$, $\alpha \uplus \gamma$ and $\alpha \boxplus \gamma$ to be umbrae such that

$$\kappa_{\alpha \star \gamma} \equiv \kappa_\alpha \dotplus \kappa_\gamma, \tag{10.23}$$

$$\eta_{\alpha \uplus \gamma} \equiv \eta_\alpha \dotplus \eta_\gamma, \tag{10.24}$$

$$\mathfrak{R}_{\alpha \boxplus \gamma} \equiv \mathfrak{R}_\alpha \dotplus \mathfrak{R}_\gamma. \tag{10.25}$$

In this way, we have

$$(\alpha \star \gamma)^n \simeq (a \star b)_n, \ (\alpha \uplus \gamma)^n \simeq n!(a' \uplus b')_n, \text{ and } (\alpha \boxplus \gamma)^n \simeq n!(a' \boxplus b')_n.$$

The umbrae $\alpha \uplus \gamma$ and $\alpha \boxplus \gamma$ were first studied in [2]. The Lagrange involution (10.6) allows us to pass from $\alpha \boxplus \gamma$ to $\alpha \uplus \gamma$ and viceversa.

**Theorem 10** *We have*

$$\mathfrak{L}_{\alpha \boxplus \gamma} \equiv \mathfrak{L}_\alpha \uplus \mathfrak{L}_\gamma \text{ and } \mathfrak{L}_{\alpha \uplus \gamma} \equiv \mathfrak{L}_\alpha \boxplus \mathfrak{L}_\gamma.$$

*Proof.* From (10.21) we deduce $\eta_\alpha{}^n \simeq -(-1.\alpha)^n$ for all $n \geq 1$. In this way $-1.(\alpha \uplus \gamma) \equiv (-1.\alpha) \dotplus (-1.\gamma)$. From (10.22) we obtain $\mathfrak{R}_\alpha \equiv -1.\mathfrak{L}_\alpha$, so that $\mathfrak{L}_{\alpha \boxplus \gamma} \equiv -1.\mathfrak{R}_{\alpha \boxplus \gamma} \equiv -1.(\mathfrak{R}_\alpha \dotplus \mathfrak{R}_\alpha)$, that is $\mathfrak{L}_{\alpha \boxplus \gamma} \equiv \mathfrak{L}_\alpha \uplus \mathfrak{L}_\gamma$. The proof of the second similarity is analogous.

Abel parametrization suggests to define a wider class of convolution umbrae.

**Definition 10.4.1** *Let $\alpha$ and $\gamma$ be two umbrae. We name Abel-type convolution of $\alpha$ and $\gamma$ each auxiliary umbra $\alpha_{(k)}\gamma$ such that*

$$A_n^{(k)}(\alpha_{(k)}\gamma) \simeq A_n^{(k)}(\alpha) + A_n^{(k)}(\gamma), \text{ for } n = 1, 2, \ldots. \tag{10.26}$$

The explicit connections among Abel-type convolutions and $\alpha \star \gamma$, $\alpha \uplus \gamma$ and $\alpha \boxplus \gamma$ are stated in the theorem below.

**Theorem 11** *Let $\alpha$ and $\gamma$ be two umbrae, $\alpha \star \gamma$, $\alpha \uplus \gamma$, $\alpha \boxplus \gamma$ and $\alpha_{(k)}\gamma$ be the auxiliary umbrae defined in (10.23), (10.24), (10.25) and (10.26) respectively. Then, we have*

$$\alpha \star \gamma \equiv \alpha \operatorname{conv} \gamma,$$

$$\alpha \uplus \gamma \equiv \alpha_{(-2)}\gamma,$$

*and*

$$(\alpha \boxplus \gamma)^n \simeq (\alpha_{(-n)}\gamma)^n, \text{ for } n = 1, 2, \ldots.$$

*Proof.* Since similar umbrae have the same sequences of cumulants (classical, Boolean and free) and viceversa, the results follow via Abel parametrization of cumulants.

If $k = -1$ the Abel-type convolution $\alpha_{(k)}\gamma$ reduces to the sum $\alpha + \gamma$. In fact, $\alpha \star \gamma$ is exactly the sum $\alpha + \gamma$ (see also [2, 3]). When $k \neq -1$, equivalence (10.10) gives

$$\alpha_{(k)}\gamma \equiv \frac{1}{k+1}.[(k+1).\alpha \dot{+} (k+1).\gamma],$$

from which

$$(\alpha \uplus \gamma)^n \simeq \left[-1.(-1.\alpha \dot{+} -1.\gamma)\right]^n,$$

and

$$(\alpha \boxplus \gamma)^n \simeq \left\{\frac{1}{1-n}.[(1-n).\alpha \dot{+} (1-n).\gamma]\right\}^n.$$

## Acknowledgements

The author thanks Prof. Domenico Senato for useful discussions and suggestions in producing this paper.

## References

1. Di Nardo, E. Niederhausen, H. Senato, D., *The Classical Umbral Calculus: Sheffer Sequences.* arXiv:0810. 3554v1.
2. Di Nardo, E., Petrullo, and P., Senato, D., *Cumulants, Convolutions and Volume Polynomial,* preprint.
3. Di Nardo, E., Senato, D. (2006), *An Umbral Setting for Cumulants and Factorial Moments,* in "European J. Combin.", 27, pp. 394-41.
4. Di Nardo, E., Senato, D. (2009), *The Problems Eleven and Twelve of Rota's Fubini Lectures: from Cumulants to Free Probability Theory,* published in this volume.
5. Doubilet, P., Rota, G.-C., Stanley, R. P. (1972), *On the Foundations of Combinatorial Theory (VI): The Idea of Generating Function,* in *Sixth Berkeley Symposium on Mathematical Statistics and Probability,* Vol. II: *Probability Theory,* Berkeley, University of California, pp. 267-318.
6. Lehner, F. (2002), *Free Cumulants and Enumeration of Connected Partitions,* in "European J. Combin.", 23, pp. 1025-31.
7. Nica, A., Speicher, R. (1997), *A "Fourier Transform" for Multiplicative Functions on Non-crossing Partitions,* in "J. Algebraic Combin.", 6, pp. 141-60.
8. Petrullo, P., Senato, D., *An Instance of Umbral Methods in Representation Theory: the Parking Function Module,* in "Pure Math. Appl.", to be published.
9. Roman, S. (1984), *The Umbral Calculus,* S. Eilenberg, H. Bass (eds.), London, Academic Press.
10. Rota, G.-C., Shen, J. (2000), *On the Combinatorics of Cumulants,* in "J. Combin. Theory Ser. A", 91, pp. 283-304.
11. Rota, G.-C., Shen, J., Taylor, B. D. (1997), *All Polynomials of Binomial Type are Represented by Abel Plynomials,* in "Ann. Scuola Norm. Sup. Pisa Cl. Sci.", 25/1, pp. 731-8.
12. Rota, G.-C., Taylor, B. D. (1994), *The Classical Umbral Calculus,* in "SIAM J. Math. Anal.", 25, pp. 694-711.
13. Speicher, R. (1994), *Multiplicative Functions on the Lattice of Non-crossing Partitions and Free Convolution,* in "Math. Ann.", 298, pp. 611-28.

14. Stanley, R. P. (1999), *Enumerative Combinatorics*, vol. 2, Cambridge, Cambridge University Press.
15. Speicher, R., Woroudi, R. (1997), *Boolean Convolution, in Free Probability Theory* (Waterloo, ON, 1995), Providence, RI, American Mathematical Society, pp. 267-79.
16. Voiculescu, D. (2000), *Lectures on Free Probability Theory*, in *Lectures on Probability Theory and Statistics*, Berlin, Springer, pp. 279-349.

# Part III
# Gian-Carlo Rota, the Philosopher

# Chapter 11
# Ethics in Thought. Gian-Carlo Rota and Philosophy
## Invited Chapter

Francesca Bonicalzi

## 11.1 Telling/Inventing the Truth

*Discrete Thoughts - Pensieri discreti*, where *"discreti"* means both discrete and discreet: the title announces, with surprising clearness, and simultaneously conceals, with ingenious ambivalence, the gigantic feat Rota intends to perform: "to tell the truth". Although it is, classically, the philosopher's task, telling the truth is proclaimed from the outset as the - difficult/impossible? - enterprise of the mathematician. "In mathematics, as anywhere today, it is becoming more difficult to tell the truth" (Rota, 1986, p. IX): these are the first words of the preface to *Discrete Thoughts,* and they ring out as a provocation and a challenge that presents itself as particularly difficult due to the complexity of the wealth of knowledge that mathematics represents, but also due to the persistence in the sciences of a reductive conception of truth. The truth is not a recording of all the facts, neither can it be reduced to an exhaustiveness of pieces of information. Rather, it requires that one rediscover "the true life of thought" (Rota, 1993, p. 11) and come to grips with the history of the problem at hand, discovering the inadequacy of previous efforts in order to find a way to see the problem's true nature. Finding the truth implies also being able to pass it on to future generations, the truth implies its transmission: "In a future that is knocking at our door we shall have to retrain ourselves or our children to properly tell the truth" (Rota, 1986, p. IX).

And thus Rota reconfirms that "in mathematics, this task will be particularly difficult. The fascinating discoveries of mathematics systematically conceal, like fleeting footprints in the sand, that analogous process that constitutes the true life of thought" (Rota, 1993, p. 11). The appeal to the necessity to tell the truth is completed by the quotation of that passage of Manzoni's text which alludes to truth as that which presents itself to the eye of the spectator when, by an erroneous maneuver, a curtain is suddenly raised: the truth is pitiless, it "surprises" reality, "offends" the

Francesca Bonicalzi
Università di Bergamo, Italy

E. Damiani et al. (eds.), *From Combinatorics to Philosophy,*
DOI 10.1007/978-0-387-88753-1_11, © Springer Science+Business Media, LLC 2009

spectator, unveils that which appearance veils, but also conceals "the true life of thought" (Rota, 1993, p. 11). Does Rota not overestimate the cognitive function of mathematics, crediting it with a function of truth? Why show mathematics the fascinating way of truth, to then belittle its consistency by individuating its sources in "offhand remarks hidden within an ephemeral essay" (Rota, 1986, p. X)? When he wants to make clear in what sense he uses the expression "telling the truth" and how far this is from "reciting a rosary of facts" (Rota, 1986, p. IX), Rota utilizes Antonio Machado and Ortega y Gasset to remind us that we tell lies only out of a lack of imagination: the truth is a question of invention (Rota, 1993, p. 11).

Telling the truth. Inventing the truth. These are truly courageous claims, both for a philosopher and for a mathematician, but are surprisingly faithful to Rota's theoretical reflection. Invention is a term that modernity has loaded with a semantic significance that tends toward discovering, imagining, technical production, and that implies the new, something that did not exist before and that is produced, either imaginatively, through narration (inventing the plot of a novel), or artificially, through technology (inventing a machine, a technical apparatus). But in its original meaning "invention" - *invenio* - is oriented towards finding, finding anew.

In Derrida's intense analysis, in *Psyche. Inventions of the Other*, of the term "invention", he emphasizes the character of novelty that accompanies it:

> To find is to invent when the experience of finding takes place for the first time. An event without precedent whose novelty may be either that of the (invented) thing found (for example, a technical apparatus that did not exist before: printing, a vaccine, nuclear weapons, a musical form, an institution - good or bad - a device for telecommunications or for remote-controlled destruction, and so on) (Derrida, 1997, p. 23).

On that occasion Derrida refers to the possibility of articulating the reflection on invention further, if one revives the - now obsolete - use of expressions such as "the invention of the Cross" (Derrida, 1997, p. 23) to indicate finding it anew and, even more, if theoretical reflection looks away from the object and turns its attention to the act of finding and of discovering - that is, to the cognitive act that generates novelty. In any event, in all its senses, "invention" refers to an event of novelty that is, par excellence, the result - the product - of a human operation:

> Invention always belongs to man as the inventing subject [...]. Man himself, the human world, is defined by the human subject's aptitude for invention, in the double sense of narrative fiction or historical fabulation and of technical or technoepistemic innovation (just as I am linking *tekhnē* and *fabula*, I am recalling the link between *historia* and *epistemē* here) (Derrida, 1997, pp. 24-5).

I refer here to Derrida's work not for the sake of distraction but, rather, to introduce a thought that I believe can act as a catalyst for questions that give resonance to Rota's theoretical proposals, bringing into play that torsion in thought represented by the philosophies of Descartes and Leibniz that institute a caesura in knowledge (but also represent the junction of the separation), where *the passage from a truth as disclosure to truth as a propositional apparatus* is consummated. Descartes individuated *in mathematics the epistemic model of the certainty of knowledge*: the Cartesian reduction of the problem of truth to the problem of method understood as order (the constitutive horizon, but also institutive of meaning) made it possible to organize scientific knowledge as an apparatus of language within which it is possible to master the causes proper to scientific discourse. Later, in Leibniz, truth will no longer be a finding anew, *invenio*, but rather the productive discovery of an apparatus that can, broadly, be called technical: for Leibniz *inventors of truth are the producers of propositions*. In this junction a transformation in knowledge is consummated and produced, whose effects bring us to the heart of the question proposed by Rota, reviving the question of truth as an event of novelty.

The truth as an event of novelty surprises and - as we saw earlier - "offends", as Rota tells us in the incipit of his other - *Indiscrete – Thoughts*: "The truth offends" (Rota, 1997, p. XIX). But which truths offend, and why?

*Certain* truths offend; namely, those that hurt because they oblige us to restructure our knowledge and demand that we leave the shores of our certainties and our myths - as Rota calls them, to underline their character of indisputability - giving up the structures taken on by certain assumptions that have proved to work ("working myths") and that we would like to preserve even when they deteriorate and begin, sooner or later, to fade.

The offense is not a theoretical response; on the contrary, it represents a renunciation of the theoretical commitment of the response in order to keep from abandoning the working myth. It is, rather, a defense (through the disqualification of the other) that allows the myth (the paradigm, Kuhn would say) to survive an event of novelty that, taken seriously, would oblige us to go on to restructure our knowledge. Rota, by contrast, invites us to commit ourselves with a strong dynamism of thought. This invitation to a movement of reason, understood as the genuine movement of a scientific knowledge that does not shut itself up in the rigor of its language but exposes itself to continual rectification, reproduces the urgency of a concern that permeates Gaston Bachelard's thought, driven by theoretical demands that are surprisingly analogous to Rota's. For Bachelard knowledge proceeds precisely through the openness to continual, profound changes that produce an effect of theoretical reorganization, overcoming that which, in knowledge itself, opposes and obstructs the scientific process of abstraction, preventing us from questioning prior knowledge; it is this continual *rectification* that broadens the horizons of knowledge (Bachelard, 1950, pp. 49-52). This process of the complication of reason keeps us from holding fast to the simplifications underlying the eighteenth-century ideal of scientific reason that, as Rota recognizes, risks being confirmed through inertia, but must rather be abandoned because it is "too narrow" and risks surviving only at the cost of operating reductionisms, confining knowledge to the sole analysis of language. How, then, shall we speak of truth?

This question addresses both philosophy and science, and mathematics in particular, insofar as it brings into play the structures of rationality interrogated in relation to a radical question on the working of reason.

## 11.2 The Winding Streets, and the Straight and Wide Avenue of Precision

The difficult relations between philosophy and science have been the object of complex reflections that, in the twentieth century, were brought together under the heading "epistemology", indicating the questions that, within scientific knowledge, interrogate philosophy: the questions of the foundation, of theories of knowledge, and of the unity of scientific knowledge. Here, then, the term "epistemology" indicates those complicated relations between philosophy and science that have marked the history of Western thought and that require critical surveillance because, even where they appear silent, they produce effects of which we are not aware, and give rise to confusion. If it is true that Rota rarely uses the term "epistemology", the question it delimits - namely, the tasks, the specificities, the reciprocal implications, but also the differences, between philosophy and the sciences - is the repeated and reasserted object of his works, which are guided precisely by the concern to wrest both philosophy and science from any form of spontaneism that gives rise to misunderstandings and ambiguity. On this terrain - or, more precisely, to escape this magmatic terrain - his contribution appears particularly significant.

On a first reading, we are struck by Rota's criticisms of philosophy, which he accuses of sclerosis, approximation, partiality. His judgment brooks no appeal:

> The assertions of philosophy are tentative and partial. It is not even clear what it is that philosophy deals with. It used to be said that philosophy was 'purely speculative,' and this used to be an expression of praise. But lately the word 'speculative' has become a *bad word.* (Rota, 1997, p. 91).

Implacable and pitiless, while sparing philosophers no criticism Rota does express a feeling of sympathy and offers a glimmer of praise in granting that philosophy, amidst so many difficulties, still manages to speak to us about our existence:

> This confused state of affairs makes philosophical reasoning more difficult but far more rewarding. Although philosophical arguments are blended with emotion, although philosophy seldom reaches a firm conclusion, although the method of philosophy has never been clearly agreed upon, nonetheless the assertions of philosophy, tentative and partial as they are, come far closer to the truth of our existence than the proofs of mathematics (Rota, 1997, p. 91).

Such a concession individuates a valid side of philosophy, but does not yet suffice to recognize the value of philosophy as an enterprise of knowledge.

This discontent with philosophy, and the judgment that stems from it, becomes more specific when Rota makes clear exactly *what sort of philosophy* he condemns: namely, the philosophy that, as a victim of the myth of mathematical rigor, under "the pernicious influence of mathematics", lost its autonomy and its distinction.

Analytic philosophy saw in mathematization a way out of the concerns with the absence of demonstrative rigor, a substitute for the nonprogressive problematization of philosophy: "Philosophy can be described as the study of a few problems whose statements have changed little since the Greeks" (Rota, 1997, p. 94). The imitation of mathematics, by way of logic, has led philosophy to deviate from its original vocation and take the way of reductionism to pursue goals and successes that distort its vocation and make it follow paths not its own.

Philosophy, charmed by the certainty of mathematics' procedures and by its practical successes, thinks it can obtain analogous results by following the same path and reproducing within itself that same axiomatic method as a guarantee of its truth. Now, it is true that, as Rota confirms, for mathematics axiomatic exposition is indispensable because it guarantees its procedure rigor and certainty:

> Every fact of mathematics must be ensconced in an axiomatic theory and formally proved if it is to be accepted as true. Axiomatic exposition is indispensable in mathematics because the facts of mathematics, unlike the facts of physics, are not amenable to experimental verification (Rota, 1997, p. 90).

But it is also true that this does not exhaust the sphere of mathematics; indeed, one may speak of a "double life" of mathematics that, like philosophy, is concerned, equally, with proofs and with facts.

The equivocal equalization of philosophical rigor with mathematical precision has led much of contemporary philosophy to tread the path of mathematization, and is at the root of the *philosophical reductionism* criticized by Rota.

In taking mathematics as its model, philosophy forgets its own most distinctive characteristic: mathematizing philosophers, philo-mathematical philosophers or normative philosophers - as Rota calls them - in order to make philosophy rigorous and objective have given up the quest for sense (Rota, 1997, p. 191) and the historical dimension of theoretical thought (Rota, 1997, pp. 93-4) which, in their opinion, have been the cause of the practical failures of philosophy. To make up for these deficiencies, they link philosophical thought to the rules of mathematics and entrust philosophical sense to the language of mathematical logic. This triggers a reductionism that is the object of Rota's incessant criticism, as is fully documented by the first monograph dedicated to his philosophical reflection, *La stella e l'intero* (Palombi, 2003, pp. 29-41). On the basis of a misunderstanding of philosophy, but also of mathematics, the mathematizing philosophers deny the heterogeneity of the two disciplines, leaving the insidious field of history to take that of pure reasoning instead, and excluding from the field of philosophy everything that cannot be mathematized. A mistaken ideal of precision drains philosophical concepts of the semantic wealth necessary to them (even if such wealth entails a certain indeterminacy) and replaces it with simplifications dictated by the rules of mathematical calculability. Rota maintains that this myth of precision (Rota, 1997, p. 94) produces in philosophy the same devastation as an urban renewal project that, to streamline the road network, razes the historic center of a city and creates an enormous avenue:

> To use an imagine due to Wittgenstein, philosophical concepts are like the winding streets of an old city, which we must accept as they are, and which we must familiarize ourselves with by strolling through them while admiring their historical heritage. Like a Carpathian dictator, the advocates of precision would raze the city and replace it with the straight and wide Avenue of Precision (Rota, 1997, p. 95).

I retrace and repropose the image evoked by Rota, both for its efficacy and the undoubted vigor of its denunciation, but also because it refers us back to the question

of the way, the path - the *odos* - that is the inaugural question of philosophical rationality, but that also represents the new way, the junction - *met-odos* - in which scientific certainty is established. In either case, whether it be the philosophical method
- the Parmenidean *odos*, the Platonic dialectical method, or Aristotelian logic - or
the scientific method that matures within the intellectual disquiet of modernity, the
question of method is closely - inexorably - linked to the question of truth.

## 11.3  The Food of Philosophical Thought and the Medicine of Axiomatics

Rota believes that the current circumstance of mathematizing philosophers stems
from the fact that philosophers - immersed in the precarious situation of a thought
that can never attain a definitive content and is rent by debates with no sure outcome - are inexorably attracted by mathematical knowledge that attains immutable
results, whose validity is attributed to the axiomatic method. Philosophers, blinded
by the successes of mathematical knowledge, skeptical about their own thought,
weakened by their experiences of failure, have placed all their trust in the axiomatic
method that they conceive to be the fundamental instrument of discovery. For Rota
this heralds philosophy's defeat: "to discover truth, they surrender to a slavish and
superficial imitation of the truth of mathematics" (Rota, 1997, p. 96).

If it is true that the situation that has arisen in mathematizing philosophy has special characteristics, it is also true that it reproposes a dependency on mathematics
that has been a constant in the history of philosophical thought. Thus this attitude
can be seen as the extreme outcome of a torsion in knowledge set in train in modernity; as the resumption, at another level, of a push given to thought centuries ago,
when Galileo indicated mathematics as the only language that makes it possible to
attain an objective knowledge of nature and Descartes posited mathematics as the
epistemic model of knowledge.

Let me dwell for a moment on Descartes' theoretical operation (I know that for
the mathematicians this is archaeology... but for the philosophers it is history, and
Rota teaches us to recognize the historicity of thought) because in some way it
brings into play questions that, growing more complicated over the centuries, have
led to those attitudes that alarm Rota by the effects of non-sense they produce in
philosophy. As I was saying, Descartes, from within philosophy, senses the urgency
to find certainties in knowledge equal to the ones attained by the sciences, and re-
sponds by giving us *method* as a guarantee for philosophy and, at the same time,
for science. Truth, the special task of philosophy, is entrusted to method also for the
sciences according to a correlation reaffirmed also in the full title of the celebrated *A
Discourse on Method*: "for the Well Guiding of Reason and the Discovery of Truth
in the Sciences". With "method" Descartes excludes logic and, with it, dialectic and
rhetoric from any cognitive claim, indicating mathematics in their place as the locus
of production of certain and sure knowledge. Whoever wishes to follow the straight
path of truth will have to be concerned solely with objects about which one can have

*certainty equal to that of arithmetic and geometric proofs*. Moreover, the Cartesian reduction of matter to extension makes it possible to renounce that sort of imperialism to which Galilean science still obliged mathematics, identifying the object of science with quantity: the uniformity of reality (order and measure) resolves the difficult relationship - already noted by Aristotle - between the mathematical disciplines and the contingency of physical reality, without having recourse to those operations demanded also by the - albeit rigorous - Galilean "book of nature", which to qualify mathematics has to refer matter back to an ideal immutability or to an impediment to be removed. Mathematical certainty releases from the demand for epistemological unity that results from architectonic unity (*connexio scientiarum*), it releases science from the multiplicity to which the plurality of objects condemns it to bring it back to the oneness of human knowledge: as the epistemic model of knowledge, mathematics shows the working of method and method universalizes the possibility that certainty be the condition and outcome also for the other sciences, independently of the object to be known. Method - we said - replaces logic and rhetoric because it produces truth and persuasion in the evident, but it replaces mathematics as well, universalizing it in the production of certainty in order. While the dialectical method of philosophy aims at a logical truth in relation to a reality defined in terms of essence, the truth of the Cartesian method is articulated as a certainty that is measured not on the character of known reality but rather as the *structure of the subject in the act of knowing*.

But in what sense is method - formulated now on the basis of other problems that reflect the fertile crisis of the sciences at the beginning of the twentieth century (Husserl, 1959), marked by the formulations of logical positivism made possible by Wittgenstein's reflections on language and by Russell's and Carnap's new elaborations of logic - in what sense, I was saying, is method that which lacerates philosophical discourse?

It is clear that philosophy cannot let itself be enslaved by the method of the sciences, as Rota agues so well. But can/must reason give up method? Method, as we have seen, does not indicate simple cognitive procedures, it does not allow itself to be reduced to methodology, it is not just a modality of investigation and research. No, more radically, we can say that method is a way on which the essence of truth is determined. Precisely because method is inherent in rationality itself, *it comes to the fore in the moments in which there are radical turning points for reason*, which, in this way, is obliged to question itself on the reasons for which it is constituted, and to recognize that the movement of truth is broader than that which any discipline of knowledge can access. The *contra* (Feyerabend) or *extra* (Gadamer) methodological positions that have so resoundingly been taken in contemporary culture are not proclamations of irrationalism or relativism, but rather represent a stance that, while it makes an appeal to rationality and vindicates the totality of its movement, *denounces the reductionism of method*. The intolerance of method, like the dependence on method, represent, in contemporary culture, a denunciation of the impossible claim of reason to exercise a domination over reality, but also of the impossible claim of method to be unique, and thus gives voice to the demand that method articulate itself in relation to the differences of the object that gives itself to knowledge.

Method interests philosophy because of its promise of truth, but is the truth the axiomatic method promises the one that philosophy seeks? How is truth to be thought, outside any temptation of normativity dictated by reflections whose origin is extraneous to philosophical practices and to mathematical activity itself? Rota unmasks the misunderstanding inherent in the use of the word "truth": the "truth" utilized by philosophy is only a synonym of the contemporary term in use in mathematics, and mathematical truth, as such, has no philosophical significance. The logical term, too, requires an analogous revision: contemporary logic has nothing to do with philosophy, logic today is a branch of mathematics, it is not philosophy because it does not indicate the structures of the procedure of reason. This is why logic no longer claims to be foundational, there is no logical foundation of mathematics, and in fact "logic has become mathematical at a price. Mathematical logic has given up all claims of providing a foundation to mathematics" (Rota, 1997, p. 92).

Logic, like topology, combinatorics or probability, represents one of the branches of mathematics; it enjoys no particular privilege, it is nothing more than a highly successful field of the practical applications of mathematics.

Mathematical logic is a trap for philosophers, deceiving them regarding the nature of mathematical truth, which is made to coincide with the correctness of an axiomatic presentation; but this truth does not speak the language of philosophy, because it gives no information on reality (Rota, 1997, p. 98), and has nothing to say about sense, i.e., is insignificant as far as the main tasks of philosophy are concerned. The pseudo-philosophical language "of mathematical logic has misled philosophers into believing that mathematical logic deals with the truth in the philosophical sense. But this is a mistake. Mathematical logic deals not with the truth but only with the game of truth" (Rota, 1997, p. 93). The question of method radically interrogates the nature of rationalism and shows the results of reductionism to be met with if one equates the *logical* with the *rational*: it is as if we pretended to feed ourselves with medications rather than with food (Rota, 1997, p. 96).

The call to life and to its historicity bursts in.

## 11.4 The Ethics of Knowledge and the Genesis of its Disciplines

The mathematical path that philosophy has taken, on the basis of nonexistent analogies and attracted by the mirage of guaranteed success, shows the full measure of its sterility: the reductionism this imposes has led it to exclude the genuine - classical - problems of philosophy, either by silencing them or by considering them to be psychological. This reductionist approach, which stems also from a false image of mathematics, hurts philosophy because it denies its autonomy and prevents it from realizing the enterprise of knowledge to which it is destined - an enterprise different from that of the sciences, and of mathematics in particular:

The fundamental error is an instance of reductionism. The *process* by which the mind works, which may be of interest to physicians but is of no help to working mathematicians, is confused with the *progress* of thought that is required in the solution of any problem. This catastrophic misunderstanding of the concept of mind is the heritage of one hundred-odd years of pseudo-mathematization of philosophy (Rota, 1997, p. 101).

Rota's work is an appeal to philosophy to break free from the situation of ghettoization and impoverishment to which it has condemned itself - a situation due to the ahistorical and acritical flattening of the formulation of its theories - and regain that cultural role it has always played and for which it is irreplaceable: in fact, "no discipline has contributed as much to our civilization" (Rota, 1999, p. 183). Philosophy must rediscover the primordial motives of its discourse, interrogated in the difference between reason and logic, between truth and knowledge. Only in this way can it repropose the philosophical question in a new formulation whose range is visible in the fact that it leads to the broadening of a question that brings the motives of rationality into play and gives rise to a genetic reading of knowledge involving the subject, reality and rationality, and seeking - in the experience of subjects engaged in the invention of the fields of rational knowledge - the movement of the production of such fields and the emergence of their norms.

In this renewed direction, Rota finds instruments and resources of thought in Husserl's phenomenological method. The strong points of philosophy, which Rota deems absolutely essential, are, on the one hand, its capacity to speak to existence (compared to science, which says nothing in relation to sense but can affect and offend us because it interferes with our existence - a judgment that Rota shares with Husserl) and to account for the "variable and inexhaustible reality of experience" (Rota, 1999, p. 94) and, on the other, its capacity to intervene effectively in the productive structures of thought. The critical vigilance and the rigor of philosophy consist precisely in its capacity to account for reality in all its complexity without simplifications or reductions:

They do everything to keep so many unpleasant aspects out of this description [...]. Thought today is the thought of open categories, of one-off cases, of ambiguous appearances, of mysteries, of problems without solution or sense, phenomena that Enlightenment reason passed over in silence (Rota, 1999, p. 95).

It is not a question, then, of falling back on extraterritorial fields from which to draw legitimacy, but rather of offering resources for those classical problems, denied as nonexistent by some philosophy, that are today on the agenda "in the forefront of science" (Rota, 1997, p. 99). Thus the situation is reversed: Rota is convinced that there are such fields of knowledge as experimental psychology, neurophysiology, informatics, etc., that can become allies of a classical philosophy that can offer them perspectives and instruments to activate a process of elaboration and criticism, offering itself as a practicable space for contemporary interrogations. The very complexity of the phenomena to be dealt with convinces scientists to valorize the contribution of philosophy in the belief that "progress in science will depend on philosophical research in the most classical vein" (Rota, 1997, p. 99), to the point of envisioning a new era of scientific thought made possible by a renewed philosophical thought. It is our conviction that the organicity of the philosophical thought of this century will definitively broach a perspective of rigor for the new sciences, thus inaugurating a new era of scientific thought (Rota, 1999, p. 116). The torsion in the question that produces these effects of novelty in the relationships between disciplines is re-examined by Rota on the basis of the Husserlian concept of *Fundierung* of which he gives an original reading, as Palombi significantly emphasized in his *La stella e l'intero* (Palombi, 2003, pp. 61-73). From my own viewpoint, linking up my reflections here, I can say that Rota's reading of *Fundierung* broaches the possibility of abandoning a reductionist methodological position to access a formulation that conceives knowledge not in logical-perceptive terms but rather as a study of the *genesis in knowledge*. In this sense philosophy offers itself as a "pathway that will lead us to broach the conditions of possibility" even of mathematics. Rota believed that new tasks are opening up for philosophy - tasks such as "laying a rigorous basis for discovering the conditions of possibility" (Rota, 1999, p. 105) and providing a foundation of the sciences now emerging, such as molecular biology, cognitive psychology and robotics. Such sciences, which "are likely to be dependent upon the disentangling of foundational concepts, are becoming dependent upon philosophical argument" (Rota, 1997, p. 183). Rota suggests that philosophy put itself to the test by working on the concept of identity after the great phenomenologies of Heidegger and Husserl.

Rota concludes his *Lezioni napoletane* with the call for a philosophical revolution that will make possible a theoretical foundation of the contemporary sciences analogous to the one that led to the triumph of the physical sciences in the time of Galileo. This reference to Galileo reminds me of how Galileo asked the Grand Duke of Tuscany to add the title of philosopher to his title of mathematician. Galileo liked to consider himself both mathematician and philosopher, yet it was precisely the scientific practice he inaugurated that was, later, to render this formulation problematic, if not impossible.

No Grand Duke is needed to credit Rota with the title of philosopher, along with that of mathematician: his ethical testimony in thought demands it and shows unequivocally that this is still possible in the genuine rigor of a nonreductionist practice of thought.

# References

Bachelard, G. (1950), *De la nature du rationalisme*, in "Bulletin de la Société de française de Philosophie", séance du 25 mars 1950, pp. 45-86.

Bachelard, G. (1953), *Le matérialisme rationnel*, Paris, PUF.

Derrida, J. (1997), *Psyche. Inventions of the Other*, vol. I, P. Kamuf and E. Rottenberg (eds.), Stanford, Stanford University Press, 2007.

Husserl, E. (1959), *The Crisis of European Sciences and Transcendental Phenomenology*, D. Carr (trans.), Evanston, Northwestern University Press, 1970.

Kac, M., Rota, G.-C., Schwartz, J.T. (1986), *Discrete Thoughts. Essays on Mathematics, Science and Philosophy*, Boston, Basel, Berlin, Birkhäuser.

Palombi, F. (2003), *La stella e l'intero. La ricerca di Gian-Carlo Rota tra matematica e fenomenologia*, Turin, Bollati Boringhieri.

Rota, G.-C. (1986), *Preface* in (Kac, Rota, Schwartz, 1986).

Rota, G.-C. (1993), *Pensieri discreti*, F. Palombi (ed.), Milan, Garzanti.

Rota, G.-C.(1997), *Indiscrete Thoughts*, F. Palombi (ed.), Boston, Basel, Berlin, Birkhäuser.

Rota, G.-C. (1999), *Lezioni napoletane*, F. Palombi (ed.), Neaples, La Città del Sole.

# Chapter 12
# Indiscrete Variations on Gian-Carlo Rota's Themes
## Invited Chapter

Carlo Cellucci

## 12.1 Introduction

I never met Gian-Carlo Rota but I have often made references to his writings on the philosophy of mathematics, sometimes agreeing, sometimes disagreeing.

In this paper I will discuss his views concerning four questions: the existence of mathematical objects, definition in mathematics, the notion of proof, the relation of philosophy of mathematics to mathematics.

## 12.2 The Existence of Mathematical Objects

Although in the twentieth century the main question in the philosophy of mathematics has been the justification of mathematics, next to it there has been the question of the existence of mathematical objects (see, for example, Cellucci, 2006).

There are four possible answers to this question: (i) Mathematical objects exist; (ii) No they don't; (iii) We don't know; (iv) The question is irrelevant to mathematical practice or meaningless.

## 12.2.1 *Irrelevance of the Existence of Mathematical Objects*

Rota's answer to the question of the existence of mathematical objects is of kind (iv).

For he states that "it does not matter whether mathematical items exist, and probably it makes little sense to ask the question" (Rota, 1997, p. 161). If "someone

---

Carlo Cellucci
Università di Roma La Sapienza, Italy

E. Damiani et al. (eds.), *From Combinatorics to Philosophy,*
DOI 10.1007/978-0-387-88753-1_12, © Springer Science+Business Media, LLC 2009

proved beyond any reasonable doubt that mathematical items did not exist", that would not "affect the truth of any mathematical statement" (*ibid.*). Discussions concerning the existence of mathematical objects "are motivated by deep-seated emotional cravings for permanence which are of psychiatric rather than philosophical interest" (*ibid.*).

Rota's answer seems a very sensible one. A similar answer was given in the Seventeenth and the Eighteenth century by such disparate thinkers as Descartes, Locke, Hume. For example, Descartes stated that "arithmetic, geometry, and other such disciplines" are "indifferent as to whether these things do or do not in fact exist" (Descartes, 2006, p. 11).

True, in the Twentieth century Brouwer attempted to develop an alternative mathematics, banning certain objects on the ground that they did not exist and replacing them by certain other objects on the ground that they did exist. His attempt, however, ended in failure, for his alternative mathematics was so awkward that nobody managed to use it for any essential purpose. Moreover, it excluded certain objects, such as discontinuous real functions, which are essential to physics. Thus Brouwer's attempt turned out to be a purely ideological one. Indeed, Brouwer explicitly stated that he did not care a bit for the applicability of mathematics to physics, because he rejected "expansion of human domination over nature" (Brouwer, 1975, p. 483). For him mathematics was a search for beauty, and in applicable mathematics "beauty will hardly be found" (*ibid.*; for more on this, see Cellucci, 2007, pp. 76-81).

While Rota's answer seems a very sensible one, what Rota positively says about the nature of mathematical objects seems less convincing.

He states that such things as "prices, poems, values, emotions, Riemann surfaces, subatomic particles, and so forth" are "ideal objects" (Rota, 1986, p. 169). The method of logical analysis inaugurated by Husserl sets itself "the purpose of construction (rather than dissection) of ideal objects to be subjected to yet-to-be-discovered ideal laws and relations" (Rota, 1986, p. 172).

Thus for Rota mathematical objects are ideal objects, and mathematical laws and relations concern such objects.

Now, an idealization simplifies certain real items by ignoring some features which make only a small difference in practice, while retaining other features which are basic. Therefore, ideal objects should retain the basic features of the real items they are said to idealize. Then Rota's statement that mathematical objects are ideal objects contrasts with the fact that several mathematical objects have nothing in common with real objects. For example, infinite sets have no basic feature in common with concrete physical aggregates, so they cannot be said to be an idealization of them.

Moreover, stating that mathematical objects are ideal objects trivializes the question of the applicability of mathematics to the physical world. For it makes all mathematical statements vacuously true of it.

For example, consider the statement: in any triangle, the interior angles are equal to two right angles. Such statement is of the form $P(x) \rightarrow Q(x)$, where $P(x)$ expresses '$x$ is a triangle' and $Q(x)$ expresses 'The interior angles of $x$ are equal to two right angles'. Now, $P(x)$ is an ideal statement which is false of the physical world for triangles do not exist in it. Therefore $P(x) \rightarrow Q(x)$ is vacuously true of the physical world. For the same reason, even the false statement 'In any triangle the interior angles are not equal to two right angles', being of the form $P(x) \rightarrow \neg Q(x)$, is vacuously true of the physical world.

Furthermore, Rota pushes his view that mathematical objects are ideal objects to the extreme. For he states that "the ideal of all science, not only of mathematics, is to do away with any kind of synthetic a posteriori statement and to leave only analytic trivialities in its wake" (Rota, 1997, p. 119). Science is "the transformation of synthetic facts of nature into analytic statements of reason" (*ibid.*).

Thus, according to Rota, not only mathematical objects are ideal objects, but mathematical statements, and all scientific statements generally, are ultimately analytic.

Holding that scientific statements are ultimately analytic, Rota seems to be attracted by the view that the laws of the world can be derived a priori. He even goes so far as saying that the mathematician "forces the world to obey the laws his imagination has freely created" (Rota, 1997, p. 70). Thus, for Rota, not only the laws of the world are ultimately analytic, but the mathematician enforces them on the world.

However, the laws of the world are not ultimately analytic, for the facts of nature are not truths of reason. Nor the mathematician enforces such laws on the world, for the laws in question are just a means by which humans make the world understandable to themselves. Science is what humans grasp of the world in their own terms, and this essentially depends on synthetic facts of nature. Moreover, the creations of the mathematician are not completely free, for they essentially depend on the mathematician's biological constitution.

## 12.2.2 Mathematical Objects as Hypotheses

A more satisfactory account of the nature of mathematical objects can be given stating that mathematical objects are hypotheses tentatively introduced to solve mathematical problems.

A mathematical object is the hypothesis that a certain condition is satisfiable. For example, an even number $x$ is the hypothesis that the condition $x = 2y$ is satisfiable for some integer $y$.

If, in the course of reasoning, the condition turns out to be satisfiable, we say that the object 'exists', if it turns out to be unsatisfiable, we say that it 'does not exist'. Thus speaking of 'existence' is just a metaphor.

That the condition turns out to be unsatisfiable typically occurs in proofs by *reductio ad absurdum*.

For example, suppose that, to solve the problem whether, in Euclidean geometry, in any triangle the interior angles are equal to two right angles, we tentatively introduce a new kind of objects: triangles whose interior angles are not equal to two right angles. We say: let $ABC$ be any such triangle. Then we draw a line through one of its vertices parallel to the opposite side and we see that the interior angles are actually equal to two right angles. We thus have a contradiction. Therefore we conclude that a triangle such as *ABC* cannot exist, and hence that, in Euclidean geometry, in any triangle the interior angles are equal to two right angles.

There is no more to mathematical existence than the fact that mathematical objects are hypotheses tentatively introduced to solve mathematical problems. Such hypotheses are in turn a problem to be solved, it will be solved by introducing other hypotheses, and so on. Thus solving a mathematical problem is a potentially infinite task (see Cellucci, 2002, Ch. 22; Cellucci, 2008b).

The view that mathematical objects are hypotheses is related to Plato's view that "those who practice geometry, arithmetic and similar sciences hypothesize the odd, and the even, the geometrical figures, the three kinds of angle, and any other thing of that sort which are relevant to each subject" (Plato, *Republic*, VI 510 c 2-5). Thus they hypothesize mathematical objects. But, in addition to them, they also hypothesize properties of such objects. Therefore mathematical hypotheses concern either mathematical objects or their properties.

Those who practice geometry, arithmetic and similar sciences, however, do not confine themselves to making hypotheses, but also give an account of them by introducing other hypotheses. For "when you had to give an account of the hypothesis itself, you would give it in the same way, once again positing another hypothesis" (Plato, *Phaedo*, 101 d 5-7). And so on. This is Plato's 'dialectical method' or 'dialectic'.

It is often claimed that "Plato thinks of the method of mathematics as one that starts by assuming some hypotheses and then goes 'downwards' from them (i.e. by deduction), whereas the method of philosophy (i.e. dialectic) is to go 'upwards' from the initial hypotheses, finding reasons for them (when they are true), until eventually they are shown to follow from an 'unhypothetical first principle' " (Bostok, 2009, pp. 13-14).

Actually, quite the opposite is true. According to Plato, the method that starts by assuming some hypotheses and then goes 'downwards' from them, that is, the axiomatic method, is a degeneration of the genuine method of mathematics, which is the same as the method of philosophy, that is, the dialectical method.

Mathematicians who practice the axiomatic method assume their hypotheses as starting-points (principles) without giving any account of them, and go downwards from them. But, giving no account of them, they do not really know their principles. And, "when a man does not know the principle, and when the conclusion and intermediate steps are also constructed out of what he does not know, how can he imagine that such a fabric of convention can ever become science?" (Plato, *Republic*, VII 533 c 3-6).

Therefore, criticizing the axiomatic method, Plato opposes the method of philosophy not to the method of mathematics but rather to a degeneration of that method.

Admittedly, according to Plato, mathematical objects in their true nature are independently existing entities, which can be known only by means of intellectual intuition. For only the latter is capable of grasping the unhypothetical first principle. But intellectual intuition is possible only if we "leave our body and contemplate the things themselves with the soul by itself" (Plato, *Phaedo*, 66 e 1-2). Until then, we may only try to state more and more general hypotheses. In this way we will get better and better approximations to mathematical objects, but will never arrive at grasping them fully. On the other hand, however, this is the only way available to us.

### 12.2.3 Hypotheses vs. Fictions

The view that mathematical objects are hypotheses must not be confused with fictionalism – the view that mathematical objects are like characters in fiction.

According to fictionalism, all there is to mathematics is that "we have a good story about natural numbers, another good story about sets, and so forth" (Field, 1989, p. 22). The sense in which " '2+2=4' is true is pretty much the same as the sense in which 'Oliver Twist lived in London' is true: the latter is true only in the sense that it is true according to a certain well-known story, and the former is true only in that it is true according to standard mathematics" (Field, 1989, p. 3).

Fictionalism is inadequate in several respects, which I cannot discuss here (see Cellucci, 2007, pp. 109-14). I will only point out that hypotheses are essentially different from fictions for the following two reasons.

1) While fictions are stated with the awareness that they are not real, hypotheses are aimed at reality. They are meant to provide an adequate approach to a still unknown or not perfectly known reality, although, as in the case of proofs by *reductio ad absurdum*, they eventually may fail to provide an adequate approach. Admittedly, the determination of reality given by a hypothesis is provisional. But one will try to make it relatively stable showing that the hypothesis is plausible, that is, compatible with the facts of experience.

2) While fictions are merely thinkable, hypotheses are supposed to be possible, so they must agree with the facts of experience. Only then hypotheses can be said to provide an adequate approach to reality. On the contrary, fictions are not supposed to be possible, and hence are not rejected if they do not agree with the facts of experience.

### 12.2.4 Existence and Identity

Hypotheses provide an only partial and provisional characterization of mathematical objects. The latter may receive new determinations through interactions between hypotheses and experience.

Rota proposes an apparently related view when he states that "a full description of the logical structure of a mathematical item lies beyond the reach of the axiomatic method" (Rota, 1997, p. 156). For example, "the real line, or any mathematical item, is not fully given by any one specific axiom system" and "allows an open-ended sequence of presentations by new axiomatic systems. Each such system is meant to reveal new features of the mathematical item" (Rota, 1997, p. 157).

On the other hand, however, Rota also states that "a mathematical item is 'independent' of any particular axiom system" in much the same way as "an idea is 'independent' " of "the words that are used to express the idea" (Rota, 1997, p. 160).

Rota acknowledges that from this one might be tempted to conclude that he assumes that "mathematical items 'exist' independently of axiom systems" (*ibid.*). But he rejects such conclusion stating that, "in discussing the properties of 'mathematical items'", he is "in no way required to take a position as to the 'existence' of mathematical items" (Rota, 1997, p. 161). He is only concerned with the question of the identity of mathematical items, and "identity does not presuppose existence" (*ibid.*).

Rota's distinction between identity and existence is an important one. What he says about the identity of mathematical objects, however, is somewhat indefinite.

On the one hand, he states that "there is no way to 'reduce' identity to any mental process" (Rota, 1997, p. 187). Thus identity is not mental. On the other hand, he states that, if I kick a stone, "it is a mistake to believe that what I kick is a material stone", rather " 'the stone' is an item that has no existence, but has an identity" (Rota, 1997, p. 186). Thus identity is not physical either.

But, if identity is neither mental nor physical, what is it? Rota does not tell us. And he could not tell us, for he claims that "identity is the 'undefined term' " (*ibid.*).

Moreover, Rota claims that "the properties of identity are the axioms from which we 'derive' the world" (*ibid.*). But this is impossible, for the properties of identity are purely logical, and from purely logical properties one cannot derive any feature of the world, even less 'the world'.

Thus on Rota's view mathematical objects have a somewhat unsettled status.

## 12.2.5 The Inexhaustibility of Mathematical Objects

This problem is avoided if mathematical objects are hypotheses. Admittedly, fixing properties of mathematical objects, hypotheses characterize their identity but say nothing about their existence. The latter requires a further investigation. But their identity is by no means, as Rota states, the undefined term. On the contrary, it is characterized by the hypothesis, and characterized differently by different hypotheses.

For example, according to Euclid, the identity of the real line is characterized by the hypothesis that it is a breadthless length which lies evenly with the points on itself. According to Cantor and Dedekind, it is characterized by the hypothesis that it is the set of singleton real numbers.

Again, hypotheses do not characterize the identity of mathematical objects completely and conclusively but only partially and provisionally, therefore their identity is always open to new determinations. It is so because mathematical objects are open to interactions with other objects, from which new properties may arise.

Gödel states that "the creator necessarily knows all properties of his creatures, because they can't have any others excepts those he has given to them" (Gödel, 1995, III, p. 311).

But this is contradicted, for example, by the fact that an artist is not necessarily a good art critic, or a mathematician is not necessarily a good philosopher of mathematics. For the creator's creatures, when put in relation with other things, may be seen to have properties not included in those the creator has given to them.

Moreover, from the interactions of the given mathematical objects with other objects, new facts may emerge which may suggest to modify or completely replace the hypothesis by which the identity of the given mathematical objects had been characterized. This is a potentially endless process, for mathematical objects are always open to interactions with other objects.

## 12.3 Definition in Mathematics

In the last century the received view on mathematical definition – that is, the view that has been taken for granted without further criticism – has been that mathematical definitions are arbitrary stipulations which serve as abbreviations and hence engender no knowledge.

For example, Frege states that "no definition extends our knowledge. It is only a means for collecting a manifold content into a brief word or sign, therefore making it easier to handle. This and this alone is the use of definition in mathematics" (Frege, 1984, p. 274). Then "it would be inappropriate to count definitions among principles. For to begin with, they are arbitrary stipulations" (*ibid.*). Even when "what a definition has stipulated is subsequently expressed as an assertion, still its epistemic value is no greater than that of an example of the law of identity $a = a$", and "one would hardly wish to accord the status of an axiom to every single instance, to every example, of the law" (*ibid.*). Therefore, "by defining, no knowledge is engendered" (*ibid.*).

### 12.3.1 Definition, Description and Analysis of Concepts

Rota criticizes the received view arguing that, while "mathematicians take mischievous pleasure in faking the arbitrariness of definition", actually "no mathematical definition is arbitrary" (Rota, 1997, p. 97).

In fact "a lot of mathematical research time is spent in finding suitable definitions to justify statements that we already know to be true"(Rota, 1997, p. 50). For example, the Euler's formula for polyedra was known to be true "long before a suitable general notion of polyedron could be defined" (*ibid.*).

Definition must not be confused with description, for "description and definition are two quite different enterprises" (Rota, 1997, p. 48). In ages past, "mathematical objects were described before they could be properly defined", and "the mathematics of the past centuries confirms the fact that mathematics can get by without definitions but not without descriptions" (Rota, 1997, pp. 48-9). For example, "the lack of definition of a tensor did not stop Einstein, Levi-Civita and Cartan from doing some of the best mathematics in this century" (Rota, 1997, p. 97).

Rota is quite right in criticizing the received view, for the fact that mathematicians often spend so much research time in finding suitable definitions to justify statements that they already know to be true would be incomprehensible if definitions were arbitrary stipulations. As a matter of fact, mathematicians often use concepts for a long time before a suitable definition is found.

When what is defined is a concept which had previously been used without a precise definition, the definition contains an analysis of the concept, and may therefore express an advance in knowledge. In such cases, a definition gives definiteness to a concept which had previously been more or less vague. Presumably, what Rota calls 'description' is the more or less vague statement of the concept on the basis of which that concept was previously used.

That definitions may contain an analysis of concepts which had previously been more or less vague, explains why mathematicians often spend so much time in looking for definitions to justify theorems which are already known. Finding an adequate definition is not the result of an arbitrary stipulation but rather of an investigation.

## 12.3.2 Definition in Mathematics and in Philosophy

Rota, however, makes some statements about definitions which seem to be in conflict with the view that mathematical definitions are not arbitrary stipulations but may contain an analysis of concepts.

For example, he asks: "Doesn't mathematics begin with definitions and then develop the properties of the objects that have been defined by an admirable and infallible logic?" (Rota, 1997, p. 97).

It would be tempting to consider this question ironic, but it is not. For Rota opposes definition in mathematics to definition in philosophy, stating that, salutary as the injunction 'Define your terms!' "may be in mathematics, it has had disastrous consequences when carried over to philosophy. Whereas mathematics starts with a definition, philosophy ends with a definition" (ibid.).

Rota would have been surprised to learn that the same view is put forward by Dummett, one of the most eminent representatives of that analytic philosophy which he criticizes.

For Dummett states that "the aims of mathematicians differ from those of philosophers. To mathematicians it doesn't matter that the definitions they devise grasp concepts as they are implicitly understood in everyday life: what matters to them is to state precise concepts" (Dummett, 2001, p. 16). Once mathematicians have done so, "their argument proceeds within the bounds established by the definitions they have adopted. On the contrary, the philosopher's reasoning depends on our pre-existent implicit understanding of concepts: it appeals to such understanding and hence, unlike the mathematical case, it does not proceed within the framework of concepts which have been previously made precise" (*ibid.* I retranslate Dummett's text from the Italian for only an Italian version of Dummett's book has been published).

Thus, like Rota, Dummett claims that, whereas mathematics starts with a definition, philosophy ends with a definition.

There is, however, little credibility in the view that, unlike philosophy, mathematics starts with a definition. If it were so, then it would be incomprehensible why mathematicians often try to find new definitions for concepts for which definitions are already available.

In mathematics, as in philosophy, definitions are not a starting point but rather an arrival point of the investigation. (In the sexteenth century Zabarella put forward a similar view; see Cellucci, 1989). Definitions are not arbitrary stipulations but rather hypotheses and, as all hypotheses, are means of discovery. Two definitions of the same concept may be not equally adequate as means of discovery (see Cellucci, 2002, Ch. 36).

### 12.3.3 The Alleged Circularity of Definitions and Theorems

Rota also claims that "the theorems of mathematics motivate the definitions as much as the definitions motivate the theorems. A good definition is 'justified' by the theorem that can be proved with it, just as the proof of the theorem is 'justified' by appealing to a previously given definition. There is, thus, a hidden circularity in formal mathematical exposition. The theorems are proved starting with definitions; but the definitions themselves are motivated by the theorems that we have previously decided ought to be correct" (Rota, 1997, p. 97).

This circularity, however, only affects definitions in the axiomatic method. If definitions are hypotheses, then there is no circularity between definitions and theorems. For a definition is not justified, as Rota claims, by the theorem that can be proved with it. As any other hypothesis, it is justified by its plausibility, that is, compatibility with the existing data. Therefore, it is justified by things distinct from the theorems which are established by means of the definition (see also Cellucci, 2008a, p. 209).

### 12.4 The Notion of Proof

In the last century the received view on proof has been the axiomatic one: a proof is a deductive derivation of a proposition from primitive premises that are true, in some sense of 'true'.

Strictly related is the notion of formal proof. In fact, according to the Hilbert-Gentzen Thesis, every axiomatic proof can be represented by a formal proof (see Cellucci, 2008b).

### 12.4.1 Proof as the Opening up of Possibilities

Rota criticizes the received view arguing that, while "some mathematicians will go as far as to pretend that mathematics is the axiomatic method", the "mistaken identification of mathematics with the axiomatic method has led to a widespread prejudice among scientists that mathematics is nothing but a pedantic grammar, suitable only for belaboring the obvious" (Rota, 1997, p. 142).

The error of the received view "lies in assuming that a mathematical proof, say the proof of Fermat's last theorem, has been devised for the explicit purpose of proving what it purports to prove" (Rota, 1997, p. 143). On the contrary, the point of proof "is to open up new possibilities for mathematics", for example, "this opening up of possibilities is the real value of the proof of Fermat's conjecture" (Rota, 1997, p. 144).

Indeed, the value "of Wiles' proof lies not in what it proves, but in what it has opened up, in what it will make possible", specifically, in "a host of new techniques that will lead to further connections between number theory and algebraic geometry. Future mathematicians will discover new applications, they will solve other problems, even problems of great practical interest, by exploiting Wiles' proof and techniques" (*ibid.*).

Rota is quite right in criticizing the received view, arguing that the point of proof is not so much to prove what it purports to prove as rather to open up new possibilities for mathematics. In fact, the main point of proof is to discover plausible hypotheses which not only prove what they purport to prove, but also establish connections between different areas of mathematics – connections that may lead to new discoveries. Briefly, the main point of proof is its heuristic value (see Cellucci, 2002, Ch. 24; Cellucci, 2008b).

On the other hand – surprisingly enough in view of his criticism of the received view – Rota also states that "a mathematical theory begins with definitions and derives its results from clearly agreed-upon rules of inference" (Rota, 1997, p. 90). For "there is at present no viable alternative to axiomatic presentation if the truth of mathematical statements is to be established beyond reasonable doubt" (Rota, 1997, p. 142).

Rota's statement is in conflict with his view that the point of proof is to open up new possibilities. For, in an axiomatic theory all possibilities are implicitly contained in the axioms, which are given from the very beginning. Therefore, an axiomatic proof cannot open up new possibilities establishing connections between different areas of mathematics. Axiomatic theories are closed systems (see Cellucci, 1998, pp. 192-203; Cellucci, 2002, Ch. 7).

### 12.4.2 Axiomatic Presentation and Gödel's Incompleteness Theorems

Rota's statement that there is at present no viable alternative to axiomatic presentation if the truth of mathematical statements is to be established beyond reasonable doubt, is also in conflict with Gödel's incompleteness results. (For a philosophically motivated treatment of these results, see Cellucci, 2007).

By Gödel's first incompleteness theorem, for each axiomatic theory satisfying certain minimal conditions, there are elementary sentences which are true but cannot be deduced from the axioms of the theory. Thus, contrary to Rota's statement, for each axiomatic presentation of a given area of mathematics, there are mathematical truths of that area that cannot be proved by means of the axioms of that axiomatic presentation.

Moreover, by Gödel's second incompleteness theorem, for each axiomatic theory satisfying certain minimal conditions, the truth of the axioms of the theory, and hence of the results which can be deduced from them, cannot be established by absolutely reliable means. Thus, contrary to Rota's statement, for each axiomatic presentation of a given area of mathematics, that presentation does not guarantee that the truth of the results proved by means of it can be established beyond reasonable doubt.

Rota's statement is surprising because it takes no account of the implications of Gödel's incompleteness results for the axiomatic method.

This, however, is not uncommon among mathematicians. Many of them believe that Gödel's results have "almost no relevance to the work of most mathematicians" (Davies, 2008, p. 88). These mathematicians, however, provide no evidence for their belief, which is then a purely emotional one – a matter of faith, not of reason. In particular, they do not explain how they reconcile the belief that the method of mathematics is the axiomatic method with the belief that Gödel's results have almost no relevance to their work.

This is due to the fact that, as Rota himself acknowledges, among scientists it is not uncommon to "believe in unrealistic philosophies of science" (Rota, 1997, p. 108).

Doubts may also be raised about the effectiveness of axiomatic presentation for teaching and learning. Such doubts were already raised by Descartes and Newton (see Cellucci, 2008b, pp. 23-24).

### 12.4.3 Are there Definitive Proofs?

Another difficulty concerning proof is raised by Rota's claim that there are definitive proofs.

According to Rota, "after a new theorem is discovered, other simpler proofs of it will be given until a definitive one is found" (Rota, 1997, p. 146). The "first proof of a great many theorems is needlessly complicated", and "it takes a long time, from a few decades to centuries, before the facts that are hidden in the first proof are understood" (*ibid.*). This "gradual bringing out of the significance of a new discovery takes the appearance of a succession of proofs, each one simpler than the preceding. New and simpler versions of a theorem will stop appearing when the facts are finally understood" (*ibid.*). Then a definitive proof will have been reached.

Rota is quite right in saying that a salient feature of mathematical practice, from ancient times to the present, has been the looking for new proofs of results for which proofs were already known. This shows that the concern of mathematicians is not truth. Otherwise a single proof would be enough for each result, and there would be no point in looking for new proofs.

Rota, however, assesses this feature of mathematical practice in terms of understanding, and assumes that new proofs of a given result will stop appearing when the facts are finally understood. This is in conflict with the fact that often finding a new proof does not lead to any improvement in the understanding of the old proofs, for the new proof is based on different ideas.

Looking for new proofs makes sense only if the main point of proof consists in its heuristic value. Specifically, it consists in the discovery of hypotheses which not only allow one to prove the result, but also establish connections between different areas of mathematics, where such connections may lead to new discoveries. The result is looked at from several different perspectives, each of which may suggest a different hypothesis and hence a different proof.

Since, at least potentially, new perspectives are always possible, this means that new proofs are always possible. Therefore there are no definitive proofs.

Rota's claim that there are definitive proofs is also in conflict with his statement that any mathematical item allows an open-ended sequence of presentations by new axiomatic systems, each of which is meant to reveal new features of that mathematical item. For example, "the real line has been axiomatized in at least six different ways, each one appealing to different areas of mathematics, to algebra, number theory, or topology. Mathematicians are still discovering new axiomatizations of the real line" (Rota, 1997, p. 156). And "the theory of groups may be axiomatized in innumerable ways" (Rota, 1997, p. 155).

Indeed, each new axiomatization is based on new hypotheses, which may allow to give a new proof of a given result. Therefore, again, there are no definitive proofs.

### 12.4.4 Proof and Evidence

Instead of saying that the main point of proof is its heuristic value, Rota states that "the purpose of proof is to make it possible for one mathematician to bring back to life the same evidence that had been previously reached by another mathematician" (Rota, 1997, p. 169). This ensures truth, for truth "is the recognition of the identity of my evidence with someone else's evidence" (*ibid.*).

Thus, according to Rota, the main point of proof is to provide evidence. In his view, "evidence is the primary logical concept. Evidence is the condition of possibility of truth. Truth is a derived notion" (Rota, 1997, p. 168). Therefore, "whereas mathematicians may claim to be after truth, their work belies this claim. Their concern is not truth, but evidence" (Rota, 1997, p. 171). Looking for new proofs of results for which proofs are already known is motivated by the search for evidence, because "a proof that shines with the light of evidence comes long after discovery" (*ibid.*).

Rota, however, provides no convincing account of evidence. He claims that evidence is aimed at understanding, for "all understanding is the fulfillment of some evidence" (*ibid.*).

But evidence cannot be aimed at the understanding of the result, for in most cases, including Fermat's Last Theorem, the result is clear enough.

As Kant points out, "nothing can be comprehended more than that what the mathematician demonstrates, e.g., that all lines in the circle are proportional. And yet he does not comprehend how it happens that such a simple figure has these properties" (Kant, 1992, p. 570).

On the other hand, evidence cannot be aimed at the understanding of proof either, because, as it has been already mentioned, often finding a new proof does not lead to any improvement in the understanding of the old proofs, for the new proof is based on different ideas.

For these reasons, saying that evidence is aimed at understanding does not seem convincing.

Moreover, Rota's view of evidence risks paving the way to irrationalism. For Rota claims that "the experience that most mathematicians will corroborate, that a statement 'must' be true, is not psychological. In point of fact, it is not an experience. It is the condition of possibility of our experiencing the truth of mathematics" (Rota, 1997, p. 168). Or rather, it is a sort of "primordial experience" which is "constitutive of mathematical reasoning" (*ibid.*).

Now, Rota's claim depends on the principle "I feel it is so; therefore it must be true", and this principle is one of the "faulty beliefs that are common among people with anxiety disorders" (Hyman-Pedrick, 2006, p. 49). In fact, to say: 'I feel it is so; therefore it must be true', amounts to saying: 'I hereby declare that this is true, and I am unwilling to discuss it further'.

Rota also claims that "after reaching evidence we go back to our train of thought and try to pin down the moment when evidence was arrived at. We pretend to locate the moment we arrived at evidence in our conscious life. But such a search for a temporal location is wishful thinking" (Rota, 1997, pp. 168-9).

In this way Rota makes mathematics ultimately depend on a ur-experience which is ineffable and cannot even be located in time.

## 12.5  The Relation of Philosophy of Mathematics to Mathematics

In the last century, the received view on the relation of philosophy of mathematics to mathematics has been that the philosophy of mathematics must be normative: it must provides a canon for mathematical practice.

This depends on the assumption that the philosophy of mathematics must provide a justification of mathematics. To assume this amounts to assuming that in mathematics there are justifiable and unjustifiable methods, proofs, practices, etc., and acceptable results are those which are obtainable by justifiable methods. The task of the philosophy of mathematics is then to determine which methods are legitimate and hence justifiably used, thus what an acceptable mathematics should be like. For that reason, the philosophy of mathematics must be normative.

### 12.5.1  The Descriptive Character of the Philosophy of Mathematics

Rota opposes the received view stating that in the philosophy of mathematics "a strictly descriptive attitude is imperative", therefore "all normative assumptions shall be weeded out" (Rota, 1997, pp. 135-6).

Rota is quite right in opposing the received view. The philosophy of mathematics must be descriptive since it must account for mathematical practice. This seems to be a necessary requirement for any philosophy of mathematics deserving the name.

From the fact that the philosophy of mathematics must be descriptive, however, Rota draws the conclusion that the question 'What is mathematics?' "requires no answer", it is "simply an inadequate expression of that wonder one feels contemplating the majestic edifice of mathematics" (Rota, 1999, pp. 104-5).

This conclusion seems unwarranted. An answer to the question 'What is mathematics?' is important because it may have an impact on mathematical practice, for example, on methods used in mathematics.

### 12.5.2  The Need for a Drastic Overhaul of Logic

Rota draws a further conclusion from the fact that the philosophy of mathematics must be descriptive: to account for mathematical practice requires an approach to logic different from mathematical logic.

Indeed, Rota states that "that magnificent clockwork mechanism that is mathematical logic is slowly grinding out the internal weaknesses of the system" (Rota, 1986, p. 180). The notions on which present-day logic is based "were invented one day for the purpose of dealing with a certain model of the world", and that model of the world is "inadequate to the needs of the new sciences" (*ibid.*). Today, "in all circumstances imaginable, including mathematical reflection (the true one, not the one of a posteriori reconstructions), logic shines for its absence" (Rota, 1999, p. 94).

In fact, mathematical logic has proved inadequate not only to the needs of the new sciences developed in the last century – from computer science and artificial intelligence to life sciences and social sciences – but also to the primary use for which Frege had designed it: to be "a tool in the philosophy of mathematics; just as other mathematics, for example the theory of partial differential equations, is a tool in what used to be called natural philosophy" (Kreisel, 1967, p. 201).

From this Rota concludes that, "if we are to set the new sciences on firm, autonomous, formal foundations, then a drastic overhaul" of "logic is in order. This task is far more complex than the Galilean revolution" (Rota, 1986, p. 180). It "will have to give the new sciences their theoretical autonomy" (Rota, 1999, p. 116).

Rota's plea for a drastic overhaul of logic seems justified. This task, however, need not be, as Rota claims, far more complex than the Galilean revolution. As Rota says, the latter "was inaugurated by a philosophical revolution" (Rota, 1999, p. 115). Like the Galilean revolution, then, the drastic overhaul of logic might 'only' require a philosophical revolution: a different outlook of the world.

## 12.6 Rota's Place in the Philosophy of Mathematics

Present-day philosophers of mathematics are currently distinguished into 'mainstream' and 'maverick' (see Kitcher-Aspray, 1988; Hersh, 1997; Mancosu, 2008).

Mainstream philosophers of mathematics view mathematics as a static body of knowledge, are mainly concerned with the question of justification of mathematical knowledge and set themselves within the analytic philosophy tradition.

Maverick philosophers of mathematics view mathematics as a dynamic body of knowledge, are mainly concerned with the question of the growth of mathematical knowledge, in particular with the dynamics of mathematical discovery, and are generally ill at ease with the analytic philosophy tradition.

The use of terms 'mainstream' and 'maverick', however, does not seem felicitous, for it suggests that the mainstream philosophers of mathematics are the orthodoxy whereas the maverick ones are the heresy – a heresy which could never become the orthodoxy. Using the terms 'static' and 'dynamic' seems preferable. For this reason, in what follows I will refer to the mainstream philosophers of mathematics as 'static' and to the maverick philosophers of mathematics as 'dynamic', and will speak of the 'static approach' and the 'dynamic approach'.

In addition to these two approaches, a 'third way' approach has also been proposed, characterized by attention to mathematical practice while remaining within the analytic philosophy tradition (see Mancosu, 2008).

However, on the one hand, mathematical practice is a main concern of the dynamic approach. Now, as soon as one begins to consider mathematical practice, it becames obvious that mathematics is a dynamic body of knowledge, and questions about mathematical discovery immediately arise. Thus attention to mathematical practice necessarily leads to the dynamic approach.

On the other hand, the third way approach pays no or little attention to the dynamics of discovery. The latter is the real discriminant between the static and the dynamic approach. Therefore, the third way approach appears to be simply an extension of the static approach to other topics, retaining the basic features of that approach.

Rota does not fit in the static approach because he does not consider mathematics as a static body of knowledge, and criticizes the analytic philosophy tradition, for which he foresees an "increasing irrelevance followed by eventual extinction" (Rota 1997, p. 103). By contrast, he shares the view of the dynamic approach that discovery should be a main concern of the philosophy of mathematics. For he points out that "heuristic arguments are a common occurrence in the practice of mathematics. However, heuristic arguments do not belong to formal logic. The role of heuristic arguments has not been acknowledged in the philosophy of mathematics despite the crucial role they play in mathematical discovery" (Rota, 1997, p. 134).

In this connection, it is worth noting that Rota states that "it is a frequent experience among mathematicians" that "they cannot solve a problem unless they like it" (Rota, 1991, p. 262). Indeed, "motivation and desire are essential components of mathematical reasoning" (Rota, 1997, p. 160).

This is a valuable remark because it stresses the importance of emotion in mathematical research, and particularly in mathematical discovery. Hadamard made a similar point, stating that "an affective element is an essential part in every discovery or invention", indeed "no significant discovery or invention can take place without the will of finding" (Hadamard, 1996, p. 31). Hadamard's statement, however, has been generally neglected, so it is all the more important that Rota reminds us of the relevance of emotion in mathematical practice. (For more on this issue, see Cellucci, 2008c, Ch. 20).

Admittedly, Mancosu qualifies the dynamic approach in a way that is somewhat different from the one suggested above. For he characterizes it not in terms of attention to the dynamics of mathematical discovery but rather in terms of the following three features: "a) anti-foundationalism, i.e. there is no certain foundation for mathematics; mathematics is a fallible activity; b) anti-logicism, i.e. mathematical logic cannot provide the tool for an adequate analysis of mathematics and its development; c) attention to mathematical practice" (Mancosu, 2008, p. 5).

Nevertheless, Rota also fits in this alternative characterization of the dynamic approach for he states: a) "The lack of certainty, the insecurity of evidence are described by Husserl as features of all evidence whatsoever. Husserl's description is confirmed by observing mathematicians at work" (Rota, 1997, p. 171); b) Mathematical logic does not provide the tool for an adequate analysis of mathematics and has even "given up all claims of providing a foundation to mathematics" (Rota, 1997, pp. 92-3); c) "The philosophy of mathematics is beset with insistent repetitions of a few crude examples taken from arithmetic and from elementary geometry, in total disregard of the philosophical issues that are faced by mathematicians at work" (Rota, 1997, p. 163).

True, Rota's views on mathematics are not always homogeneous. As I have pointed out in this paper, sometimes they are even inconsistent with each other, probably because they were stated at different times. Moreover, his claim that evidence is the primary logical concept is unconvincing. Nevertheless, his criticism of the static approach and his penetrating remarks on some aspects of mathematical practice are an important contribution to the dynamic approach.

## Acknowledgements

I am grateful to Miriam Franchella, Donald Gillies and Andrea Reichenberger for useful comments on an earlier draft of this paper, to Fabrizio Palombi for helping me to locate a quotation from Rota's MIT lectures, and to the audience of the Rota memorial conference for useful remarks.

## References

Aspray, W., Kitcher, P. (eds.) (1988), *History and Philosophy of Modern Mathematics*, Minneapolis, University of Minnesota Press.

Bostock, D. (2009), *Philosophies of Mathematics. An Introduction*, Madden, MA., Wiley-Blackwell.

Brouwer, L. E. J. (1975), *Collected Works*, vol. I, A. Heyting (ed.), Amsterdam, North-Holland.

Cellucci, C. (1989), *De conversione demonstrationis in definitionem*, in G. Corsi, C. Mangione, and M. Mugnai (eds.), *Atti del convegno internazionale di storia della logica. Le teorie della modalità*, Bologna, CLUEB, pp. 301-6.

Cellucci, C. (1998), *Le ragioni della logica*, Rome-Bari, Laterza.

Cellucci, C. (2002), *Filosofia e matematica*, Rome-Bari, Laterza.

Cellucci, C. (2006), *Introduction* to *Filosofia e matematica*, in R. Hersh (ed.), *18 Unconventional Essays on the Nature of Mathematics*, Berlin, Springer, pp. 17-36.

Cellucci, C. (2007), *La filosofia della matematica del Novecento*, Rome-Bari, Laterza.

Cellucci, C. (2008a), *The Nature of Mathematical Explanation*, "Studies in History and Philosophy of Science", vol. 39, pp. 202-10.

Cellucci, C. (2008b), *Why Proof? What is a Proof?*, in R. Lupacchini and G. Corsi (eds.), *Deduction, Computation, Experiment. Exploring the Effectiveness of Proof*, Berlin, Springer, pp. 1-27.

Cellucci, C. (2008c), *Perché ancora la filosofia*, Rome-Bari, Laterza.

Davies, E. B. (2008), *Interview*, in V.F. Hendricks and H. Leitgeb (eds.), *Philosophy of Mathematics: 5 Questions*, New York, Automatic Press/VIP, pp. 87-99.

Descartes, R. (2006), *Meditations, Objections, and Replies*, R. Ariew and D.A. Cress (eds.), Indianapolis, IN., Hackett.

Dummett, M. (2001), *La natura e il futuro della filosofia*, Genoa, Il Melangolo.

Field, H. (1989), *Realism, Mathematics and Modality*, Oxford, Blackwell.

Frege, G. (1964), *The Basic Laws of Arithmetic. Exposition of the System*, M. Furth (ed.), Berkeley, University of California Press.

Frege, G. (1984), *Collected Papers on Mathematics, Logic, and Philosophy*, B. McGuinness (ed.), Oxford, Blackwell,

Gödel, K. (1995), *Collected Works*, S. Feferman et al. (eds.), Oxford, Oxford University Press.

Hadamard, J. (1996), *The Mathematician's Mind: the Psychology of Invention in the Mathematical Field*, Princeton, Princeton University Press.

Hersh, R. (1997), *What is Mathematics, really?*, Oxford, Oxford University Press.

Hyman, B. M., Pedrick, C. (2006), *Anxiety Disorders*, Minneapolis, MN., Lerner.

Kac, M., Rota, G.-C., Schwartz, J.T. (1986), *Discrete Thoughts. Essays on Mathematics, Science and Philosophy*, Boston, Basel, Berlin, Birkhäuser.

Kant, I. (1992), *Lectures on Logic*, J.M. Young (ed.), Cambridge, Cambridge University Press.

Kreisel, G. (1967), *Mathematical Logic: What has it Done for the Philosophy of Mathematics?*, in R. Schoenman (ed.), *Bertrand Russell, Philosopher of the Century*, London, Allen and Unwin, pp. 201-72.

Mancosu, P. (ed.) (2008), *The Philosophy of Mathematical Practice*, Oxford, Oxford University Press.

Rota, G.-C. (1986), *Misreading the History of Mathematics*, in (Kac, Rota, Schwartz, 1986), pp. 231-4.

Rota, G.-C. (1991), *The End of Objectivity. The Legacy of Phenomenology. Lectures at MIT*, Cambridge, MA., MIT Mathematics Department.

Rota, G.-C. (1997), *Indiscrete Thoughts*, F. Palombi (ed.), Boston, Basel, Berlin, Birkhäuser.

Rota, G.-C. (1999), *Lezioni napoletane*, F. Palombi (ed.), Neaples, La Città del Sole.

# Chapter 13
# On the Courage Needed to Do Phenomenology. Rota and Analytic Philosophy
## Contributed Chapter

Albino Lanciani and Claudio Majolino

The philosophical discourse of the twentieth century is often pictured as a sort of trench warfare between two opposite factions: Analytics vs. Continentals.

Some might be tempted to see in this conflict a renewed version of the traditional *querelle des anciens et des modernes* – the role of the moderns being played by dynamic Anglo-American analytical philosophers, trained in mathematical logic and acquainted with the more recent results of scientific research, while old fashioned continental classicists (mostly German or French) would be clearly recognizable by their deep fascination for text interpretation and history of philosophy. But, in spite of bold claims, things are clearly far more complicated, and the careful observer can't help but notice that the composition of the two factions harbors a huge heterogeneity.

On the one hand the "guiding thread" relying Neo-Kantianism, Phenomenology, Critical Theory, Psychoanalysis, Structuralism, French Post-Modernism and Hermeneutics etc. is manifestly too thin to be revealing. On the other hand, it is safe to say that the criteria usually applied to identify the members of the analytic camp, as even M. Dummet had to recognize, seem to be quite loose.

Rota has always been aware of the rather artificial – one may even say "ideological" – nature of such an oversimplified picture. As a consequence he never championed the idea of "continental philosophy" as such, and in spite of some vague references to the Western philosophical tradition, he preferred to simply portray himself as a "phenomenologist". Yet Rota's most judgmental and controversial papers are openly devoted to a critique of what he insists on calling "analytic philosophy". The aim of this paper is not to provide a systematic reconstruction of Rota's views on contemporary philosophy, nor to study in details the academic and historical context of his thought. We will rather try to spell out, as much as it can be done in the limited space allowed, some of the reasons according to which Rota *claims himself to be a*

Albino Luciani
École Normale Supérieure, Lyon, France.

Claudio Majolino
Université Lille III, France.

E. Damiani et al. (eds.), *From Combinatorics to Philosophy,*
DOI 10.1007/978-0-387-88753-1_13, © Springer Science+Business Media, LLC 2009

*"phenomenologist"* and, conversely, conceives his critique of "analytic philosophy" as a consequence of his endorsement of phenomenology.


## 13.1 Twilight of an Idol

First, as to analytic philosophy. There is no doubt that Rota's idea of analytic philosophy is neither historical nor philological. There is certainly a distinction to be made between those thinkers who are currently *considered* as analytical philosophers and those who *claim* the existence, the virtues and even the intellectual superiority of such a thing as "analytic philosophy". While Rota may pay a great deal of respect to some of the former, he attempts to dismantle in a quite devastating way the practices and the presuppositions of the latter.

His papers are in fact scattered with sympathetic references to the late Wittgenstein or Austin – philosophers that one might be keen to label quite easily as "analytics" – and he even comes to state that G. Ryle's book *The Concept of Mind* (1949) "would serve as a good introduction to phenomenology" (Rota, 1986, p. 18). On the other hand, every now and then we bump into the names of Frege and Russell, we recognize a reference to Tarski, and learn that Wittgenstein's *Philosophical Investigations* (1953) are nothing but the last attempt to correct the huge "gaffe" of the *Tractatus* (1922) (Rota, 1997, p. 95). It is even useless to mention Rota's caustic apothegms on Quine, Passmore and Mackie – who allegedly stated that "all philosophy, to be any good, must be analytic" (Rota, 1997, p. 257). Yet the critique of such illustrious thinkers or colleagues is not Rota's main concern. At least in his published texts the controversy with actual analytical philosophers is somehow incidental. What seems to be at stake in this critique is a more general view of the meaning and the scope of philosophy as such. That is precisely the reason why, as we will see, Rota's idea of "analytic philosophy" is more complex than it looks.

What is analytic philosophy, then?

If we start from the beginning one of its basic features is, according to Rota, the mistrust in the inner resources and peculiarity of philosophy as such.

History is there to prove that, unlike mathematics, philosophy has only been able to provide theories that are hard to consider as conclusive or definitive. Notwithstanding, each philosopher claims boldly to have reached the truth. Such a situation is rather disturbing. Not only philosophical concepts are vaguely defined and philosophical statements are approximate and hard to prove, but, worse than that, neither the object of philosophy, nor even its tasks seem to be clearly identified. At best one could say that philosophy, from its very beginning, seems to struggle in vain with the same handful of problems: the relationship between soul and body, the question of reality, the nature of truth and things alike. And in spite of the efforts, it appears to be doomed to what Rota calls, quoting Ortega y Gasset, a "constant shipwreck" (Rota, 1997, p. 94).

This somewhat disconcerting picture looks even more discouraging once it is compared with the official records of mathematics, whose "success" are commonly accepted as taken for granted. Not only mathematical problems get solved (sooner or later) but its truths are everlasting and, once established, impossible to disprove; more than that, what is currently considered as the very method of mathematics, the axiomatic method – whose study is carried out by mathematical logic –, appear to be extremely powerful and its foundational possibilities far-reaching. Once the quite peculiar situation witnessed by the history of philosophy is recognized, two possible attitudes can be taken.

According to the first, the "constant shipwreck" of philosophy is precisely a *necessary* feature, a structural element of the philosophical inquiry itself. It cannot and should not be taken as a sign of weakness or as a proof of its lack of rigor. In a sort of reversed version of the Pascalian bet, philosophers choose to play a game they will necessarily loose. But then again, once the bet accepted, part of the game is precisely to try to find a way to understand the structural reasons of such a constant shipwreck and, eventually, its logic. To be a philosopher one has to show some courage.

The second attitude considers such a failure as *contingent*. If philosophers are still running round in circles, it is because they have failed to provide a rigorous method and a clear set of plainly defined concepts. Such an error should be corrected following the example of mathematics. Mathematical and philosophical truths share the same defining mark: they cannot be empirically verified. And the fact of their possible "applications" cannot be considered as a sign of truth. However, mathematical truths can be demonstrated, expressed within the framework of a formal theory that proves them true in a strictly formal way. No courage is needed to do philosophy then. The only thing required is being as honest as to recognize the success of mathematics and take all the consequences from this fact.

In order to picture this second attitude Rota uses the somewhat oxymoronic syntagm "mathematizing philosophers" (Rota, 1997, p. 92 *passim*). There is nothing wrong in being a philosopher *and* a mathematician or a philosopher *of* mathematics, since in both cases the specificity of mathematics and philosophy is preserved. But as soon as the idea of a "mathematics *of* philosophy" is introduced, in order to avoid the shipwreck of philosophy thanks to the compass of the axiomatic method, we lose both philosophy *and* mathematics. Philosophy, because of its "loss of autonomy". Mathematics, because of a double reduction (1) of the mathematical *practice*, whose main task is to discover and investigate theorems, to the mathematical *method* employed to prove them formally, and (2) of the mathematical method to the axiomatic method (Rota, 1997, pp. 91-2).

Now, the official biography of analytic philosophy runs roughly like this: at the end of the Nineteenth Century, thanks to the enormous success of mathematics, and especially of mathematical logic applied to the analysis of language and meaning, philosophers were about to recognize at last the immense powers of formalization and symbolization. Frege's introduction of quantifiers provided a new tool to solve old philosophical and metaphysical problems. Later on, Russell's distinction between logical proper names and definite descriptions, which ends up becoming evident in the use of quantifiers in his formalism, put some order in the jungle-like universe of the meinongian ontology. Exit "round squares" and "actual Kings of

France". Such a job will be carried out and accomplished by Quine, who will finally "clear the slums" of ontology pounding with the hammer of bounded variables.

A new conception of philosophy comes to the forth. Following Wittgenstein, logical positivists start to defend the idea that, under the guidance of the logical analysis of language, philosophy should become a sort of therapy. On the one hand, its *positive* mission is to study the basic features of a discourse adapted to the expressiveness needed in contemporary science. On the other hand, its *critical* mission is to fight against pseudo-philosophical concepts and propositions whose truth conditions are out of reach. Concerning the first point, philosophy has to strive for a language (and that was already implicit in Frege's idea of a *Begriffschrift*) which is able to "picture" in the clearest way possible a fragment or a state of the world (or, at least, the way in which we conceive it). Accordingly, philosophical concepts have to be univocally defined, organized in well formed truth functional propositions, and connected in argumentative patterns that clearly follow the limpid outlines of mathematical logic. Armed with such a symbolic, formal language, empowered with a set of univocal definitions and inferential laws, it is finally possible to reach the second point and distinguish *real and valuable* philosophical concepts and "true" problems from *pseudo-concepts or pseudo-philosophical problems* – unclear, confused, and hence useless. Carnap's critique of Heidegger is a good example of this situation.

Rota's version of the story is quite different. The introduction of mathematics, through mathematical logic, in the field of philosophy as a true philosophical method is not due to the achievement of mathematics itself, but to its collapse; it has nothing to do with the success of the axiomatic method, but rather with its failure.

Here is the secret story of the "mathematizing philosophers". At the end of the Nineteenth Century the "working myth" of mathematics as the key to reach a full rigorous comprehension of the world blossomed in the Renaissance and grown up in the Enlightenment, begins to wilt. "Working myth" is not a bad word. Mentioned quite incidentally in his "Introduction" to *Indiscrete Thoughts* (1997) this term describes what Rota calls "the bedrock of civilization" (Rota, 1997, p. xix), a set of unquestioned and silently operating assumptions that shape the way in which things are gathered together and understood to be. If the world appears in such and such a way that things "make sense", if we feel ourselves "at home in the world" – as Kant said in the quite different context of the third *Critique* –, that is because a myth is at work.

Every working myth is a *necessary* ideal formation – and it is necessarily unthematized. As Rota puts it "We could not function without the solid support that we get from our working myths. We are not aware of our working myths" (Rota, 1997, p. xix). But working myths are also necessarily *historical*. Their contingency comes to the forth when they unexpectedly come to language.

> Sooner or later every working myth begins to wilt. We can tell that a myth is wilting as soon as we are able to express it in words. It then turns into a belief, to be preserved and defended (Rota, 1997, p. xix).

Interestingly, Myth is not opposed to Reason (as in the classic rivalry between *mythos* and *logos*), but to Belief. Once a myth has lost its silent power to make things meaningful, two options are still available. One may summon the courage to discard it, and question the world that the myth was supposed to disclose, or turn it into an idol in order to prevent its disappearance.

Hence Rota's powerful suggestion: behind the fact that certain contemporary philosophers "have become great believers in mathematization" (Rota, 1997, p. 92), there is not the *raising power of contemporary mathematics*, but the *fading myth of modernity*, i.e. the myth according to which mathematics is (as claimed by Gauss, using a definition once reserved to metaphysics) the "regina scentiarum"; a discipline entitled to disclose what Husserl called in the *Krisis* "being as such" (*ontos on, vere ens*), the world as it really is.

The so called crisis of the foundations of mathematics, perceived with anxiety by many mathematicians of the end of the Nineteenth Century, was nothing but the fact of a working myth coming to language. Mathematics risked to become a science like any other science: with its own peculiar objects, picked up from a very specific "eidetic domain" of ideal objects (Rota, 1986, pp. 167-73; Rota 1997, p. 118); dealing with facts, discovered as simply being-there by observation and experimentation and whose truth is recognized independently from any formal proof (Rota, 1997, pp. 117-8); hypothetical and exploratory (Rota, 1997, p. 166); using a wide range of different techniques of proof in order to secure in a formal way the previously accepted truth; developing proof techniques that are not just heterogeneous and loosely related, but also historical and contingent (Rota, 1997, pp. 93-4). Mathematics is just *another* way to talk about the world.

Frege's, Russell's and Whitehead's, Zermelo's efforts rest on the *belief* that mathematics had to be rescued from decline through a reform of its "method"; in spite of the fact that, *de facto*, such a "decline" wouldn't have affected (and it didn't) the activity of the working mathematician or the truth of its theorems.

Rota's conclusion is striking: the raise of the *idol* of mathematical logic as the true method of philosophy is a consequence of the fall of the *myth* of mathematics as "regina scientiarum". The axiomatic method was supposed to avoid the shipwreck of mathematics (even if it never really worked, as Gödel's incompleteness theorem has unmistakably proven), *now* is going to "rescue" philosophy as well. Like those street-corner preachers stalking people to save their souls, the worshippers of the idol of mathematical logic, after having attempted in vain to rescue the soul of mathematics, turn to philosophy in search for a better chance to do their compulsory duty.

There would be many reasons to agree with Rota's reconstruction. Some are clearly historical, others more philosophical and have to do with the role played by ontological concerns in the early analytic inquiries. Flown away form traditional metaphysics ("regina scientiarum") ontological concerns are rescued by modern mathematics; they now jump on the philosophy of language bandwagon. It is especially noteworthy that the first attempt to perform such a "foundational transfer" of mathematical logic from mathematics to philosophy (in order to free the latter from its "speculative" assumptions) has to be found in the connection between theory of meaning and ontology. The old problems of metaphysics – or, at least, those problems that survive the foundational transfer – are finally exposed thanks to the logical analysis of language. We have already mentioned the task, shared by Russell and Quine, to renew the ontological discourse with the support of a "regimented language". But the best example is probably Wittgenstein's idea, defended in the *Tractatus*, according to which propositionally expressed "*Gedanke*" are nothing but "logical images of facts": the structure of the world is mirrored by the structure of language, which in turn, once formalized, will make the latter *appear*. The once metaphysical problem of the "*ontos on*" has been modified and, at the same time, legitimized.

Rota's reconstruction of the rise of the idol of "analytic philosophy" can also be summed up as the joint venture of (1) the traditional *fear of the shipwreck* (philosophers' pessimistic view of their own field) and (2) the more recent *wilting myth* of mathematics as "regina scientiae", which brought the hypostasis of the axiomatic method. Than we have: (3) the first attempts (failed) to "rescue" mathematics through the axiomatic method; (4) the foundational transfer of the axiomatic method from mathematics to philosophy; (5) the choice of the theory of meaning as the promised land where classical ontological problems may be solved thanks to the first order logic formalism.

In one single blow, Rota gives a historical-philosophical account of the origins of analytic philosophy and a non-ideological explanation of the widespread success of analytic philosophy as a ruling trend. The idol of analytic philosophy is a man-made product whose shape is venerated in academic circles as if it had a divine nature precisely because, on the one hand, *it diverts the attention from the aporetic and uncomfortable situation of philosophy*, which still needs to be understood in its full scope; on the other, *it allows the wilting myth of modernity to survive its own death*.

## 13.2 Three Senses of Phenomenology

It should be clear by now that if Rota decides to take issue with the idol of "analytic philosophy", it is not because of a previous profession of faith in the "continental" true god. Rota is rather an iconoclast – precisely because he is a phenomenologist. Once again: there's nothing wrong with myths. Working myths are there to be debunked and dropped by, sooner or later. Idols are nothing but myths that try to survive their own death. The working myth still operational behind the rise of an idol is most likely the myth of immortality.

Let us turn now to phenomenology. If "analytic philosophy" is the name given by Rota to one of the most powerful – and hard to die – wilting myths of modernity, finally changed into an idol, it is obvious that "phenomenology" cannot be simply the name of a philosophical movement or a method. Yet, it is well known that Rota has never really defined what phenomenology is all about. And we won't certainly take the risk to answer on his behalf. However, it is our belief that one might isolate at least three "senses" of "phenomenology" in Rota's work. The list, of course, is open and other alternative configurations would be as legitimate.

1) On a quite general level, the word "phenomenology" names *a certain relationship to time*. To be a phenomenologist means to be able to acknowledge the historicity of any object- formation, be it explicit and thematic (like concepts, theories, methods, patterns but also values, esthetic criteria and institutions) or implicit and unthematized (like working myths and naïve assumptions).

If we take the word "object" in a rather broad sense, as Rota does, this position may be summed up in the following formula: "Every object carries within itself the seed of its own irrelevance" (Rota, 1986, p. 170). Accordingly, the expression *genetic phenomenology* names the attempt to disclose what Rota calls the "original drama" that leads to the constitution of every single object or concept-formation, and that, at the same time, points to its further destitution. As Rota puts, it, referring to Levinas:

> anything that has been made into an object *eo ipso* begins to conceal the original drama that led to its constitution. Most likely, this drama followed a tortuous historical path, through things remembered and things forgotten, through cataclysm and reconstructions, pitfalls and lofty intuitions, before terminating with its objective offspring, which will thereupon believe itself to be alien to its origin (Rota, 1986, p. 168).

Such a way to conceive historicity revolves around the twin notions of "genesis" and "crisis". The "loss of the origin" is the *terminus a quo* of every object-formation while the "loss of evidence" represents its *terminus ad quem*, i.e. the way in which its historical and genetic constitution comes into sight. It would be easy to show that this first sense of "phenomenology" is not unrelated with Rota's broad reflection on Husserl's *Krisis* and its notion of *"sedimentation"*. The most striking example of phenomenology as an attempt to disclose the inner logic of change concealed in object-formations is to be found in Rota's genetic account of the ideal object-formations of science, sketched in the last part of his *Husserl and the reform of logic* (Rota, 1986, pp. 172-3). A study, even cursory, of this texts, and especially of the idea of "enlargement of an eidetic domain" would be the task for another paper.

In section I. we have extensively shown to what extent, according to this first sense, "phenomenology" can be opposed to "analytic philosophy" and its hypostasis of the axiomatic method.

2) On an even broader level, "phenomenology" names an *act of resistance*: resistance against a philosophy proudly resigned to define itself "from outside", as it were, ready to dismiss its own philosophical peculiarity. In this second sense phenomenology is nothing but a plea for philosophy itself.

In spite of their "continuous shipwreck", philosophers keep on trying to reformulate and redefine the same old questions, uncovering the genesis of their object-formations (as we will see, philosophical concepts too are ideal object-formations), bringing hidden presupposition out into the open, and taking the risk of describing "phenomena that lie at the other end of the spectrum of rationality that science will not and cannot deal with" (Rota, 1997, p. 90). The courage of philosophy blends into the courage of description, a description of the world in all its complexity.

> The reality in which we live is constituted by a myriad contradictions, which traditional philosophy has taken pains to describe with courageous realism. (…) The real world is filled

with absences, absurdities, abnormalities, aberrances, abominations, abuses, with *Abgrund* (Rota, 1997, p. 102).

Yet this peculiar complexity of the world can be either reduced or described. The second option is not an easy one, and the simple *will* to describe does not necessarily mean that we have the right *tools* to accomplish this task. Philosophy requires fine-grained descriptive devices, able to capture the *abysmal* (*abgründig*) sides of the world. In other words, philosophy needs to supplement the rational tools of language, logic and argumentation – which are necessary but not sufficient philosophical tools – with brand new descriptive strategies.

Rota gladly quotes Machado's sentence "truth too is a matter of invention" (Rota, 1986, p. ix), and there is a precise reason for that. To be truly descriptive, a philosophical theory needs to display new concept-formations, created *ad hoc* in order to describe all the complexities of a *"abgründige Welt"* the best way possible. Such concept-formations that every philosopher ends up to conceive (let us thing about Husserl's eidetic variation, Hegel's dialectics etc.), knocked together in a rather pragmatic way, are not to be conflated with irrational or arbitrary procedures. They are precisely empirical attempts to define *alternative patterns of descriptive rationality*. Compared to standard logic they may be labeled as illogical, but they are certainly not irrational.

It is indeed remarkable that Rota clearly talks of a *"spectrum of rationality"* (Rota, 1997, p. 90). The idea is quite intuitive: rationality may be pictured as an ideal field locating at one end, say, the "less *abgründig* phenomena" (those that can be handled within the established forms of rationality proper to science) and at the other end those "fringe" phenomena – usually "kept in the background" and repressed (Rota, 1997, pp. 98-9) – that call for new and unconventional rational patterns. The "otherwise rational patterns" required to describe such phenomena could be considered irrational or merely subjective only if measured with the scale of a normatively conceived form of standard rationality (Rota, 1997, p. 102). Yet such a normative viewpoint, conflating the "otherwise rational" (required by the very logic of the phenomena) with the "irrational" is twice as misleading: it prevents any comprehension of the tortuous paths of rationality and nips in the bud all understanding of the truly irrational forces acting behind the curtains of the philosophical discourse. As Rota puts it, "Behind every question of philosophy there lurks a gnarl of emotional cravings" (Rota, 1997, p. 91), no need to deny it. And because philosophy aims to describe phenomena "at the limits of rationality", there is no way to avoid the risk that philosophers end up to describe what they want to be seen to be *personally* satisfied by a description of the world.

In recognizing the role played in philosophy by unarticulated passions, needs and desires, Rota completes his description of what we may call the "threefold layer of philosophy". Philosophy is (1) partly *rational*, because it necessarily rests, in a quite loose way, upon the basic principles of logic (mostly on classical logic); (2) partly *otherwise rational*, because it rearranges, assembles and makes up new descriptive tools in order to think the structurally blurred form of the "fringe phenomena" (full-fledged worldly phenomena, as worldly as those "ordinary phenomena" commonly accepted and broadly studied by the science); (3) partly *irrational*, because it is practiced by philosophers, true battlefields of conflicting forces, emotional beings with interests, fears, desires running trough their words and actions.

Now, put in a nutshell: point (2) constitutes precisely the *phenomenological core of every philosophy*. Phenomenology can be opposed to analytic philosophy, precisely because of its radically descriptive task and unrestricted philosophical ambition (and that explains why Ryle and Austin are often considered by Rota as "phenomenologists"). Conversely, Rota's "analytic philosopher", as noted above, chooses not to take the risk of complexity, which for Rota is nothing but the risk of philosophy itself ; lacking of any "courageous realism", holding back the "*abgründig*" spots of the world and looking for a methodological supplement, he finally meets the wilting myth of mathematics. Joining the normative hypostasis of (1) – with the only noteworthy difference that classical logic has been replaced by predicate logic – and the conflation of (2) and (3) – both equally relegated to the predictable realm of psychology – Rota's analytic philosopher can finally deal with philosophical *questions* without having to struggle with philosophical *problems*.

3) One may have the feeling that so far, the limits between phenomenology and philosophy are quite blurred. Be it as "*a certain relationship to time*" or as "*a certain act of resistance*", the concept of phenomenology may give the impression to represent, say, a non-independent part (and even the core) of the whole of philosophy itself. A world made of timeless and quite simple object formations, revealed by a finite set of well defined and logically articulated concepts, is the furthest thing from the world described by philosophy (and experienced by human beings). But is that all?

That brings us to the third, narrower, sense. "Phenomenology" names now what Rota calls *a certain "readjustment of perception"*, a rearrangement "that would enable us to gaze at, say, moral or esthetic values, inaccessible cardinals or quarks with the same objective detachment we adopt in gazing at the starry sky above us" (Rota, 1986, p. 170).

This third sense of "phenomenology" is deeply related with a second wilting myth: *the myth of the physical object*. As Rota puts it:

> one of Husserl's basic and most firmly asserted themes is that physical objects (such as chairs, tables, starts and so forth) have the same 'degree' of reality as ideal objects (such as prices, poems, values, emotions, Riemann surfaces, subatomic particles, and so forth). Nevertheless, the naïve prejudice that physical objects are somehow more 'real' than ideal objects, remains one of the most deeply rooted in Western culture (Rota, 1986, p. 169).

The working myth of a world made exclusively of physical objects and properties thereof, is indeed one of the most powerful myths of the Western tradition. In a way, it is even more deep-rooted and tenacious than the modern myth of mathematization. As a matter of fact, the two are quite independent. One can easily admit that the "real" world is made of individual physical objects without claiming *eo ipso* that mathematics is a firmly established and sturdy formal discipline free of contradictions and apt to picture this world as it really is. On the other hand, the hypostasis of first order logic is not only rooted in the everlasting "fear of the shipwreck", but, more deeply, in this preliminary and unquestioned restriction. As Rota claims:

> A consequence of this belief – which until recently was not even perceived as such – is
> that our logic is patterned exclusively upon the structure of physical objects (Rota, 1986, p.
> 169).

This third sense gives a more complete picture of Rota's analytic philosopher. According to the first myth, analytic philosophers are "mathematizing philosophers"; we have already seen that this hypostasis of the axiomatic method rests on a conflation of mathematical *practice* and mathematical *method*, reinforced by an unhistorical comprehension of this method itself (first sense of "phenomenology). Mathematizing philosophers,

> are convinced that the axiomatic method is a basic instrument of discovery. They mistak-
> enly believe that mathematicians use the axiomatic method in solving problems and proving
> theorems. To the misunderstanding of the role of the method they add the absurd pretense
> that this presumed method should be adopted in philosophy. Systematically confused food
> with medicine, they pretend to replace the food of philosophical thought with the medicine
> of axiomatics. This mistake betrays the philosophers' pessimistic view of their own field.
> Unable or afraid as they are of singling out, describing and analyzing the philosophical rea-
> soning, they seek help from the proven technique of another field, a field that is the object of
> their envy and veneration. Secretly disbelieving in the power of autonomous philosophical
> reasoning to discovery truth, they surrender to a slavish and superficial imitation of the truth
> of mathematics (Rota, 1997, p. 96).

Eventually, analytical philosophers are "normative metaphysicians": reducing the scope of the philosophical inquiry, they tell us how the world should be, rather than how it is (second sense of "phenomenology").

> Philosophy should be precise, it should follow the rules of mathematical logic; it should
> define its terms carefully; it should ignore the lessons of the past; it should be successful at
> solving problems; it should produce definitive solutions (Rota, 1997, p. 102).

But the third sense of "phenomenology" sketched above introduces a new feature: analytic philosophers are "ontological reductionists". Physical objects represent, at the same time, the core elements of reality and the paradigms of objectivity in general. Everything else that is neither a physical object nor is reduced to a physical object is not in the world. It is just in our minds. Accordingly, first order logic and extensional semantics, operating within an ontological framework, become the powerful tools to disclose the inner structure of such a world. Psychology will do the rest.

Phenomenology rests on a quite different view. First of all "nothing whatsoever is 'purely psychological'" (Rota, 1986, p. 172). They way in which things appears tell us something about their constitutive features. In other words, phenomenology "depsychologizes" the crucial notion of appearing, and gets rid of the quite old adagio "being is objective, appearing is subjective". Consequently, if the mode of appearing is not extrinsic to the object as such, phenomenological descriptions are about things as they appear. But how do things appear? Of course, guided by the myth of the physical object, we might be tempted to answer: real objects appear as existing (having a certain mode of being, as bundles of actual properties belonging to actually existing beings), psychological items as lacking of any existence (being merely "in our heads"). However, at close sight the situation is far more complex.

Things as ordinary as tools, pencils, newspapers or decks of cards, do not primarily appear as existing things but as *identities*. More precisely, they appear as *functional identities* through *factual manifolds* (Rota, 1997, p. 185). Rota's favorite example is today's issue of the *New York Times*, which is experienced as *the same*, despite the fact that it may be read on different printed copies or even on a computer screen (Rota, 1997, pp. 184-5. But the very same feature, identity in a manifold, is present in many other quite common experiences of objects, such as the perception of my friend Pierre *as the same* Pierre I met yesterday (Rota, 1986, p. 176), or the recollection of a stone I saw yesterday as the *same* stone I am seeing right now (Rota, 1997, p. 184). Once appearing depsychologizied and embedded in the very core of the object, an object is nothing but the identity of a manifold, say, a *recurring sameness*.

It would be useless to reproduce all the examples used by Rota in order to make his point. Let us just insist on stressing the core idea of phenomenology according to which "*Identity precedes existence*" (Rota, 1997, p. 186).

> Speaking esoterically, all physical, ideal, or psychological presentifications of any items are secondary to the one primordial phenomenon which is called *identity*. Exoterically, the world is made of objects, ideas or whatever; esoterically, it is made of items sharing one property: their permanence throughout each presentification (...). The problem of existence is exoterically motivated by or cravings for a physical basis of the world. Esoterically, this problem is subordinate to the understanding of that fundamental happening that is the miracle of identity. (Rota, 1997, p. 185).

That brings us to the last point: if "things as given" are not *beings* but *functional unvarying identities amid varieties of possible or actual presentifications*, it goes without saying that the world of a phenomenologist is neither the set of existing objects with their individual properties, nor the totality of facts shown in the formal structure of a propositional language. It is rather a complex layer of *Fundierungen*.

> *Fundierung* is a *primitive relation*, one that can in no way be reduced to simpler (let alone to any "material") relations. It is the primitive logical notion that has to be admitted and understood before any experimental work on perception is undertaken. Confusing function with facticity in a *Fundierung* relation is a case of *reduction*. Reduction is the most common and devastating error of reasoning in our time. (Rota, 1997, p. 176).

*Fundierung* is a primitive relation, irreducible to any other relation and to any of its *relata*. It is made up of a couple: function and facticity. Facts are function-bearers and functions are fact-dependant. Together they constitute the stuff that the world is made of. But *Fundierung* is also a "logical concept" (Rota, 1997, p. 172). In an "ontological" world, made of objects and properties, the fundamental relation is that of "inherence", mirrored on a propositional level by the functional link of predication. But in a "phenomenological" world, made of identical functions and multiples facts, the key relation is precisely that of *Fundierung*. Facts alone do not constitute reality. They just represent the *factual layer* of reality. "Facticity is the essential support, but it cannot upstage the function it *founds*" (Rota, 1997, p. 176). A logic of the world should therefore formalize the different *Fundierung* relations occurring between functions and facts and the contextual nets of functions, what Rota also calls the "context-dependence of *Fundierung*" (Rota, 1997, p. 180).

According to the third sense, phenomenology is a "reajustment of perception" in a rather radical sense. In order to picture a world which is more resemblant to the world in which we live, phenomenology diverts the focus of description from physical objects to phenomena, from things to recurring items, from existence to identity. Beyond the myth of the physical object, and thanks to the concept of *Fundierung*, phenomenology gives up *ontology* for a *genealogy of identity*, which is nothing but a logic of the "real" world.

## 13.3 Coda

Rota's "analytic philosophy" is a descriptive and a genetic concept. It describes the raise and fall of an idol, built upon the fray pillars of a *normative metaphysics* and *ontological reductionism*, praised by the prophets of a "new philosophy" shaped at image and likeness of a fetichized – and hence misunderstood in its own nature – mathematical science. And the funny thing, that Rota never gets tired of remembering, is that those prophets are neither scientist nor mathematician, but simply not-so-courageous philosophers. What is phenomenology then? In Rota's view it is just a way to face the oddities of a quite complex world with a bunch of courage, a robust sense of time, a different sight and a brand new logic.

## References

Rota, G.-C. (1986), *Misreading the History of Mathematics*, in (Kac, Rota, Schwartz, 1986), pp. 231-4.

Kac, M., Rota, G.-C., Schwartz, J.T. (1986), *Discrete Thoughts. Essays on Mathematics, Science and Philosophy*, Boston, Basel, Berlin, Birkhäuser.

Rota, G.-C. (1997), *Indiscrete Thoughts*, F. Palombi (ed.), Boston, Basel, Berlin, Birkhäuser.

Ryle, G. (1949), *The Concept of Mind*, London-New York, Hutchinsons University Library.

Wittgenstein, L. (1922), *Tractatus logico-philosophicus*, London, Kegan-Trench-Trubner.

Wittgenstein, L. (1953), *Philosophische Untersuchungen*, Oxford, Basil Blackwell.

# Chapter 14
# Rota's Philosophical Insights
## Invited Chapter

Massimo Mugnai

## 14.1 Introduction

There is a story, told by Fabrizio Palombi, about Gian-Carlo Rota and the philosopher Willard van Orman Quine, according to which, during a meeting at the American Academy in Boston attended by both, Quine invited the audience to take Rota's mathematical inquiries very seriously, but to ignore his philosophical investigations (Palombi, 2003, p. 40). After reading Rota's philosophical papers, however, and thus disregarding Quine's advice, I have the impression that Quine's judgement was unfair. First of all, Rota belonged to the philosophical tradition of *phenomenology* and this, as a long honored branch of contemporary philosophy, is far from deserving the fate of being ignored. Secondly, Rota's philosophical work, even though naive for some aspects, contains very interesting considerations and, compared to the more recent developments in philosophy and logic, some 'prophetic' insights. Clearly, Rota was not a professionally trained philosopher but, as one may learn from the history of science and from the history of culture in general, self-taught persons sometimes look at a given discipline from an unusual point of view, thus contributing to its development with important discoveries or, at least, recognizing problems and questions which otherwise would escape the attention of a well-trained specialist of the field.

## 14.2 What Kind of Reductionism?

Rota identifies *neopositivism* and *analytical philosophy* and attributes to both a strong reductionistic attitude. 'Reductionism' and 'objectivism' are two words that

Massimo Mugnai
Scuola Normale Superiore di Pisa, Italy

E. Damiani et al. (eds.), *From Combinatorics to Philosophy,*
DOI 10.1007/978-0-387-88753-1_14, © Springer Science+Business Media, LLC 2009

recur very often in Rota's philosophical writings, but the conceptual content that they express is not clearly defined. In many cases Rota seems to think of reductionism in terms of 19th (or even 18th) Century materialism, i.e. as the doctrine which identifies any mental activity with physiological and physical processes. He disparagingly characterizes this kind of reductionism as "Billiard ball materialism where everything is reduced to billiard balls banging against each other according to the laws of classical mechanics" (Rota, 1991, p. 138). But he doesn't attempt to explain what is reduced to what. That Rota's ideas about reductionism are not completely clear emerges from many passages of his *magnum opus*, *The End of Objectivity*; amongst them, the following seems to me particularly striking: "For example, suppose that we try to employ reductionism by trying to reduce reading to some physical or mechanical act. We might do so by saying that reading is the physical act of seeing certain written symbols and then interpreting these symbols to have a particular meaning" (Rota, 1991, p. 38). Here an example of reductionism is given by appealing to "some physical or mechanical act", substantial parts of which are the interpretation of symbols and the determination of 'meaning'. A reductionist of the sort suggested by Rota, however, cannot speak without any further qualification of the activity of interpreting some written symbols and grasping their meaning: for a reductionist, meaning is the main phenomenon to be explained or 'reduced', it cannot be given for granted.

Another example of ambiguity with the use of the term 'reductionism' emerges from a paper devoted to Husserl (Rota, 1986, pp. 175-81). Introducing Husserl's philosophy to the general reader (to 'consumers of philosophy'), Rota discusses a case of recognition of someone in the street: "I meet someone in the street. As he walks towards me, I recognize him:"It is Pierre". The Pierre I am meeting now is *the same* Pierre I met yesterday. How do you know he is *the same*? This question is usually given a mechanistic answer, by translating the event into physiological terms. My eyes register Pierre's image and transmit it to the brain, where, in turn, by another physiological process, it is identified with yesterday's image" (Rota, 1986, p. 176). Rota calls this explanation 'mechanistic' and states that "the mechanistic explanation of any phenomenon of perception is faced with an impasse that cannot be evaded without intellectual dishonesty" (Rota, 1986, pp. 176-7); thus he concludes: "Husserl's debut in philosophy is a head-on attack on this problem. Several hundred pages of *Logical Investigations* (1900-1901) and about half the text of *Krisis* are dedicated to it and its variants, under the condemning epithet "psychologism" (a more common word nowadays is "reductionism")".

It is absolutely true that in his *Logical Investigations* Husserl launches a tremendous attack against psychologism, but psychologism cannot be identified with a "mechanistic explanation of any phenomenon". In particular, the kind of psychologism which Husserl had in mind was a doctrine grounded on introspection, i.e. a doctrine which considered the first person's point of view as the privileged access to mental phenomena. From this point of view, a typical advocate of psychologism was John Stuart Mill, according to whom the validity of logical laws was based on induction performed on the data of 'internal experience'. As is well known, Husserl vehemently reacted to this thesis, arguing that experience can neither explain nor justify the alleged universality and necessity of logical laws. Probably, Rota assimilates psychologism and reductionism because he interprets psychologism as an attempt to 'reduce' logical laws to mere psychological processes, but the great majority of the advocate of psychologism made recourse to psychological processes to *justify* the validity of logical laws, whereas a reductionist is supposed to aim to some sort of identity between what is reduced and what reduces. A typical 19th-century psychologist believed that the laws of logic were empirical generalizations of psychological phenomena, in the same way as a physicist believed that physical laws were empirical generalizations obtained observing physical phenomena. Rota, however, insists on assimilating psychologism and reductionism, thus employing the term 'psychologism' in a rather idiosyncratic way, attributing to it the aim of explaining every mental phenomenon in terms of processes of the brain: "Earlier this century, one of the common forms of reductionism was 'psychologism' - that is, saying everything was explained by the workings of the brain" (Rota, 1991, p. 119).

In a more general sense, Rota employs the term 'reductionism' to qualify any attempt to identify a given 'function' with its 'material basis' or, to use the jargon of phenomenology, its *facticity*. This is the case, for instance, when one plainly identifies a chair with a particular set of atoms, without carefully distinguishing the 'abstract' social role that a given object plays in a social context (and which, to some extent, is independent of the materials that compose it), from the concrete, physical structure of this very object. Rota's favorite example to illustrate this point is that of the Bridge Game: "when we describe the game of bridge, we use entirely sense words like 'trump' and 'cross-ruffing'. We don't think of cards as being made of cardboard unless it is absolutely necessary if, for instance, we make playing cards and sell them" (Rota, 1991, p. 51). This example, according to Rota, is aimed at showing that we don't live exclusively in a world of physical objects (the cards made of cardboard), but that we live in a plurality of worlds in which 'things' assume different roles or functions: thus, for instance, in the world of the Bridge Game, there are 'trumps' and not merely pieces of cardboard (even though, without pieces of cardboard or of some other material, it is impossible to play Bridge). Rota, borrowing both expressions from phenomenology, calls *facticity* the pieces of cardboard and *function* the role played by the cards in the game. A *function* cannot exist without a kind of 'embodiment' in something which offers to it the concrete possibility of being performed or of being recognized as existing: this 'something' is *facticity*. Thus, a sequence of letters written on paper (Rota's example) constitutes the material condition, i.e. the facticity for the expression of a meaning (the function); another example may be taken from economics:"Prices are [. . . ] functions. The price of an item at the drugstore is given in a dollar amount. The dollar amount is the facticity of the price" (Rota, 1997, p. 177).

## 14.3 Rota's Objectivism and Husserl's Naturalism

*Reductionism* and *objectivity* are strongly connected in Rota's philosophical perspective: the Western scientific tradition, considering the physical world as if it were the only existing world and attempting to reduce any kind of functions to physical things and processes, has created the illusion of a unique world of objects. On the basis of this illusion, physics has become the leading science and a physicalistic point of view now pervades all philosophical accounts of reality. What Rota calls *objectivism* is analogous to what Husserl calls *naturalism*, i.e. the doctrine according to which nature is considered "as a domain of spatio-temporal being governed by exact natural laws, a notion which was born during the scientific revolution in the seventeenth century. According to the naturalist, everything is nature in this sense, primarily physical nature. The mental is regarded at best as a dependent and secondary epiphenomenon. In the eyes of the naturalist, philosophy should follow the natural sciences, broadly conceived. It should become empirical psychology, for instance, or cognitive science, as we would say now" (Philipse, 1996, pp. 242-3).

Rota's *objectivism*, however, is more general than Husserl's naturalism. Rota defines *objectivity* as "the understanding of our world as made up of 'things'" (Rota, 1991, p. 124), but this does not directly imply that we have to think of things exclusively in terms of *physical* things. As we have seen, Rota distinguishes *facticity* from *functionality*, and in a passage of *The End of Objectivity* he emphasizes that facticity is not synonymous with physical world: "In translating a text, the sense becomes what in ordinary speech was taken for granted as facticity. One has to focus on features that were formerly facticity. So, we should guard against confusing facticity with the physical object. Facticity is not always the physical object. For example, in the project of translation, the facticity is the facticity of words, grammar, syntax, and of meaning considered as such - it is not the facticity of anything physical" (Rota, 1991, p. 111). Thus, 'objectivity', in general, seems to be produced when someone tries "to explain a phenomenon by reducing it to its factical components" (Rota, 1991, p. 112). In other words, 'objectivism' is the attitude to render a particular aspect absolute and dominant over the others: it is a kind of narrow-mindedness attempting to reduce to only one the multiple layers which constitute what we call 'reality'. According to Rota, this narrow-mindedness limits in an essential way even of the most basic facts of our cognitive activity as, for example, the understanding of a simple declarative sentence: "So objectivism is the error we persist in believing that we can understand what a declarative sentence means without a possible thematization of this declarative sentence in one of endless variety of possible contexts" (Rota, 1991, p. 155). Rota here implicitly refers to what, amongst phenomenologists, is known as *eidetic variation*, i.e. the change of perspective, imposed by experience or performed voluntarily, from which to look at things, facts or sentences of the world. A typical example, proposed by Heidegger in *Sein und Zeit* (1927) and repeated many times by Rota, is that of the hammer. A hammer is a tool that we usually employ to hammer nails, but if it gets broken, then suddenly it looses its function as a tool and we are forced to change our way of looking at it. The 'eidetic variation' shows that to view a hammer and "see it as a hammer requires the contextual knowledge of the possible uses of a hammer as a tool, without which the hammer will be incomprehensible" (Rota, 1991, p. 100). The hammer is a hammer only if there is a "recognition of its possible functionalities" to hammer and if a particular purpose accompanies our looking at the hammer. Without this and "some recognition of the everyday, past-accepted, unthematized notion" of a hammer, it is impossible to view a given object as a hammer (Rota, 1991, p. 161).

From all this Rota argues that in our experience of the 'surrounding' world we are not confronted directly with 'things', but with 'functions': the world we are living in, is a world made of functions. Thus, a proper way to describe our perceiving activity is not to say that we perceive (see, hear, etc.) 'that', but that we perceive (see, hear, etc.) 'as': we do not perceive directly a 'chair', but something 'as a chair'. In Rota's own words: "the table is a function viewed" (Rota, 1991, p. 123); "you see an object, and it's really not the object at all, you see what that object is for" (Rota, 1991, p. 129). Any act of knowing is an act of interpreting; there are no 'mere' facts which are not contextual facts, where 'contextual' means 'given in a context of functions'. At the semantical level, this means, according to Rota, that meaning cannot exist independently of possible contextualizations.

## 14.4 Beyond Classical Logic

A strong link connects Rota's criticism of what he calls *objectivism* to the need for a reform of logic. 'Logic' for Rota means *classical logic*, and as objectivism is the result of 'reducing' the surrounding world to a unique aspect (that of physical things), classical logic derives its fundamental concepts from "the structure of physical objects": "[...] our logic is patterned exclusively upon the structure of physical objects. Mathematical logic has done us a service by bringing this pattern to the fore in unmistakable clarity: the basic noema is the set, and all relations between sets are defined in terms of two of them: $a \subseteq b$ ($a$ is contained in $b$) and $a \in b$ ($a$ is a member of $b$). The resulting structure was deemed sufficient for the needs of the science up to a few years ago" (Rota, 1986, p. 169). For someone who presents himself as a champion of anti-reductionism, conceiving the entire logic as exclusively based on the notion of set and on the two relations of membership and containment sounds quite paradoxical. Surely if one considers only this statement it would be very difficult to retrace the fact that Rota was trained in logic by Alonzo Church! At any rate, Rota does not explain in what sense the basic set-theoretical notions derive from "the structure of physical objects". This, however, seems to be a firm belief of Rota, as clearly shown also by the following passage: "The foundations of our logic were set by Aristotle and have remained unchanged since. That magnificent clockwork mechanism that is mathematical logic is slowly grinding out the internal weakness of the system, but it will never revise its own foundations without an impulse from outside. Present-day logic is based upon the notion of *set* $(A, B, C, \ldots)$ and upon two relations between sets: "*A* is an element of *B*" and "*A* is a subset of *B*" (Rota, 1986, p. 180).

The statement that "the foundations of our logic were set by Aristotle and have remained unchanged since" echoes Kant's well-known (and unfortunate) opinion, expressed in the preface to the *Critique of Pure Reason*, according to which logic had improved poorly since Aristotle and had attained its perfect form in Kant's times. Rota seems to ignore deliberately the authentic revolution represented by Frege's work in the development of logic; in fact, he even explicitly criticizes Frege's semantical notions, which he judges as "a quagmire of contradictions" (Rota, 1991, p. 53, footnote 7).

Of (classical) logic, Rota criticizes the assumption that every sentence is true or false (the principle of bivalence), which he considers as a consequence of an 'absolutist' attitude to thinking: "The myth of canonized logic refers to the tendency of many to evaluate the truth value of a statement very grossly - that is, sentences are unquestionably true or unquestionably false. In actuality, it must be admitted that virtually all sentences are not the unalloyed essence of trueness or falseness of the absolutist thinker, but consist often of inseparable strands of conditionality, uncertainty, or contextual dependency" (Rota, 1991, p. 19). Classical logic is the product of an attitude to ignore eidetic variation. Because eidetic variation is strongly associated with the interplay between *function* and *facticity*, and because *function* and *facticity* are two basic notions on which the concept of *Fundierung* rests, it is quite obvious that Rota should consider this latter as a kind of cornerstone of the new logic. Indeed, Rota explicitly expresses the hope "that the concept of *Fundierung* will one day enrich logic, as implication and negation have done in their time. That is, *Fundierung* is a connective which can serve as a basis for valid inferences and for the statement of necessary truths. However, *Fundierung* is not just one more trick to be added to the baggage of logic. Quite the contrary. It is likely that the adoption of the concept of *Fundierung* will alter the structure of logic more radically than Husserl might have wished".(Rota, 1997, p. 172)

According to Rota, *Fundierung* is a primitive relation which usually expresses the link subsisting between a given function and the 'support' or the conditions on which the function itself is grounded (i.e. facticity). Interpreting the Fregean notion of sense as a function, Rota argues that *sense* and *facticity* would be much more pregnant and rich notions than *sense* and *denotation* to give rise to a coherent semantics: they would "be formalized much better than sense and denotation" (Rota, 1991, p. 53, n. 7). This is clearly considered by Rota as a first step towards the creation of a new logic, which is assumed to be a priority task for philosophers and mathematicians.

Rota does not deny the important achievements of contemporary mathematical logic, but he considers the present form of this discipline inadequate to give an account of situations that imply development or processes of any sort. Again, to build a new form of logic, one has to look at the doctrines developed by Husserl and the phenomenological movement (Merleau-Ponty included) (Rota, 1986, p. 171). Rota, however, fully recognizes that the main results achieved by Husserl and his followers attempting to construe a kind of 'genetic' logic "admirable as they are, came before the standard of rigor later set by mathematical logic, and are therefore insufficient to meet the foundational needs of present-day science" (Rota, 1986, p. 171). As Rota vividly writes: "In contemporary logic, to be is to be formal"(Rota, 1986, p.171): hence, even if one aims at creating a new logic suitable for explaining genetic processes, one has to maintain the formal structure and the 'spirit' of classical logic.

As we have seen, however, Rota's critical attitude towards what he calls 'objectivism' involves even the two basic set-theoreticalnotions of *inclusion* and *membership* on which, according to him, contemporary logic is built. An obvious consequence of this is that, to reach a status of formal presentation comparable to that of present-day mathematical logic, the new logic needs tools quite different from those employed in set theory. A considerable advancement towards the construction of the new 'genetic logic' would be to formalize, in Rota's own words, "the ontologically primary relations of being which had to be veiled as soon as the two relations $\subseteq$ and $\in$ were constituted". (Rota, 1986, p. 171) In *Husserl and the Reform of Logic*, Rota incidentally mentions some of these relations: "*a* lacks *b*, *a* is absent from *b* (one could describe in precise terms how this differs from the classical "$a \notin b$"), *a* reveals *b* (as in "the possibility of error haunts the truth") *a* is implicitly present in *b*, "the horizon of *a* (Rota, 1986, p. 171)". Another list of 'ontological' relations which only in part overlaps the previous one is presented in a text devoted to explaining Husserl's philosophical perspective to the general reader: "Husserl begins his attack by digging out of ordinary experience a number of crucial relations-of-being which mathematical logic ignores. He then attempts or at least proposes their formalization. Examples: "*A* is absent from *B*", "*A* is *already* contained in *B*", "*A* anticipates *B*", "*A* is a perspective (*Abschattung* of *B*)" (Rota, 1986, p. 180). These, however, are scattered and rather vague suggestions that cannot offer, themselves alone, a solid ground to constitute a new kind of logic. Rota was well aware of this and considered as a very difficult task that of building a system of logic that aims at being more powerful than the one corresponding to classical logic: "If we are to set the new sciences on firm, autonomous, formal foundations, then a drastic overhaul of Aristotelian logic is in order. This task is far more complex than the Galilean revolution. Pre-Galilean physics was by large a failure. But mathematical logic is a wildly successful enterprise in its avowed aims, and it is ingrained as second nature in our minds" (Rota, 1986, p. 180).

## 14.5 Concluding Remarks

Rota gives an oversimplified account of phenomenology, centering it on the inter-play of few notions as *Fundierung, function, eidetic variation* and *facticity*: the com-plexity of Husserl's philosophy is evoked all the time in *The End of Objectivity*, but not a serious attempt is made to develop an authentic phenomenological inquiry. Even Rota's celebrated paper on the concepts, respectively, of truth, beauty and proof in mathematics (Rota, 1997, pp. 108-150), as it has been remarked (Leng, 2002), contains *descriptions* of how the working mathematician usually operates rather than *phenomenological analyses* in the proper sense. Moreover, Rota seems to be more attracted by the interpretation that Heidegger gives of phenomenology rather than by the genuine Husserlian doctrines. This may contribute to explain Rota's hostility towards formal logic.

Even though Rota's views about philosophy and logic have not been very influ-ential, some of Rota's main desiderata, however, have been realized since the last two decades of the past century. Different sorts of non-classical logics, for instance, have been developed, and some of them do not accept the principle of bivalence: the investigation of these logics has become at present a flourishing business. Husserl's theory of part and whole, which constitutes the core of his *Third Logical Investiga-tion* and to which Husserl attributed "a fundamental role in the entire development of his phenomenology" (Casari, 2000, p. 1), has been deeply analyzed and recon-structed (Simons, 1982; Simons, 1987; Fine, 1995; Casari, 2000).

Concerning philosophy, ironically, starting from the second half of the past cen-tury, a strong interest for phenomenology began to arisein the field of analytical philosophy, i.e. amongst authors that Rota would have considered as completely extraneous to the 'spirit' of the phenomenological school. Dagfinn Føllesdal, for example, with his dissertation on Frege and Husserl (Føllesdal, 1958), and with his several essays on Husserl's philosophy, has given a decisive contribution to the studies on phenomenology in the Anglo-saxon philosophical tradition. Several years after Føllesdal's essay, Barry Smith's miscellaneous book (Smith, 1982) paved the way for other relevant contributions in the same direction by Peter Simons and Kevin Mulligan.

From the present-day point of view, even Rota's unfair attacks to the Anglo-saxon analytical tradition appear as a legitimate reaction to a kind of scholasticism from which many representatives of this tradition are not completely exempt.

# References

Casari, E. (2000), *On Husserl's Theory of Wholes and Parts*, "History and Philosophy of Logic", 21/1, pp. 1-43.

Fine, K. (1995), *Part-whole*, in (Smith, Woodruff Smith, 1995), pp. 463-85.

Follesdal, D. (1958) *Husserl und Frege: Ein Beitrag zur Beleuchtung der Entstehung der phänomenologischen Philosophie* (Thesis for the degree of Magister artium, Oslo, 1956), Aschehoug.

Heidegger, M. (1927), *Being and Time*, New York, Hagerstown, San Francisco, London, Harper & Row, 1962.

Husserl, E. (1900-1901), *Logical Investigations*, London, New York, Routledge, 2001, 2 voll.

Kac, M., Rota, G.-C., Schwartz, J.T. (1986), *Discrete Thoughts. Essays on Mathematics, Science and Philosophy*, Boston, Basel, Berlin, Birkhäuser.

Leng, M. (2002), *Phenomenology and Mathematical Practice*, in "Philosophia Mathematica", 3, 10, pp. 3-25.

Palombi, F. (2003), *La stella e l'intero. La ricerca di Gian-Carlo Rota tra matematica e fenomenologia*, Turin, Bollati Boringhieri.

Philipse, H. (1995), *Transcendental Idealism*, in (Smith, Woodruff Smith, 1995), pp. 239-322.

Rota, G.-C. (1974-1991), *The End of Objectivity. The Legacy of Phenomenology. Lectures at MIT*, Cambridge, MA., MIT Mathematics Department.

Rota, G.-C. (1986), *Misreading the History of Mathematics*, in (Kac, Rota, Schwartz, 1986), pp. 231-4.

Rota, G.-C. (1997), *Indiscrete Thoughts*, F. Palombi (ed.), Boston, Basel, Berlin, Birkhäuser.

Rota, G.-C. (1999), *Lezioni napoletane*, F. Palombi (ed.), Neaples, La Città del Sole.

Simons, P. M. (1982), *The Formalization of Husserl's Theory of Wholes and Parts*, in (Smith, 1982), pp. 113-59.

Simons, P. M. (1987), *Parts: A Study in Ontology*, Oxford, Clarendon Press.

Smith, B. (ed.) (1982), *Parts and Moments: Studies in Logic and Formal Ontology*, Munich, Philosophia Verlag.

Smith, B., Woodruff Smith, D. (eds.) (1995), *The Cambridge Companion to Husserl*, Cambridge, Cambridge University Press.

# Chapter 15
# "A Minority View". Gian-Carlo Rota's Phenomenological Realism
## Invited Chapter

Fabrizio Palombi

> Artists [...] fail to give an accurate description of how they work, [...] scientists [...] believe in unrealistic philosophies of science.
> Gian-Carlo Rota.

In 1997 Rota's second anthology was published, entitled *Indiscrete Thoughts*. The theses put forward in the book were "minority" positions in a Unites States cultural context that, after long having attempted to replace philosophy with logical analysis and the analysis of language (Hersh, 1997, pp. IX-X), was preparing to interpret it also in neuroscientific terms. Rota intended to show that he did not fear uncomfortable positions and chose the phrase "a minority view" as a provocative title for the philosophy section of the book, inspired by phenomenology. We cannot understand the importance of Rota's intellectual figure, within the American cultural context of the end of the twentieth century, and the importance of his heritage if we interpret it in terms of architectonic of philosophy. In fact, Rota did not aim to create parts of a philosophical system or to realize complete phenomenological analysis. His goal was another, and we may assimilate it to the aims of the polemist and the antidogmatic intellectual, that have an important role in the tradition of the philosophy. I believe this is the right interpretation of Rota's works, which allows a filigreed reading of his intellectual heritage. We see Rota's precise choice of sides also in *The End of Objectivity*, the text of his course of lectures (Rota, 1991), which is an anthology of examples drawn from everyday life and scientific practice compiled in order to update phenomenology and transform it into an efficient instrument of critical culture. One of his principal philosophical references is represented by *Being and Time* (Heidegger, 1927), which is examined and commented on in detail (Rota, 1991, p. 2). Let us recall the fact that Rota always tends to valorize the elements of continuity of the phenomenological movement to the detriment of its internal fractures and of its splits. He applies this interpretative perspective in particular to the relationship between Husserl and Heidegger, relegating the reasons for their disagreement to a

Fabrizio Palombi
Università della Calabria, Italy

E. Damiani et al. (eds.), *From Combinatorics to Philosophy*,
DOI 10.1007/978-0-387-88753-1_15, © Springer Science+Business Media, LLC 2009

muted background (Rota, 1997, pp. 189-90). This work of re-elaboration and up-dating of Heideggerian philosophy was part of the intellectual battle that Rota was fighting in proclaiming himself, simultaneously, antireductionist and realist. Let me provisionally define as "phenomenological realism" this dual reference that consti-tutes the synthesis of his "minority view" and shows the topicality of his reflection. My paper is dedicated to the analysis of this theoretical hallmark.

## 15.1 Considerations on the Problem of Realism

It was Kant who codified the contemporary opposition between idealism and real-ism. Kant based idealism upon an ontological interpretation that makes the totality of what exists depend on thought, in which, however, the meaning of "thought" is not univocal. Realism, by contrast, holds that reality possesses an existence that is independent of thought (see Costa, 2007, p. 27). Realism, moreover, has an episte-mological variant, fundamental for our interests, which asserts the independence of the things known with respect to the knowing subject.

The celebrated philosophical topology proposed by Gàston Bachelard (1884-1962) graphically refers the most important contemporary epistemological options (formalism, conventionalism, positivism and empiricism) back to the opposition be-tween realism and idealism (see Bonicalzi, 2007, pp. 29-30). I also connect with this schema the opposition between objectivism and psychologism, examined at length by phenomenological research. Furthermore, I shall also use the terms naturalism, reductionism and scientism in a shaded and correlated way, as different perspectives that fore-shadow the same vision of the world.

At the conclusion of this paper I shall return to these Bachelardian coordinates that - while taking no part in Rota's philosophical constellation - can help us to take our bearings and to individuate the place of phenomenology in a historiographic problem that is herewith too complex for me to discuss exhaustively.

For the moment, let me point out a significant theoretical custom: phenomenol-ogy is, in fact, usually presented as a sort of third way capable of getting through an insidious philosophical bottleneck and gaining a terrain that surpasses any form of realism or idealism (Heidegger, 1927, pp. 57-8). In most cases Rota, too, followed this tendency, yet there are some passages that constitute interesting exceptions. One of them emphatically describes phenomenology as "an extreme form of real-ism" (Rota, 1991, p. 1; see also pp. 51, 374). This is an important reference since it comes right at the beginning of *The End of Objectivity*, in the section bearing the programmatic title "Foundations of Phenomenology". His position is not an illegit-imate one: Heidegger himself declares that he pursues

> a more precise characterization of the concept of Reality in the context of a discussion of the epistemological questions oriented by this idea which have been raised in realism and idealism (Heidegger, 1927, p. 228).

Rota's intention, however, is different, since his vein is essentially polemical. He harbors no prejudices against the idealist tradition, which, on the contrary, played a major role in his cultural development (see Palombi, 2003, p. 17). Rota, rather, wished to take a position against a widespread interpretation of realism that sees reality in ingenuously materialistic terms and reduces all the other philosophical positions to a sort of futile reverie. Maintaining that phenomenology is a form of realism is equivalent, for Rota, to asserting its value and concreteness. Asserting phenomenological realism means, first of all, criticizing scientism, reductionism and naturalism.

## 15.2 Phenomenological Realism

Rota opens his fullest examination of this theme with a sort of oxymoron, declaring that "our ideal of realism is taken from the phenomenology of Edmund Husserl" (Rota, 1997, p. 135). Comparing the abstractness of an ideal with the claimed concreteness of realism is a precise rhetorical strategy that shows the paradoxicality of contemporary scientism that often, drunk on technological progress, seems to lose its points of reference and to provoke some short circuits that I shall touch on later. Let us examine the phenomenological rules of "realistic description" summarized by Rota in the following four points:

1. A realistic description shall bring into the open concealed features [...].
2. Fringe phenomena that are normally kept in the background should assume their importance [...].
3. Phenomenological realism demands that no excuses be given that may lead to the dismissal of any features of mathematics, labeling them as psychological, sociological or subjective.
4. All normative assumptions shall be weeded out (Rota, 1997, p. 135).

The first point refers us to the hermeneutic circle that in all aspects of existence, from everyday existence to the existence characteristic of mathematics, makes it possible to shed a new light on phenomena considered banal and obvious and to grasp new aspects of them (see Rota, 2007, p. 25). Rota believes that logic and the axiomatic method are sufficient to weed out incorrect proofs but are only necessary for the complex genesis of those that are correct (Rota, 1997, p. 134).

On the basis of these convictions Rota re-assesses the importance of heuristic lines of reasoning founded on fore-shadowing and on the hermeneutic circularity that makes it sensible to interchange theorems and proofs (Rota, 1997, p. 143). The sense of this circularity can be understood if we assume that the value of a proof consists not in the logical verification of a thesis but in "open[ing] up new possibilities for mathematics" (Rota, 1997, p. 144). This is a statement that derives from the Heideggerian rereading of his experience as a scholar and, in particular, of his practice of mathematical research.

The mathematician is, first and foremost, someone who, as such, relates to certain possibilities shared by the scientific community to which he or she belongs. Not every possible proof is significant (Palombi, 2004, p. 36) and there are no mathematicians who know all the fields of research of their discipline in a certain historical period. The opening up of new possibilities brought about by a proof reveals unexpected connections between different and distant sectors of mathematics (see Cellucci, 2008). Their meaning can be Heideggerianly interpreted first of all as the emerging of new references between known mathematical entities that sometimes reveal other references that were unknown until that moment. The structure of the reference, in everyday as in mathematical existence, reveals the phenomenon of sense.

Mathematicians too, like everyone else, find themselves before an infinity of possibilities that cannot all be realized, and therefore the preliminary selection of the roads to be taken is fundamental. Rota believes that this choice is made in the light of hermeneutic preunderstanding and the hermeneutic project. A conjecture is such only when mathematics, as a tradition of scientific research and community, wagers on and invests in its truth before this truth can be logically proved. The conviction here is neither easy nor painless, since it almost always involves years of study and of research that may prove to be sterile.

Rota finds no better term than the word "faith" to describe that anticipation of truth, logically untenable, that often becomes the irreplaceable keystone of a proof. The proof of Fermat's theorem obtained by Wiles and Kostant's proof regarding the exceptional Lie groups represent a valid example of this type of proof (Rota, 1997, pp. 139-44).

In this sense heuristic reasoning installs itself in one of the crucial points of Heideggerian circularity constituted by the term *Auslegung*, which I render as "explicitation" (see Costa, 2003, pp. 256-9) instead of the usual term "interpretation" used in the English translation of *Sein und Zeit* (see the translators' note and the glossary drawn up by Macquarrie and Robinson in Heidegger, 1927, p. 19, note 3, and p. 545).

The term "explicitation" avoids the subjectivistic misunderstandings sometimes implied by the word "interpretation", and better expresses the phenomenological rereading of mathematical proof proposed by Rota. We can consider the emphasis attributed to the hermeneutic circle and to the function of explicitation as a sort of circumscribed repetition, or reproposition, of the Heideggerian operation to describe phenomenologically the Being-in-the-world of the mathematician (Palombi, 2003, p. 77). In this way Rota intends to prepare the ground to graft onto the traditional description of mathematical proof not only the themes of understanding and of project but also that of state-of-mind.

## 15.3 Fringe Phenomena and State-of-Mind

"Fringe phenomena" is a term Rota uses to indicate all the aspects of experience that scientific practice tends to marginalize because they are considered inessential, subjective or judged to be epiphenomena of underlying material structures (Rota, 1991, pp. 75-6, 410). The reductionist and naturalist approaches tend to disembody the fact from all those parts that present it in its concreteness, neglecting (by epistemological statute) the marginal phenomena. By contrast, Rota considers fringe phenomena to be part of that which we are describing without any reference to possible or hypothetical imperfections or explanations (Rota, 1991, pp. 73-6). This statement provides us with two important phenomenological points of reference. The first is represented by the particular Husserlian mereology that considers proper parts of a whole also some of its subjective qualities such as color (Husserl, 1900-1901). This is a theme that Rota extrapolates from Husserl's text and re-elaborates in a number of interesting writings dedicated to the theory of *Fundierung* (Rota, 1997, pp. 172-181, 188-191) and which I have examined in depth elsewhere (Palombi, 2003, pp. 61-73, 78-80).

The second consists in the intention to recover the relation between the cognitive and the emotive values of fundamental aspects of human existence that scientism has ostracized. Rota considers the interpretation of human behavior in strictly rational terms to be nothing else than philosophical mythology and deems it indispensable to develop a theory of knowledge that also includes human state-of-mind (Rota, 1991, p. 28). The recovery of these residual aspects at times leads Rota to compare phenomenology with psychoanalysis and to integrate it into a Sartrean perspective (Rota, 1991, p. 425). In fact he believes that

> We are describing the intricate network of relevancies or relevant functions [...]. Part of a phenomenological description is precisely bringing up everything. Just like a psychoanalyst would bring up a circumstance that we don't like to mention, similarly, in a phenomenological description, we dig up emotional, mystical [relevance], or anything else that gives sense to the meaning (Rota, 1991, p. 224).

Nevertheless, his principal reference is represented by the pages of *Being and Time* (Rota, 1991, pp. 1, 224) dedicated to the analysis of state-of-mind (Heidegger, 1927, sec. 29, 30, 40, 68) that are examined in detail in some sections of *The End of Objectivity* (Rota, 1991, pp. 252-67, 327-30, 373-83). In the phenomenological perspective affectivity is not something that is simply added to understanding but, rather, the two "existential structures" must be considered in their equiprimordiality (Heidegger, 1927, p. 182).

Rota believes that a phenomenological description must be distrustful of all the forms of binary logic that interpret reality in the form of an exhaustive opposition (Rota, 1997, p. XXI). On the contrary, such description must always take into account shadings and graduality (Rota, 1991, p. 264). This norm holds also for the analysis of the role of understanding and of state-of-mind in a given phenomenon. Rota writes that

> There are phenomena of disclosure where the actual grasp in the context fits the major role and the mood component fits the minimal role - for example, our approach to solving a mathematical problem. This is not saying that we have to like the problem, but the minimum of mood lets us get involved in it (Rota, 1991, p. 264).

He examines some examples that demonstrate the importance of state-of-mind for phenomena that are seemingly cool and distant from the emotions, such as mathematical practice, that contribute to the proof of a theorem or to the opening up of new fields of the discipline (see Rota, 2007, p. 32).

This is the significance also of Rota's short unauthorized biographies of some great twentieth-century mathematicians that he knew and associated with. His analysis does not possess clinical or moral value but intends to examine some of the ways in which the state-of-mind that accompanies creativity is manifested. He gives us a sort of blow-up of some aspects of the psyche that thematizes aspects common to all human beings that emerge, with particular force, in a number of famous scholars.

It is in this perspective that we have to understand the detailed descriptions of the bizarre work habits of MIT hackers (Rota, 1993, pp. 137-40), of Stan Ulam (1909-1984) (Rota, 1997, pp. 63-85), and of some famous Princeton professors (Rota, 1997, pp. 3-85). Rota considers certain traits to be particularly widespread among these scholars, such as continual application, the scrupulous attention to certain details, constant practice, and the indifference to the clock and the calendar that leads them to frequent their university offices and libraries at all hours of the day or night.

These attitudes seem to manifest the passion for and urgency of their research, the emerging of ideas that demand to be in some way tested, deposited, transcribed, and that cannot be grasped in their existential concreteness unless state-of-mind is taken into consideration. They are secondary aspects that reveal certain deep structures of thought. In fact, for Rota,

> Our project of viewing is guided by mood phenomena such as curiosity, liking, fascination and desire [...]. What directs our viewing is some sort of liking for what we want to view. Without that liking, however minimal, we cannot view. It is a frequent experience among mathematicians and scientists [...] that they cannot solve a problem unless they like it (Rota, 1991, p. 262).

## 15.4 Psychologistic Misunderstandings

The attention to hermeneutic understanding and to state-of-mind must not be misunderstood. This is an aspect of crucial importance, upon which I intend to dwell, in order to understand the topicality of Rota's reflection and to respect its authenticity. I would do my master wrong if I sought to compromise his thought with scientistic or naturalist drifts that are foreign to his work.

Let us recall that Rota, on several occasions, warned his readers about the danger of confusing hermeneutic understanding with the physiological and psychological substrates on which it is founded. "Founding", in this case, refers to the Husserlian *Fundierung*, and must not be confused in any way with a causal relation. Rota insists on emphasizing that

> "Founded" does not mean "causes" [...]. That is a misuse of the term "caused", and it is probably because of this misuse that the word "founded" is misunderstood (Rota, 1991, p. 224).

It is his peculiar way of taking in the phenomenological criticism of psychologism and up-dating it in the light of the philosophical interpretation of scientific progress that, at the time of the publication of *Indiscrete Thoughts*, was mainly concerned with Artificial Intelligence and that, today, looks above all to the neurosciences. Nowadays it is important to remember that Rota analyzed the reductionist interpretation of Nuclear Magnetic Resonance Imaging in his last course on phenomenology at MIT. He criticized any kind of identification between human thought and images produced by new medical imaging devices which represent the workings of the brain (Rota, 2007, p. 47). He underlines that "the brain alone is not the thinking. It's a logical error to think that the thinking is the same as the brain" (Rota, 2007, p. 21; see Franzini, 2007, p. 161).

Rota dedicates interesting pages to the phenomenological analysis of the genesis of scientific concepts and to the two - opposite and complementary - directions that characterize it (Rota, 1973). Let me attempt to summarize this dual perspective briefly, using a generic example taken from neuroscientific research. In the first direction the phenomenon of human understanding is made possible by our brain, our body and by its nervous system that, in the current phase of history, are explained scientifically by means of neurons and by their complex organization. This scientific perspective has produced extraordinary conquests and great new knowledge in the field of human cognitive mechanisms.

We must not forget, however, that it is possible to interchange the viewpoint and individuate a second direction in which it is the phenomenon of human comprehension that, in a logical and historical sense, makes the discovery of the neuron possible. This is the philosophical perspective that considers the nerve cell to be a derived concept that belongs to science and not to Husserlian *Lebenswelt* (see Franzini, 1991, p. 87) or Heideggerian Being-in-the-world.

Rota is convinced that the philosophical reflection on understanding cannot be entirely replaced by scientific research due to the different objectives, interests and times that characterize the two enterprises. The former is situated on the plane of sense, avails itself of models of argumentation that are rigorous but imprecise (Rota, 1999, p. 94), and deals with questions that are age-old (Rota, 1997, p. 94). The latter studies causes, uses exact procedures, and develops concepts that change and age quickly, driven by scientific progress.

The Husserlian criticism of naturalism and psychologism lead Rota to believe that philosophical analysis of the phenomenon of human understanding cannot be reduced to, replaced by, or confused with some of the scientific concepts that are extraordinarily topical today: mental modules (Rota, 1997, p. 187), Intelligence Quotient (Rota, 1991, p. 24; see Gould, 1981) or genetic inheritance (Rota, 1986, p. 232).

I believe that this reflection is of great help in dealing with some recent short circuits between philosophical positions held to be traditionally opposed and that, in my view, recall the previously mentioned Bachelardian topology.

Let us consider the specific question of the ontological status of mathematics. Robert Sokolowski, in the preface to *Indiscrete Thoughts*, emphasizes how the phenomenological reflection on mathematics must beware of the symmetrical dangers of ingenuous objectivism and ingenuous subjectivism or psychologism (Sokolowski, 1997, pp. XIII-XIV). The current cultural context influenced by the neurosciences has transformed this philosophical geography and produced a situation that is (at least partially) unprecedented. The psychologistic interpretation attributes mathematical entities to human intellectual activity but this activity, in its turn, is attributed by the philosophical reflection on the neurosciences to neuronal activity. In this way, mathematical objects are made to sublimate into the evanescence of thought to then solidify suddenly in the matter of the brain.

Rota understands the philosophical sense of this process, emphasizing that everyday human existence and, in other forms, scientific practice, is articulated on "the forgetting of constitution" (Rota, 1991, p. 374). In our epoch such forgetting takes the form of a repression of the complex historical genesis of scientific concepts and of their phenomenological constitution. In this way, in the current philosophical climate, ingenuous objectivism is no longer set against psychologism but is capable of including it as a particular case of its own domain.

## 15.5 Descriptions, Not Prescriptions

How does a mathematician, a scientist, an artist or a philosopher work? What makes their work creative? Rota refuses to answer these questions with one of the usual attempts at psychological investigation of the context of scientific discovery, of philosophical reflection and of artistic inspiration. The fourth rule of phenomenological description prevents him from having recourse to the normative replies provided by epistemology or by psychology.

Rota prefers to describe significant aspects of his own research experience, divided between combinatorics and phenomenology, and to compare them with others taken from the biography of some famous scholars in order to individuate the components that I have illustrated in this paper. On another occasion I showed that the detour, the sidelong glance and lateral thinking are, for Rota, fundamental parts of scientific activity and, in particular, of the complex relation that connects mathematics with technological applications (Palombi, 2005). Here, I would like to conclude by indicating a few philosophical and cultural consequences of the four points that articulate his phenomenological realism.

First of all, they manifest his concern about the excessive specialism of research that constitutes a serious danger for mathematics, but also for science and knowledge in general, as he demonstrates in his superb article "Ten Lessons for the Survival of a Mathematics Department" (Rota, 1997, pp. 204-8). The problems regarding the education of the new generations, for Rota, cannot be solved by giving courses in creativity (Rota, 1997, p. 127) but rather by keeping the cultural tradition alive (Rota, 1985, p. 94) and by allowing young people to stir up their curiosity without forcing them into excessively rigid study programs.

Nevertheless, his greatest fears were related to the future of the philosophical tradition (Rota, 1997, pp. 89-103), which he believed could be saved by remaining faithful to its history and to its inexhaustible demand for sense. The attention to hermeneutical circularity, to state-of-mind, to the shadings of Being-in-the-world cannot be replaced by scientific results - on the contrary, they must find in scientific results material to delve further into phenomenological research. The lesson of phenomenology thus becomes an irreplaceable resource to resist the enticements of contemporary naturalism (Lanciani, 2003, pp. 69-75).

If we keep these demands in mind we may, perhaps, be able to find a term more suitable than the overused, and implicitly prescriptive, word "realism" to synthesize Rota's complex reflection. Inspired by Bachelard's lexical research, I venture to propose the term "surrealism".

Think of a deeper analysis of realism that phenomenologically meets with flesh and blood things, of a sophistication of realism that does not claim to construct ontological hierarchies. Think of surrealism's research on double images that can find some significant counterparts in Rota's writings, but also of the provocative force of the surrealist works. Think, finally, of Rota's smile and of the sophisticated irony of his writings and of his reviews. For all these reasons I'd like to think he would have been happy with my description of his life's work as "phenomenological surrealism".

# References

Bonicalzi, F. (2007), *Leggere Bachelard. Le ragioni del sapere*, Milan, Jaca Book.

Cellucci, C. (2008), *Why Proof? What is a Proof?* in (Lupacchini, Corsi, 2008), pp. 1-27.

Costa, V. (2003), *La verità del mondo. Giudizio e teoria del significato in Heidegger*, Milan, Vita e Pensiero.

Costa, V. (2007), *Il cerchio e l'ellisse. Husserl e il darsi delle cose*, Soveria Mannelli, Rubbettino.

Franzini, E. (1991), *Fenomenologia. Introduzione tematica al pensiero di Husserl*, Milan, Franco Angeli.

Franzini, E. (2007), *Immagine e pensiero* in (Lucignani, Pinotti, 2007), pp. 137-164.

Gould, S. J. (1981), *The Mismeasure of Man*, New York, Penguin.

Heidegger, M. (1927), *Being and Time*, New York, Hagerstown, San Francisco, London, Harper & Row, 1962.

Hersh, R. (1997) in (Rota, 1997), pp. IX-XI.

Husserl, E. (1900-1901), *Logical Investigations*, London, New York, Routledge, 2001, 2 voll.

Kac, M., Rota, G.-C., Schwartz, J. T. (1986), *Discrete Thoughts. Essays on Mathematics, Science and Philosophy*, Boston, Basel, Berlin, Birkhäuser.

Lanciani, A., (2003), *Phénoménologie et sciences cognitive*, Beauvais, Association pour la promotion de la Phénoménologie.

Lucignani, G., Pinotti, A. (eds.) (2007), *Immagini della mente. Neuroscienze, arte, filosofia*, Milan, Cortina.

Lupacchini, R., Corsi, G. (eds.) (2008), *Deductions, Computation, Experiment. Exploring the Effectiveness of Proof*, Berlin, Springer.

Palombi, F. (2003), *La stella e l'intero. La ricerca di Gian-Carlo Rota tra matematica e fenomenologia*, Turin, Bollati Boringhieri.

Palombi, F. (2004), *La forma della mente matematica: contesto della scoperta e bellezza*, "Ou. Riflessioni e provocazioni", Vol. XV, pp. 33-8.

Palombi, F. (2005), *Une application mal interprétée: une contribution de G.-C. Rota á la philosophie des mathématiques*, in G.-C. Rota, (2005), *Mathématique et Phénoménologie*, Beauvais, Association pour la promotion de la Phénoménologie, pp. 129-139.

Rota, G.-C. (1985), *Mathematics, Philosophy and Artificial Intelligence*, "Los Alamos Science", no. 12, pp. 92-104.

Rota, G.-C. (1986), *Misreading the History of Mathematics*, in (Kac, Rota, Schwartz, 1986), pp. 231-4.

Rota, G.-C. (1991), *The End of Objectivity. The Legacy of Phenomenology. Lectures at MIT*, Cambridge, MA., MIT Mathematics Department.

Rota, G.-C. (1997), *Indiscrete Thoughts*, F. Palombi (ed.), Boston, Basel, Berlin, Birkhäuser.

Rota, G.-C. (1999), *Lezioni napoletane*, F. Palombi (ed.), Neaples, La Città del Sole.

Rota, G.-C. (2007), *Lectures on Being and Time* 1998, M. van Atten (ed.), "The New Yearbook for Phenomenology and Phenomenological Philosophy", pp. 1-99.

Sokolowski, R. (1997), *Foreword*, in (Rota, 1997), pp. XIII-XVII.